*The Best American Science
and Nature Writing 2019*

The Best American Science and Nature Writing™ 2019

Edited and with an Introduction
by Sy Montgomery

Jaime Green, Series Editor

MARINER BOOKS

HOUGHTON MIFFLIN HARCOURT

BOSTON • NEW YORK   2019

hmhbooks.com

ISBN 978-1-328-51900-9 (print) ISSN 1530-1508 (print)
ISBN 978-1-328-51901-6 (ebook) ISSN 2573-475X (ebook)

Printed in the United States of America
DOC 10 9 8 7 6 5 4 3 2 1

Contents

Foreword

EARLIER THIS YEAR, I received an email from a self-declared longtime reader of *The Best American Science and Nature Writing*. The reader noted that the anthology had been getting more political over the last few years, and he asked me to please keep the selections for 2019 focused on science and nature.

My first thought was: *Sorry, guy. Tough luck.* (I hadn't even yet read our guest editor, Sy Montgomery's, introduction at the time.) My second thought was: *Tony Kushner.* Yes, the playwright. There are many thinkers who have argued that *everything* is political, that an apolitical stance is inherently political, too, but Tony Kushner is the one I think of first. In his essay "Notes About Political Theater," Kushner writes, "In life, as in art, much energy is devoted toward blurring the political meaning of events, or even that events *have* a political meaning . . . When theater artists assiduously avoid politics, we deny the existence of the political and are making a political statement, committing a political act."

You can easily swap *science* in for *events, readers of this book* for *theater artists.* Here:

> In life . . . much energy is devoted toward blurring the political meaning of [science], or even that [science] *has* a political meaning . . . When [readers of this book] assiduously avoid politics, we deny the existence of the political and are making a political statement, committing a political act.

We may desperately want science not to be political, because that seems simpler or more pure. But if science teaches us anything, it is that simplicity is an illusion, and that ignorance cannot be a resting place.

Science has not gotten more political in the last few years, and neither has this anthology. What has changed is the sense of urgency, a rechanneling of energy from blurring political meaning to, as a planet and as people, admitting that we are fighting for our lives.

It isn't only about climate change, although of course that crisis is top of mind (and features in many of the pieces in this book). This is where science is political because politicians hold the power to make meaningful change. We can't reduce/reuse/recycle our way out of the wildfires so meticulously and vividly evoked in Apricot Irving's "The Fire at Eagle Creek," or the looming insect apocalypse described by Brooke Jarvis. All the individual actions in the world can't save us. Science is political because it demands action from power.

Readers who don't want science to be political share much with people who say *I don't see race!* It is a privilege to be able to live your life in a way that race is not a factor. And it is a privilege to be able to turn off your TV or close your computer and think that politics has no bearing on your life. That means that you feel safe, and it means you're not worrying about the people who aren't. Kushner, again: "The act of wishing away the world of political struggle is a deeply reactionary gesture." It is a privileged ignorance—a privileged ignoring—to think that science is, can be, or ever was apolitical.

Say *Science isn't political* to the women profiled in Linda Villarosa's "The Hidden Toll: Why Are Black Mothers and Babies in the United States Dying at More Than Double the Rate of White Mothers and Babies? The Answer Has Everything to Do with the Lived Experience of Being a Black Woman in America" or to the uninsured patients conjured by Molly Osberg in "How to Not Die in America." Say *Science isn't political* with respect to Jarvis's insects and the vaquita in Ben Goldfarb's "The Endling," the rhinos in Jeremy Hance's "The Great Rhino U-Turn," and the cetaceans of J. B. MacKinnon's "You Really Don't Want to Know What It's Like to Be a Right Whale These Days." MacKinnon's title supposes you might not want to know, but you *do* want to, also, because stories like these are elegy and alarm, mourning what we're too late to save so that we might open our eyes to the next victims, hopefully in time.

I don't know if the author of the email imploring me to focus on science and nature would say that these stories aren't political, or if he would rather not read them at all. But they're not in this book because of political arguments or despite them. They're in this book because, as some of the best science and nature writing of the year, they each tell the story of a slice of the Earth, a story that, for better or for worse, humanity is writing.

The essential question of politics, I believe, is: *What do we owe each other?* Or, as the writer Kayla Chadwick titled an essay appearing in *HuffPost* in 2017, "I Don't Know How to Explain to You That You Should Care About Other People." That is the divide in American politics today. Would you, as Chadwick cites in her piece, pay an extra 17 cents for your Big Mac so a McDonald's employee can earn a living wage, or do you just want your cheap burger? Would you sacrifice an oil company's profits to preserve a national monument's untrammeled space? We don't, individually, get to answer that second question, so we rely on elected officials to care beyond themselves, too.

Science writing is political because it shows us how to care about other people, and the world. In "This Sand Is Your Sand," Chris Colin illuminates the tensions inherent in the idea of public lands with a paddling trip down a river. Rebecca Mead's "The Story of a Face" is political not just because it's about trans people, whose civil rights have been relentlessly politicized and endangered; Mead's story is political because it draws you, the reader, into empathy. I'm sad that empathy is political, but it obviously is.

It is political to call attention to looming dangers—as Ed Yong does in "When the Next Plague Hits," highlighting the vulnerabilities in America's epidemic-response infrastructure. It is also political to celebrate the beauty in the world and beyond it, which Rebecca Boyle does so powerfully in "The Search for Alien Life Begins in Earth's Oldest Desert." There are political forces that oppose this view of nature as valuable unto itself, apart from its commodification, or the view of pure scientific research as valuable unto itself apart from applications. Under capitalism, it is political to seek knowledge for its own sake, to wonder with no goal other than to learn.

I want to focus on the political implications of the writing in this book, but I should also mention that scientific research is often funded by the federal government; so are the agencies that would protect our food, water, and natural resources. Politicians are constantly seeking to insert themselves into scientific and medical decisions, imposing their personal beliefs on the lives and bodies of women and trans people. And sexism and racism have gifted us scientific institutions disproportionately dominated by white men.

"Nature" is hardly apolitical, either, considering that our American sense of wilderness was—as environmental historian William Cronon writes in his 1995 essay "The Trouble with Wilderness" —born out of a fantasy sired by romanticism and manifest destiny. From the sublime and the frontier we conjured a vision of pristine nature untouched by humans, where, obviously, indigenous humans had been living before. Every piece of nature writing, every celebration of prairie or park, grapples with this history.

We shouldn't be embarrassed that writing about science and nature is political. This foreword is neither a defense nor a 1,600-word response to one cranky email (though I thank that writer for the spark of inspiration). If anything, I want this foreword, and this book, to be a celebration of the power of telling stories of research and discovery, of human ingenuity and dedication and wonder and hope. And the possibility of change.

Science has not gotten more political, but politics has changed. Kushner, whose husband, Mark Harris, once joked that his drag name could be Eara Lee Prescient, wrote, back in 1997, "We have entered into an age the politics of which I like to call neobarbaric, in that the previously unassailable fundamentals of civilization, of community, are under attack, and logic, causality, even coherent narrative are gone." This sounds, to many people, at least to me, a lot like 2019. Politics sucks really bad right now. But there is hope in politics, too, hope for progress and change. What species can we write the story of saving next year?

This is my first year as series editor of *The Best American Science and Nature Writing*. It was a gift and a pleasure to spend part of my year reading as widely as I could in this genre. I learned so much. I am so grateful to all of the writers doing the hard work of writing about the world. (I'm also grateful to the writers and editors who nominated these pieces. You can make nominations for future editions at jaimegreen.net/BASN.) Many thanks, also, to Sy for her selections, and to you, reader—as if writing about politics in a science and nature book weren't enough, I've been using the words of a playwright to do so. I realize that's a big ask.

Allow me one more Kushner quote: "The *Dies Irae* is looming as the fate of the poor, the marginal, the immigrant, the refugee, the disenfranchised, the vulnerable. We are in terrible trouble. We must watch out." Reading the pieces collected in this book may not be a grand work of activism, but it is keeping our eyes open. Which is the first step of watching out. Which is the first step of whatever comes next.

Jaime Green

Introduction

SEVERAL YEARS AGO I was invited to speak to kids and teens at the Boston March for Science. On a cold, rainy day in early April, I looked out at a sea of young faces framed by dripping umbrellas and the hoods of ponchos, and spoke to them about tree kangaroos.

Tree kangaroos sound made-up, I told them, but they're real. There are 12 species of kangaroos who really live in trees—and can jump down 60 feet and bound away, hopping just like the "regular" kangaroos we know from Australia. The tree kangaroos I was speaking about lived in a cloud forest clad in orchids and moss on the island of Papua New Guinea. They had orange and yellow fur, pink noses, and, like other kangaroos, belly pockets for the babies —but other than that, scientists knew very little about them. Until one scientist, Dr. Lisa Dabek, decided to go find out more, mounting expedition after expedition into the trackless, high-altitude cloud forest of northeast Papua New Guinea's Huon Peninsula . . .

Few of these kids had ever met, or might ever meet, a tree kangaroo, even in a zoo. Fewer still had ever been, or ever would go, to Papua New Guinea. And I wondered how many of them had met a scientist. Yet these kids were standing, spellbound, in the rain, riveted by the story of Dr. Dabek's quest, transfixed by wonder about this beautiful and enigmatic animal.

That first Boston March for Science could have been a pretty depressing day for everyone, and not just because of the cold and wet. Why did we need a March for Science in the first place? Because science was—and is—under siege in America.

Never mind the proven adage "science is the engine for prosperity." Never mind that science has immeasurably improved human health and longevity, or that using science to discover and invent new ways to protect the environment, save endangered species, and curb climate disruption has never been more urgent. Even amid fires, floods, and record temperature swings, our president denies climate change is real. Fewer than a third of Americans understand evolution and natural selection; and so many parents refuse to vaccinate their children that diseases we thought we had conquered through science are now surging back to kill our kids.

Yet I looked out at all those kids braving the cold and the rain, many wearing costumes and carrying signs, and felt not just hopeful, but joyous. In Boston alone, 70,000 people turned out for that first March for Science—and it was but one of the more than 600 cities across the world where people rallied to show how much science mattered to them.

Just why is science so important? Well, there are the reasons mentioned before: Science improves our lives. Science makes money. Science can help us heal our wounded world.

But there's another reason, and this is why those kids were so spellbound listening to the story of Dr. Dabek's work that day: our thirst for discovery.

Science is important because this is how we seek to discover the truth about the world. And this is what makes excellent science and nature writing essential. (*What is nature?* "The phenomena of the physical world collectively," begins the definition that came up first in my Google search—which makes all science writing nature writing.) Science and nature writing is how we share the truth about the universe with the people of the world.

Truth—not fake news. Not the fake news that two centuries of carbon emissions are harmless; not the notion that our planet is merely thousands of years old; not that vaccines cause autism, a lie based on fabricated data published in the *Lancet*, which the journal retracted in 2010 when it discovered the fraud.

Like science itself, science writing has also faced serious obstacles in recent years. In 1989, 95 newspapers had science sections. Today only 19 newspaper science sections remain. With a double-digit decline in print advertising, print outlets of all kinds are shrinking as Americans turn instead to online sources. And while there are notable exceptions—*Mongabay* is one; it originally published "The Great Rhino U-Turn," which you will read in these pages, one of a four-part series—web-based media don't often offer science writers the resources that newspapers and magazines once did to fund serious science journalism, which may require months of research as well as extensive travel.

And yet: On the first day of 2019, Jaime Green, the series editor of this volume, sent me 100 longform journalism articles, gathered from print and online sources, on science and nature topics published in 2018. I read every one of them. All were excellent. From these, as well as others, I selected—only with great difficulty —those in this book.

All of these selections, disparate as they are, illustrate why telling the truth about the world is important. Some of these stories are powerful calls to action. "The Hidden Toll: Why Are Black Mothers and Babies in the United States Dying at More Than Double the Rate of White Mothers and Babies? The Answer Has Everything to Do with the Lived Experience of Being a Black Woman in America" outlines an outrageous crisis: black mother and baby mortality in the United States is now worse than a quarter century ago. Education and income disparity are not to blame. A black woman with an advanced degree is more likely to lose her baby than a white woman with a less than eighth-grade education.

"The Endling: Watching a Species Vanish in Real Time" records what is probably the last days of the vaquita—the world's smallest porpoise, so appealing that it's known as "the panda of the sea." It's going extinct on our watch, in the backyard of one of the richest countries in the world. We have had the power to save it for decades. But will we? Have we waited till it's too late?

Some of these stories are important because they illuminate the surprising way that science itself works. "The Brain, Reimagined" asks: Are some discoveries made because of accidents of history? Hardware limitations may influence what discoveries are made in the first place, discoveries that sculpt thought for decades.

"What If the Placebo Effect Is Not a Trick?" challenges the thinking behind how we evaluate how medicine works. What if the thing that heals us is not a particular drug or therapy, but a physiological response to an act of caring?

And some of these stories simply delight us by sharing the joy of discovery—even when that discovery is an ordinary, everyday miracle of a tiny piece of prairie in Iowa, as we see in the evocative writing in "Little Golden Flower-Room: On Wild Places and Intimacy."

This book is a cabinet of wonders. You'll learn why, despite leaps in sophisticated technologies, paper jams are still such a plague—and why fixing them offers "the ultimate challenge," demanding a working knowledge of physics, engineering, computer programming, and interface design. You'll learn that the answers pronounced by the Oracle of Delphi were probably induced by natural, psychoactive gas. You'll discover that cell membranes are like quartz watches, able to convert mechanical forces to electrical pulses, and vice versa. Wow! Facts like these make me love our world even more.

And that love, I think, is why I'll never forget those kids at the Boston March for Science. Theirs were faces shining with what Albert Einstein called "holy curiosity"—the natural inclination to question and to probe, searching for the truth about the way the world works.

At a moment in time when many forces conspire to suppress science—particularly about climate disruption and pollution—the stories in this book are part of the antidote. Excellent science and nature writing has the power to spark our holy curiosity, and to ignite our love—for our planet, for its creatures, for each other. My hope is that you emerge from reading these pages transfixed by wonder, bolstered by knowledge, and inspired to help heal our fascinating, fractured world.

SY MONTGOMERY

*The Best American Science
and Nature Writing 2019*

PHILIP BALL

A Compassionate Substance

FROM *Lapham's Quarterly*

WATER ATTRACTS TROUBLE. Time and again this ubiquitous and vital substance becomes the subject of controversial claims. The latest is about "raw" or "live" water, consumed directly from natural springs with no treatment or purification.

It's largely a Silicon Valley thing. About $30 will buy you five gallons from the Oregon-based company Live Water.

Sure, raw water might be full of other stuff like bacteria, algae, and minerals. But these, say devotees, are good for us—unlike the antimicrobial agents and additives in tap water or the plastic additives leached into bottled water. Fluoride, added to tap water for dental health, has a particularly long history of health scares and conspiracy theories; in the 1950s some said fluoridation was a communist plot to undermine the health of Americans. Raw-water advocates contend that fluoride is neurotoxic even at very low levels, although there's no evidence of that.

I'll happily attest that spring water straight from the source can be splendid—if you have the right source. But no law of nature prescribes it free from nasty pathogenic bacteria such as shigella or parasites like giardia. "I can't stress enough how many lives have been saved because of functioning water and wastewater treatment facilities," writes Gail Teitzel, editor of the journal *Trends in Microbiology*.

At any rate, raw water hardly qualifies as the "purest substance on the planet," as it has been advertised. If that were so, whither the assertions that water is "so much more than just H_2O molecules" and that purified water (here called "refined H_2O," which is entirely harmless) is "dead"?

It's all so confusing. Notions of *purity* seem to have come un-stuck from any conventional definition of that word. But where water is concerned, such dissonances shouldn't surprise us. While clothed in scientific-sounding language, the rhetoric of the raw-water movement taps into an ancient discourse about the values of water. It is indeed "much more than just H_2O molecules," because it has nurtured human civilizations throughout time. It is a sacred substance, and in their confused and sometimes misleading way, raw-water enthusiasts are at least attempting to honor that legacy.

When science advances in ways that challenge old beliefs, what results is often not simply the displacement of one by the other. Some beliefs don't persist just because they haven't yet been dis-proved. When they serve a deep psychological purpose, they're likely to evolve to accommodate new information. Water demon-strates this again and again.

It is the original purifier. By baptism it cleanses the believer for a life reborn; blessed by a priest, it confers protection from evil. The tensions of the Israeli-Palestinian conflict have been exacer-bated by water issues not only via restrictions placed on Palestin-ian access to aquifers in the West Bank but also by the collision of water values: some Muslims, for whom the ritualistic use of water is of profound importance, regard Israeli use of water resources as profligate and negligent. (Of course, Jewish tradition also advises respect for water.) In 1871 the anthropologist Edward Burnett Tylor described how water's sacredness seemed to make it invul-nerable to contamination in the Middle East: the faithful Persian "may be seen by the side of the little tank where scores of people have been in before him, obliged to clear with his hand a space in the foul scum on the water, before he plunges in to obtain ceremo-nial purity." It would take a powerful belief in water's self-cleansing propensity to keep Hindus taking purifying dips in the dreadfully polluted Ganges.

Science, you might think, can clear (if not clean) these things up. A good chemical and microbiological test will reveal exactly which noxious substances a sample of water might contain and in what doses. But scientific studies have sometimes supplied an apparent rationale for sustaining the mystical attributes of plain old H_2O —such as the "memory of water."

At a rather strange conference on water in 2004, I met the

French immunologist Jacques Benveniste. He was charming, witty, and seemingly comfortable in his role as scientific pariah. Benveniste was convinced that he had been "punished" for the embarrassment he had caused the editor of *Nature*, the august science journal that published his paper on the so-called memory of water in 1988.

I had joined the editorial team at *Nature* just weeks after its publication, and so had a ringside seat for the entertainment that followed. Far from opposing Benveniste's paper, *Nature*'s editor, John Maddox, had pushed it into print against the instincts of his staff. If Maddox had a feud with Benveniste, it was certainly not an establishment-versus-outsider conflict but was more a spat between two mavericks. Maddox's "investigation" of Benveniste's research after the paper was published was unconventional, to say the least: he traveled to the French lab to see attempts (ultimately unsuccessful) to replicate the findings, accompanied by a specialist in scientific fraud and the professional magician and "pseudoscience debunker" James Randi.

Benveniste's team had reported studies of one component of our cells' immune response: the release of the chemical histamine from certain types of white blood cells when they encounter an allergen. The researchers could provoke this response using a protein molecule called an antibody that attached itself to the cells' surface. But because the cell response sometimes seemed out of proportion to the amount of antibody present, they tried systematically diluting the antibody and seeing how the cells reacted.

What they found made no sense according to the standard principles of chemistry. As dilution steadily reduced the amount of antibody, the activity of the cells didn't fall off but instead kept resurging. At first glance this rise and fall looked random, but Benveniste and colleagues discerned a pattern within it: an approximately regular oscillation of activity reappearing at certain intervals of dilution. What's more, this behavior persisted up to an extremely high level of dilution: so dilute, in fact, that there should not have been a single antibody molecule left in the solution. The French team concluded that somehow the water was "remembering" the presence of the antibody long after it had been diluted to nothing.

The controversy caused by the paper's publication in the world's most prestigious scientific journal wasn't just about the apparent clash with what all scientific experience would lead one to expect. Biological activity in a solution of some ingredient diluted to the vanishing point seemed to vindicate the claims made by homeopathy, which employs ultrahigh dilution of the "active" compound.

Ever since its introduction in the mid-18th century, homeopathy has defied conventional scientific wisdom. Its inventor, the German physician Samuel Hahnemann, claimed that it cured disease by treating "like with like." The cure comes from an ingredient that, if administered in more than a "homeopathic" dose, would cause the disease or something like it. Homeopathic remedies were prepared from some highly toxic substances, such as arsenic and belladonna, but such was the dilution that they caused no ill effects. No study has ever shown convincing evidence that homeopathic medicines work beyond the placebo effect. But in Hahnemann's time, many conventional medicines were also useless—and sometimes so positively harmful that giving patients nothing but water may have seemed effective in comparison.

Some homeopaths still seek vindication in Benveniste's "memory of water." But his results were never repeated in a careful and convincing experiment. Unless the results were fabricated (which I do not believe) or bungled, something odd really did happen in Benveniste's experiments. But cell biologists know how unpredictable living cells are. The research is a textbook case of what the American chemist Irving Langmuir in 1953 christened "pathological science": it found a phenomenon that was barely detectable, hard to repeat, conflicts with all previous experience, and is explained with arbitrary invention and defended with ad hoc excuses.

I believe Benveniste was sincere in his convictions. But he made it hard to respect that sincerity. Rather than trying to systematically pin down how his strange findings had come about by simplifying his experimental system, he investigated ever-more complicated and uncontrollable manifestations of the alleged memory of water, testing his high-dilution solutions on rabbits and plants. He ended up claiming he could use radio waves to program biological activity directly into water according to what he called "digital biology." Water, he said, acts as a "vehicle for information"—a receptive medium of infinite capacity and versatility.

There is something wonderfully alluring about the idea of a universal medicine in unlimited supply. Water is, after all, essential for life, and a shot of it can be deliciously reviving. It's no wonder that it should be reimagined as a panacea. "We attribute to water virtues that are antithetic to the ills of a sick person," wrote the French philosopher Gaston Bachelard in his 1942 treatise on the poetics of water, *Water and Dreams.* "Man projects his desire to be cured and dreams of a compassionate substance."

There is no better encapsulation of water's sacred mythology than that conjunction: a compassionate substance. Chemists might balk at this juxtaposition: as though mere chemical compounds could have such attributes! But chemists do it themselves, recognizing the mythic dimension of gold when they call it a "noble metal" (they quietly redefine *noble* to mean "unreactive," but don't be fooled). Some material components of our world have, through their constant presence in history, acquired psychic associations. Bread and grain are more than sustenance, air more than a mix of gases, diamond more than an optically transparent solid. Water is indeed not just H_2O.

That's true even in chemical terms—and therein lies water's potential for spawning pseudoscience. For one thing, water is a fabulous solvent; things dissolve in it so readily that it's all but impossible to keep them out. Water soaks up oxygen and carbon dioxide from the air, and that makes plant and animal life possible below the ocean surface. Even the atoms in water itself aren't all united as H_2O molecules. At any moment, a small proportion of them have fallen apart into electrically charged fragments, hydrogen and hydroxide ions; an excess of one or the other turns the water acidic or alkaline, respectively.

Besides, the simple chemical formula H_2O hides a formidable complexity at the molecular scale. Water molecules may form weak chemical links with one another, called hydrogen bonds, and via these unions—which are constantly being made and broken—the molecules in liquid water become joined into a gigantic network that is constantly shifting, breaking, and reforming, arranging the molecules into temporary structures and clusters with less randomness than is seen in many other liquids.

Water isn't unique in forming hydrogen bonds, but it is the only common substance that can be joined by these gentle, frangible molecular handclasps into a three-dimensional network. Most liquids are little more than a disorderly scrum of jostling molecules.

But water is delicately poised between order and disorder, constantly adopting a defective version of the framework structure that, in ice, immobilizes the water molecules into crystalline regularity.

It's extremely hard to find the right scientific language to describe this dynamic state, and molecular scientists who study water still debate how to do that. This leaves an ambiguity about the structure of water that allows superficially plausible pseudoscience —and its relative, pathological science—to slip in.

In the mid-1960s a team of scientists in the Soviet Union led by the distinguished physical chemist Boris Deryagin reported they had discovered a new type of water that formed when it was contained in very narrow glass capillary tubes. This stuff didn't flow like the ordinary liquid but had a gummy consistency. They called it "anomalous water."

Soon other groups reported experiments that seemed to verify the claim, and it was suggested that this new form of water might arise from a more robust form of hydrogen bond linking the H_2O molecules into a kind of polymer. It became known as polywater.

Yet these ideas didn't fit with standard thinking about what water should be like, and they were hotly contested over the next several years. Eventually most researchers accepted that what the Soviets had seen was a solution contaminated by impurities, perhaps silicate material leached from the tubes' glass walls—"polycrap," as one critic put it. By 1973 Deryagin himself had capitulated to this interpretation, and polywater was dismissed as bad science—or another example of pathological science.

The furor about "cold fusion" is yet another pathological episode that hinged to some degree on the real mysteries of water's properties. In 1989 two chemists claimed to have conducted energy-releasing nuclear fusion on a laboratory workbench in a beaker of dissolved salts, drawing strength from old ideas about water as fuel, which also sustain perennial tales of vehicles and engines that run on water alone.

These theories are part of an old tradition. In some ways it harks all the way back to the transmutation of water by holy blessing to make it a potent prophylactic against the devil and a solvent for purifying and washing away sin.

In the 18th century, the German physician Franz Anton Mesmer offered "cures" involving the insertion of magnetic rods into baths in which his patients were immersed. Water seemed to become a conduit for an "animal magnetism" that Mesmer claimed to be able to command and channel. The belief that water can be altered by magnetism persists. You can still buy magnetic devices that allegedly counteract the effects of hard water and prevent scaling of kettles. There's no reason to suppose they work and no scientific argument for why they should. Nevertheless, I was once assured by a respected physicist that stirring a glass of water with a magnetic rod changes its viscosity and mouthfeel.

Water energized to some peculiarly "vital" state is an old dream, too. In the 1920s and 1930s, an Austrian forest warden named Viktor Schauberger claimed he could make what he called "living water" by sending it down specially constructed channels to activate it with the energy of vortices created by the flow. He developed a theory for how these vortices could release energy by inward bursts he called implosions, which could develop a thrust that might power flying devices. It was rumored that Schauberger was forced to work on secret weapons for the Nazis during World War II to exploit this principle. After the war, entrepreneurs brought him to the United States, but he returned to Austria, convinced that the military was trying to steal his ideas.

New scientific understanding simply creates new niches where these old notions may flourish. I once spent a couple of hours with a very famous pop singer hosting a group of Russian scientists in her elegantly minimal London residence while the scientists sought to persuade me that they had developed some manner of treating water—I think it involved electromagnetic fields —to "neutralize radioactivity" and thereby decontaminate the lakes around Chernobyl. When I explained to the assembled company that radioactive decay does not really respond to this sort of intervention, I was offered an explanation based on quantum mechanics. I have heard nothing of this solution for radioactive decontamination since, but the technique ("quantum resonance technology") resurfaced for the production of yet another brand of health-giving, and expensive, bottled water, Kabbalah Water. I sense a pattern here.

Supercharging water to confer restorative virtues is a recurring theme in the mystical pseudoscience associated with "hydration." The company behind Penta Water—a bottled water produced in Southern California—originally claimed it had been "restructured" at the molecular level, which somehow produces better hydration. They no longer advertise that idea but instead boast of a formidable series of purification steps that, for fans of raw water, must leave Penta Water as dead and sterile as could be imagined. (Ultrapure water, used for industrial processing of semiconductors, is in fact said to taste bitter, and to leach minerals from the body to a degree that won't do your health any favors, even if it's not as lethal as it is sometimes claimed.)

But high tech is not always required to "improve" water. The Japanese writer Masaru Emoto claimed in his best-selling book *The Hidden Messages in Water* (released in English translation in 2004 and popularized the same year in the film *What the Bleep Do We Know!?*) that human consciousness can influence water at the molecular level thanks to the transmission of emotional "energies" and "vibrations." The effects, he said, can be seen in the transformative influence of Buddhist prayer on the forms taken by water as it crystallizes into ice. Music has this effect too: crystals exposed to Elvis singing "Heartbreak Hotel" split in two.

This is indeed industrial-strength water pseudoscience. But it has traction because—whether knowingly or not—it was planted in such rich soil.

Nothing better illustrates the coexistence of old beliefs and new science in the study of water than the way many of today's water companies still employ dowsing, with handheld divining rods, to locate underground sources such as pipes or aquifers. When the British science blogger Sally Le Page revealed the ongoing practice in 2017, it sparked a furious debate about the value of dowsing. Perhaps the most revealing aspect of the affair, however, was the merry insouciance with which water companies and their engineers admitted to sometimes taking up the bent rods—as though this was simply a good, old-fashioned, and cheap alternative to satellite-surveying methods. They use dowsing, they explained, because it works for them.

But as the old adage goes, the plural of anecdote is not data, and the most extensive tests of dowsing under scientific conditions, conducted in Germany in the 1980s, showed that it located water sources no more reliably than chance. This is hardly surprising, for there is no known physical force or mechanism that could explain why water flowing underground or under a floor should cause rods clasped in the hand to move. The only plausible reason why water might be found this way with better than chance efficiency is that the dowser has developed a good intuition for signs of possible water sources (the nature of vegetation, maybe?) and subconsciously causes the rods to cross.

The tradition of water divining is ancient; the Roman writers Pliny the Elder and Vitruvius mention it. But it was not until the 16th century that Georgius Agricola, a German expert on mining, mentioned the use of handheld rods or twigs for earth divining; he offered a skeptical appraisal of how miners used it to locate mineral and ore veins. There are records of the use of twigs to find water from around the same time.

Dowsing, like astrology, retains aspects of a quasi-animistic vision of the natural world, with roots in the early Renaissance tradition called natural magic. The world of the natural magician was a web of hidden (occult) forces woven by God with design and purpose. This design was thought to be imprinted on nature and could be "read" by those who knew how—so that, for example, herbs with particular healing powers resembled the parts of the body they could be used to treat. In such ways, the cosmos is imbued with properties that make human habitation possible. Our existence within it is no mere accident, for our very presence awards it meaning.

That was doubtless a consoling thought when Copernicus's heliocentric theory threatened to dislodge the Earth from a central position in all creation. But this notion also has a modern scientific equivalent, and you might not be surprised to hear that it makes use of water. In 1913 the American biochemist Lawrence Henderson argued that water seems uniquely "fit," in the Darwinian sense, to act as the solvent of life. "Water, of its very nature," he wrote in his book *The Fitness of the Environment,* "as it occurs automatically in the process of cosmic evolution, is fit, with a fitness no less marvelous and varied than that fitness of the organism which has been won by the process of adaptation in the course of organic evolution." In water's unusually high capacity to hold heat, keeping lakes and oceans at a stable temperature through the seasons; its large "latent" heat of evaporation and condensation, which helps redistribute heat across the globe as the sea surface evaporates and rain clouds condense; its power as a solvent delivering minerals to organisms; its large surface tension, which lets it rise high above the ground in the capillary systems of plants; and other properties that mark it out from other liquids, water seemed, in Henderson's view, just what is needed to sustain life on our planet.

There's a clear invitation to see this as an "argument from design" for the divine creation of the cosmos. Henderson refrained from any direct statement about whether water is indeed "purposeful" in this sense. But scientists today continue to ask: Is water indeed somehow uniquely suited to acting as life's matrix—or does it just look that way to beings adapted to living in a watery world?

Here again the old idea that water's agency is somehow bound up with human existence finds new apparel from the science of the times. If water had an anthropocentric "purpose" in the 20th century, it would have to be expressed in the languages of physical chemistry and Darwinism. That Darwinian theme is echoed today in the claim of raw-water enthusiasts that we are "adapted" to the waters of natural springs, not those artificially purified of all other ingredients. There's no factual basis for that assertion—there would be no adverse consequences from drinking only filtered water, but you'd suffer plenty from drinking water that natural selection has populated with microorganisms—although it makes sense against the backdrop of a belief in a teleological universe.

How dismayed should we be by the survival of mystic ideas about water? While there can be a hectoring edge to the advertisements warning that you will not be properly hydrated without an expensive brand of bottled water, most people are unlikely to be taken in by it. Rather, such fads and fashions are typically just one expression of a lifestyle choice. Beliefs about water here serve a symbolic function: they express, in what might sometimes be a clumsy and garbled way, an ancient sense of why water is important beyond its chemical formula.

And it is, of course. All cultures have a tradition of respect for water because that is how they survived. Sacred springs exist in many cultures, and our stories and philosophies encourage care of water resources. "The highest good is like water," wrote Lao-tzu in the foundational text of Chinese Taoism. Water offers a metaphorical model for good conduct both in that tradition and in Confucianism. The Hani, Deang, and Naxi ethnic minorities in Yunnan Province in China fetch water at the dawn of the new year to bring good luck and prosperity. Pools and wells all over Southeast Asia are said to be guarded by dragons, as though to say, *You abuse water at your peril.*

The price we pay, though, for these important and indeed sacred associations of water is that they may distort, even traduce, our scientific understanding of this decidedly strange substance. This is not, as is often implied, because New Age mystics, health gurus, and Gwyneth Paltrow are drowning out the voices of sober scientists. For as we've seen, science is itself vulnerable to our deep-seated intuitions about water. We haven't seen the last of strange "water science."

The fact is that, when we study water, we study dreams and myths. As Bachelard discerned, we conjure up reveries of birth, of milk and blood, of change and flux, of healing and succor, but also Poe's nightmarish maelstrom and the deathly tug of the Styx. "One drop of powerful water," he wrote, "suffices to create a world and to dissolve the night." No microscope or spectrometer can contain all of that; it leaks off the slide and out of the frame, and flows everywhere.

REBECCA BOYLE

The Search for Alien Life Begins in Earth's Oldest Desert

FROM *The Atlantic*

MARY BETH WILHELM'S right arm shot out of the passenger window, and the convoy of vehicles behind her crunched to a stop. Spread among two trucks and an SUV, her four colleagues and I squinted and craned our necks, wondering what had caught her eye in the colorless wasteland of the northern Atacama Desert. She opened the door of our SUV and hoisted herself out.

The temperature was somewhere around 90 degrees Fahrenheit, and the midday gales were already picking up. Nothing beckoned. Nothing but tan spread out before us in every direction. I steeled myself against the wind, got out, and looked around.

Behind our convoy and down a hillside was the great salt flat called Salar Grande, a scorching, parched expanse about the size of San Francisco Bay. Ahead of my silver SUV, I saw a rolling Martian world of sand-colored rock spread beneath a blazing blue sky. No mosquitoes buzzed near our ears; no birds flew overhead. Wilhelm walked a few dozen feet away from the convoy, stopped, and stooped. Then everyone saw it.

A pebble field roughly the area of a two-car garage was dappled with chartreuse flakes: lichen. The first life we'd seen in days. Wilhelm crouched in the heat and squinted, flashing her hot-pink eyeshadow. She scooped some rocks into a sterile canister.

Later, Wilhelm would ship the rocks to her lab at NASA's Ames Research Center, in Silicon Valley, where she works as an astrobiologist. She would scrape bits of lichen from the rocks, liquefy them, and sequence their DNA. She would do the same for microbes she collected from ravines, and preserved cells she scraped from salt rocks. In many places in the Atacama, such hardy creatures are the only life forms, and Wilhelm and other scientists think that they might be similar to the last surviving life on Mars—if Martian life ever existed.

The odds seem pretty good that we will find extraterrestrial life, someday. Everything Earth's history has taught us suggests that once life takes hold, it is really, really hard to snuff out. Almost everywhere that astrobiologists like Wilhelm look—from sunless caves in the Italian Apennines to hydrothermal vents on the seafloor—they find something respiring and reproducing. Life is everywhere on this planet. Life can handle a lot. And yet we remain, so far, alone. No one has visited. Nobody has called. Nobody has turned up in our telescopes, our robotic instruments, or our collected space rocks.

In the Atacama, whose waterless, wind-worn landscapes mimic the surface of Mars, Wilhelm and her sunburned crew are mapping the edges of life's domain, learning what life needs to survive, form communities, and produce future generations. And they are approaching the question they most long to answer: Are we really alone in the universe?

On every leg of my journey to the Atacama, the density of life dropped as the heat intensified. I traveled from wintry St. Louis; to Dallas; to late-summer Santiago, Chile; and then to Antofagasta, a small coastal city creaking under growth fueled by lithium and copper mining. From there, loaded with groceries, outback supplies, and Chilean wine, Wilhelm's team and I drove north up the coast toward Salar Grande.

The Atacama Desert stretches 600 miles south from the Peruvian border, nestled between the Pacific Cordillera and the Andes, "a cross extended over Chile," in the words of the Chilean poet Raúl Zurita. Some parts of it are so devoid of life that their microbe-per-inch count can compete with near-sterile hospital surgical suites. Some areas of the Atacama, Earth's driest nonpolar desert and the oldest desert anywhere, have been rainless for at least 23 million years, and maybe as long as 40 million years. Carbon cycling happens on timescales of thousands of years, comparable to Antarctic permafrost and places deep within Earth's crust; the Atacama contains some of the most lifeless soils on the planet. The Atacama is one reason that Chile has become a haven for astrobiologists and astronomers: its pristine dark skies offer an unparalleled view of the stars, and its depleted desert offers a peerless lab for studying the dry limits of life, including how life might survive among those stars. And honestly, it just *looks* a lot like Mars. It is the closest that these astrobiologists will ever get to the planet that occupies their grant proposals and their imaginations.

I'm neither an astrobiologist nor a professional astronomer, but I spend a lot of time thinking about Mars. I keep tabs on the robots spread across its surface and in its orbit, and sometimes I check their nightly photo downloads. The Atacama is not a giant leap from the Mars of my mind. As I drove up the coast, I found the view so much more like Mars than Earth. There are no palm trees or tourists or bleating gulls. There is nothing but brown, tumbling tanly down the hills, darkening to chocolate inside shadowy ravines and runnels, bleaching to an impoverished shade of cardboard, and crumbling into fine white beach before being swallowed by the cobalt hues of sea and sky. With no trees or succulents or even a blade of grass—not a smidge of green—the only disruption in the brown is a strip of asphalt, Ruta 1. With my cruise control set and David Bowie blaring, I pictured myself driving through Meridiani Planum, a vast equatorial Martian plain, en route to visit the *Opportunity* rover. The only reminders of other humans were the grim commemorations of car-wreck victims: almost every mile of Ruta 1 is marked with roadside shrines to the dead.

Four hours north of Antofagasta, our convoy peeled off Ruta 1 and turned east onto a dirt road. Alfonso Davila, a soft-spoken NASA astrobiologist who also works at Ames, was in the lead vehicle. He was a calming presence, quintessentially Spanish in his affect and manners. He gently translated the language and the landscape for me after I arrived, and I found him easy to trust. As he approached a featureless hill, I looked in vain for a tunnel. I followed his cherry Toyota Hilux up the hill, then watched him wrench the truck 180 degrees around a bend and up an even steeper incline. The switchback was dense with sand and the loosest gravel I have ever tried to clear—and I grew up in Colorado. I was certain I would get stuck, and I hung back. Davila saw me hesitate and reversed, compressing the dirt for my SUV. He hopped out and ran over to my window. He was wearing a black San Francisco Giants hat and, improbably, a tan corduroy shirt. I thought, *He must be really hot,* and also, *I am definitely going to die.* "Just, um, gun it," he told me, in his Catalan accent. I waited for him to clear the path. I gunned it.

The hill was so steep that no road was visible past my dashboard; my SUV might as well have been a rocket ship launching me off the Earth. I felt my back left wheel slide and start inching toward the edge of the roadway, such as it was. Wilhelm, cool as a cucumber in my passenger seat, commented that the view was incredible. My palms were soaking wet. I wiggled the steering wheel and gunned it again, and the SUV lurched forward, finding purchase. Clear! I still couldn't see the road beneath my wheels, but I could see the hillside to my right, and I hugged it until we cleared the summit. I let out a half-sigh, half-cheer "Whoo!" as we turned east, over the cordillera, and down into the shimmering Salar Grande.

From a distance, Salar Grande looks like a sea of pebbles, glistening underneath the bluest possible sky. Up close, the pebbles resolve into a field of fissures and slabs called polygons, which form only in the Atacama's ultra-dryness. They are edged with countless creamy, windswept nodules, as smooth and round as carved marble. They are made of pure salt, but are deceptively slippery, so the six of us moved slowly after leaving our trucks at camp.

We could not see it, but we were walking on a forest. The nodules, called halites, harbor a community of microbes. They include a member of Archaea, the third domain of life, named *Halobacterium salinarum.* The salt also contains a few species of cyanobacteria adapted to ultrasalty conditions. The bacteria eke out a living by drawing energy from sunlight, which they absorb through the translucent salt knobs. The archaea and other bacteria feed on those cells.

They might be the only permanent residents of Salar Grande. "These are the last outposts of life in the Atacama," Davila said. "These are the last outposts of life on Earth, in terms of dryness." And they might be similar to the last permanent residents of Mars. Today, the fourth planet has environments like the Atacama, with similar polygons and salt flats and humidity levels.

Salar Grande was once a coastal inlet, much like today's San Francisco Bay. It dried up between 1.8 and 5.3 million years ago, leaving behind a salt flat between 225 and 300 feet thick. The salar is therefore an analogue for the last time Mars was habitable, after Mars's oceans, if there were any, dried up, when Martian ecosystems became concentrated in smaller places. And, like Mars itself, the Atacama is a glimpse into Earth's own future. One day, billions of years from now, all of Earth may resemble this parched land of fissures and knobs, after our own oceans boil away, after the last trees fall, after the algae are all that is left of us.

"In the beginning," Davila said, "there was bacteria. And at the end, there will be bacteria."

We stopped after walking the length of a football field or two into the salar. Wilhelm, wearing a Sia concert T-shirt and yoga pants, began laying red rope in a straight line. Davila helped, wearing his long-sleeved corduroy shirt. He didn't seem to mind the heat.

"He's Mr. Atacama," Wilhelm reassured me, and laughed. He travels here twice a year, at least.

"I'm trying to be," he replied.

As they worked, Davila compared notes with Jocelyne DiRuggiero, a French biologist at Johns Hopkins. I chatted with Kathryn Bywaters, a biologist at the Ames-affiliated SETI Institute. I kept my gaze high to keep sweat from pouring into my eyes. I started to worry I would contaminate the site. In this part of the Atacama, fog is the only water source keeping the microbes alive. It rolls over the coastal range at night and temporarily moistens the salt, which awakens as a shape-shifting blob in a phenomenon called deliquescence. On a previous expedition, Davila and others had wrapped a rock in gauze, and returned to it a year later to see what had happened. The gauze had been swallowed, as the salt rock continually melted and resolidified around it. "The salt is very dynamic," Davila said, cracking one rock open with a hammer.

Inside were bright layered stripes, in every shade of pond scum: pine, jade, pea soup. DiRuggiero hypothesizes that the pigments are chromatic adaptations, essentially a form of sunscreen that protects the microbes from DNA-stripping solar radiation. She thinks that the halobacteria are changing their color based on the wavelengths of light they receive within the rock. Closer to the surface, they are lighter; down below, they are darker. Or it could be that the layers are made of different organisms entirely. Only DNA will tell, and to get that, the scientists have to bring samples home.

DiRuggiero also planned to collect viruses, and later, to drill into a rock and load it with sensors to track what happens inside it over one year.

The researchers wanted to figure out what lives in the Atacama kiln, how the communities work together, how they signal to one another using molecules such as RNA, and how they are adapting to the Atacama's changing climate. Their story may echo on Mars: *Here lie the last leftovers, and this is what happened to them.*

DiRuggiero passed me her rock hammer, and I used it to smack my own Seussical tuft of white. The meringue broke in two, and I saw small open pockets within the salt crystal, which were colonized in green. The other half of the rock stayed attached to the ground, its insides now exposed. I felt guilty for cleaving this teeming microbial city. "You are a destroyer of worlds," Bywaters joked.

As our first night approached, we set up camp and cooked dinner: beef tacos with the best avocados I have ever had. While we ate, the golden hour finally brushed the brown away, and suddenly the Atacama was all color. Hilltops glowed pink and mauve. The far horizon darkened from cobalt to purple. The salt rocks of the salar became amber jewels. Stars started emerging well before they had any right to, well before the sun had vanished into the Pacific. To the west, to my relief, I found the waxing crescent moon, but it looked wrong: I had forgotten that it hangs backward in the southern hemisphere.

After sunset, the wind shut off as abruptly as if someone had closed a door. By the time everyone turned in for the night, the air was utterly still. The emptiness seemed to press down on me. I thought of home, where even on the darkest, quietest night, sensory reminders of life are ever present. I see airplane lights blinking 30,000 feet above my yard. Leaves rustle in the wind or under animal feet. In my house, I hear my family; I see dog fur piling up in corners. But I had said goodbye to all that to sleep on Mars, where there is nothing.

I lay awake waiting for the moon's scythe to sink below the western horizon. The Milky Way crossed my open-ceilinged tent, and I found that I could see by its light. To my right was the Southern Cross. The Chilean poet Gabriela Mistral instructs: "Lift up your face, child, and receive the stars."

If there were ever life on Mars, and if it could see, it would have witnessed a similar sky. I looked for the Centaurus constellation, but it had not risen yet. The centaur's raised front hoof contains the nearest star to our own: Proxima Centauri. Astronomers think that star has a planet. Each of the other numberless, anonymous glitterings might have one, too. Some of those planets might be warm, nestled close to cool stars or safely distant from fiery ones, and some of those warm planets might have water. Add an atmosphere, and one of those warm, wet planets might be hospitable to life.

I am comforted by this possibility, but I know that for others, the certainty of life elsewhere would be world-rending, faith-shattering. While some people of faith question the possibility of extraterrestrial life, others may question their faith in the face of its discovery. Some may associate its discovery with doom. Growing up in the Catholic Church, I was never taught that aliens were doctrinally impossible, though theologians have long debated how a discovery would affect church teaching. Catholicism's relatively neutral stance may well be informed by its institutional embrace of astronomy; in 1891 Pope Leo XIII founded the Vatican Observatory so that "everyone might see clearly that the Church and her Pastors are not opposed to true and solid science, whether human or divine." The Vatican's current chief astronomer, Guy Consolmagno—an MIT graduate, a Jesuit, and a practicing planetary scientist—assured me that Scripture accommodates aliens. He cautioned against imposing our own understanding of religion onto the science that might find those aliens.

"The only theological point is to remember it's always a mistake to think there's something God can't do," Consolmagno told me, "including making parallel universes, much less other worlds with other inhabitants."

I woke up chilled and thirsty, and eager for the sun—"the bitter sun, Lord Burn-All-Things," as Mistral wrote. DiRuggiero's graduate student, German Uritskiy, volunteered to hike another couple of miles into the wastes to retrieve a GPS device that had recorded the salar's temperature, humidity, and other vitals for the past year. He volunteered in part because he is color-blind, meaning he was hard-pressed to find our quarry in the salt forest: he is an algae-hunting biologist who cannot see green. His absence meant DiRuggiero needed my help. I slathered sunscreen over every inch of my skin, swaddled myself in a white linen scarf, and stepped into the salar.

DiRuggiero gave me a quick lesson in how to scrape bits of compressed salt into a sterile sample bag, and handed me her hammer. I pounded on a nodule until it cracked apart, then donned purple sterile gloves, spritzed my knife with alcohol solution, and picked up a chunk. I looked for the stripe of color that would indicate life within. I saw nothing. I turned the snowy sample toward the sun, then cupped it in my palm to shade it. In different light, with my sunglasses on and off, I still saw no evidence of life. I wanted so badly to see a line, even a faint one, but this rock, it seemed, was barren. I put it down.

DiRuggiero and the others had found life in Salar Grande before. Other researchers had even grown Atacama microbes in the lab. They expected to find life again, but also knew that difficulty was part of the point. They were trying to find the limit beyond which life becomes hard or impossible to sustain. It might be a temperature, or a moisture level, or a rate of change. If they could pinpoint some threshold, that might be a clue to life's minimum requirements.

This liminal state, between finding and not finding, is characteristic of astrobiology in general. We don't know whether we're alone, but we don't know whether we aren't. We do know that there is one planet with life, and on this planet, life is everywhere; because of us, we can be sure life in the universe is possible. If we don't find life on Mars pretty soon, or on Enceladus or Titan or Europa or Trappist-1b, that doesn't mean we should stop trying. But it's also possible that life has happened only once. We might be it.

I took DiRuggiero's hammer, stood up, walked a few feet, and thwacked another nodule. The hammer twanged in my hand and a foamy chunk of salt fell apart beneath. Dark green! I scrambled back to the sterile bags and started scraping. Uritskiy had told me that as long as I got a tiny bit of green, he could get to work; in the lab later, he would pour the salt into a vial, liquefy it, and purify the sample before running the contents through a DNA sequencer, to find out what was living in the rock and what it was doing. We collected samples from eight rocks, scraping both light and dark stripes, so DiRuggiero could test her chromatic-adaptation hypothesis.

DiRuggiero packed my samples into a cooler for their journey home before we left for Salar Grande's northern reach. After the lichen pit stop, we arrived at the second field site, which welcomed us with a dust devil. Everyone took iPhone videos as I imagined my rental car tossed like a tumbleweed. The scientists all seemed at home in the desert, despite its hostility. A few weeks before the trip to Chile, Wilhelm and Davila had returned from Antarctica, home to the world's driest desert, the McMurdo Dry Valleys. But Davila, who spends four weeks a year in the Atacama, finds the frozen continent a breeze by comparison.

"You set up a much more comfortable camp. There are lab tents, heated tents," he said. "Outside, it's extreme, but inside it's fine. Here, there is no inside."

As Lord Burn-All-Things beat down on us, DiRuggiero set up a tabletop light experiment. She took a rock from the salar, placed it under a sawed-in-half milk crate, and used a modular spectrometer to measure the colors of the sun. Then she got out the drill. Brushing her straw-colored bangs from her face, DiRuggiero set her jaw and squeezed the trigger. The rock fought back; like my halite earlier, this one was very dry, and firmer than usual. She pressed on, and drilled a hole halfway through the rock.

"We want to know what the organisms are seeing, the wavelengths of light and the intensity," she explained. A spectrometer probe nestled deep in the rock would tell her.

Later, she told me that the light inside the rock was red-shifted, meaning toward the infrared range of the electromagnetic spectrum. In her lab this spring, she and Uritskiy isolated cyanobacteria and sequenced their DNA and RNA. In April they identified genetic markers for an adaptation called FaRLiP, for far-red light photoacclimation. This summer, she grew the cells in white light and near-infrared, and in still-unpublished research she found that they are using a newly discovered pigment called chlorophyll f. This pigment, only isolated in 2010, allows some species of microbes to harvest energy from far-red light. On Earth, this light penetrates salt rocks in the Atacama. On a distant planet orbiting a faraway sun, this wavelength of light may dominate. So chlorophyll f, just like the salt-loving microbes of the salar, offers tantalizing hints of life's possibilities beyond this planet.

By the end of our stay in Salar Grande, my face looked like it was red-shifted, too, crisped by sun and wind. It was time to go.

The wind buffeted my SUV, keeping me company as I drove seven hours solo down the coast of Chile. "The hurting, healing, flying wind," Mistral wrote. The next morning, as I drove to the beach in Antofagasta, I passed a park and thought my eyes were tricking me. The green looked somehow off. I wondered, *Is that an accurate grass rendering, or is the simulation failing?* I had forgotten what plants should look like.

While DiRuggiero, Davila, Wilhelm, and the others collected fresh samples, their NASA colleagues were busy at another Mars stand-in a day's drive south, in a ghost town called Yungay. For a few weeks every year—it would be longer if the team had more funding—NASA engineers and scientists descend on a ramshackle five-room building near the ghost town's cemetery, on land owned by a copper mine. Their five-year project, the Atacama Rover Astrobiology Drilling Studies, is the space-robot version of the work in Salar Grande.

The decibel level was shocking after my desert baptism. It was like entering a sports bar during the Super Bowl after a weeklong silent meditation. In Salar Grande, I knelt like a penitent and quietly scraped green crud into sample bags. In Yungay, a laser-equipped metal box on wheels trundled around, noisily spinning its three-foot drill. In Salar Grande, a handful of us shared quiet dinners around a campfire; in Yungay, two dozen engineers—mostly, but not all, young men—passed tortillas down a loud, long table as they tried to make one another laugh. In Salar Grande, I basked in a silent shower of stars and saw my shadow under their light. In Yungay, an engineer stood in front of the rover and dabbed, making sure his friends could see his jagged lidar shadow on the laptop screen.

Yungay was hotter than Salar Grande, and less beautiful, though no less like Mars. The grit still covered every inch of me. The wind consumed me, and the sunlight was starting to make me feel manic. I felt as if I were hallucinating. But there were some upsides. A watercooler meant we could actually enjoy staying hydrated, and the University of Antofagasta Desert Research Station even had a shower. At the close of the NASA mission, Yungay would fall silent, and the building would soon be picked clean by local mine workers scavenging for home-improvement materials.

If the goal in Salar Grande is to uncover life's last holdouts, the goal in Yungay is to teach a robotic emissary to do it for us. Wilhelm's keen eye spotted the yellowish-green speckles of lichen from a moving truck. But no human scientists are going to Mars anytime soon. A robot geologist would be able to identify the algae-fungus composite lichen flakes only after drilling into the planet, bringing up a deep sample of dirt, liquefying and purifying it, and running some kind of peptide or DNA-sequencing test. This is tricky even for a human; from 100 grams of soil, Wilhelm extracts just 40 microliters, four-tenths of one gram, of organic material. From that minuscule sample, she can detect life and its makeup, but a robot will not be able to pipette.

"You have to turn your dirt into liquid, and liquid into data. The liquid is the technology gap we're trying to bridge," Davila told me.

The rover, nicknamed K-Rex, was designed to do all this. In Yungay, its mission was to drive where it was told, drill, scoop, run an analysis, and beep "Eureka!"

Like K-Rex, Mars rovers planned for 2020 and beyond will look for chemistry that cannot be explained without the busyness of life. But this is much more complicated than it sounds. Amino acids are abundant in space, so that would not be enough. Same with hydrocarbons. It's the type that matters: some organic molecules come in mirror varieties, like our hands; on Earth, life uses all left-handed amino acids and right-handed sugars to build proteins and DNA. If we sift the sands of an alien world and find an abundance of certain hydrocarbons, amino acids, and lefty sugars, then things could get interesting.

And yet—will that be enough? Chiral molecules are quite a long way from little green microbes, let alone little green humanoids. Getting to "Eureka! We are not alone!" is a lot to ask of a miniature car driving alone on a world 300 million miles from home.

The thing is, everything we know about Earth suggests that we should keep trying. When you take away almost all the water, add copious heat, eliminate all vegetation, and turn up the bitter sun, it is still possible to find something alive, even a whole community of living things. Yes, we're on Earth, a rock that has spent 4 billion years crawling with creatures desperate to survive and make copies of themselves. Nowhere else that we know of has such a history. If life took hold on Mars in its watery past, or if microscopic life persists on the moons of Jupiter or Saturn, it is not going to be easy to find its exhalations, or its remains. It will be even harder for the robots we send in our stead. But we have reason to hope. We know life arose in the universe once. We just don't know whether it can happen again. The only way to be certain, to end this state of unknowing, is to find life somewhere else, and to answer *yes*. Until then, the search will continue.

PETER BRANNEN

Glimpses of a Mass Extinction in Modern-Day Western New York

FROM *The New Yorker*

I BEGAN VEERING INTO the rumble strip somewhere near Binghamton. Traffic engineers are lucky that few people know what's on the side of the highway, but I had been spending time in the dangerous company of geologists—and now I was a rubbernecking menace. This time, I was slack-jawed and lurching on I-86, disoriented by a road cut that placed me somewhere at the bottom of an ancestral ocean 385 million years ago.

I had already had quite the trip. A short drive on the highway that morning had dropped me from the airless, snowcapped peaks that once towered over New York City, down through a sprawling delta in the tropics where the planet's first trees rose from the edge of the sea. But the Hudson Valley roughly marked land's end, and, by now, I had pushed off this secret coastline to head west, and offshore. The red earth that earlier bracketed the highway—rumors of ancient rivers on land—now gave way to gray, banded rocks filled with seashells, where stacks of seafloor piled up, millennia-thick.

"The farther west you go in New York, it's all marine fossils," a paleontologist told me before I left. "New York would have been facing into a great continental sea. All the way out to Ohio, it's all marine."

This upstate ocean poked out from under farmland, and crumbled from rock walls behind gas stations. In the Devonian period —hundreds of millions of years ago—it was filled with sea lilies, sea scorpions, armor-plated monster fish, forests of glass sponges, and patch reefs of strange corals. At night, these reefs were cast in shimmering chiaroscuro, inviting moonlit patrols of sharks and coelacanths. Where the water met land in eastern New York, dawn revealed fish hauling ashore on nervous day trips—slimy, gasping astronauts under a withering sun.

In the ages since, the tropical inland sea drained away, the continents merged and rifted, and the seafloor turned to stone. As fish conquered the land at last, the ocean was buried and forgotten. Era stretched to eon, and, where there were once croaks of stranded lungfish, sail-backed dimetrodons now groaned confidently across an arid supercontinent. A hundred million years later, these groans gave way to the wails of dinosaurs. And a hundred million years after that—still more dinosaurs. But, all along, in New York, the old seabed lay buried in darkness. Another hundred million years passed.

Then, not long ago, continents of ice planed off the state, removing these untold histories. As the ice retreated, meltwater carved gorges through the ancient seafloor, and sunshine fell on its depths once more. Humans arrived. The elephantine fanfare of the Ice Age went silent, and the highway department chiseled out a few more crags with dynamite. Now shells of a strange vintage tumble from the side of the road across the entire state.

This is the surprising inheritance of New York, where, from Albany to Buffalo—and just beneath the thin, photosynthetic rind of our world—an alien ocean planet, still groping toward the land, is frozen in stone.

In the eastern half of the state, and especially in the Catskills, the pulsing shoreline of this tropical sea wafts back and forth over millions of years—a collage of floodplains, lagoons, deltas, estuaries, and beaches, arrested in the strata and invisible to all but stratigraphers. Three hundred eighty-five million years ago, the first trees in the history of life sprouted along this forgotten coast. One day, after a storm, some of these coastal trees were buried by floodwaters, cast in sandstone, and packaged for safe passage to the far future. A century ago, these fossil tree stumps were uncovered by construction workers in Gilboa, New York, who were quarrying stone for a dam that would eventually drown the Catskills town. Now the village of Gilboa is under 120 feet of New York City tap water, and the planet's first forest is behind the high-security cordon of the New York City Department of Environmental Protection. But branches of these first trees, some of which washed out to sea, can be found elsewhere in the state: behind a Binghamton strip mall, in a retaining wall next to a Starbucks drivethrough, I found the ageless twigs embedded in boulders that had been quarried nearby.

The rise of these plants and their invasion of the land ended more than 4 billion years of continental desolation. The trees invited fish to come ashore and consider becoming tyrannosaurs, humans, and hummingbirds. But this lifestyle, severed from the sea, was still a daydream: the pioneering fish of the Devonian, and the halfway world they inhabited, claim more distance from the beginning of the age of the dinosaurs than the dinosaurs do from our own time.

This onshore world, then, was still a novelty—an absurd one even, for what had always been an ocean planet. And, in the Devonian, it was still provisional. I was interested in that much older, more confident planet long rioting under the surf.

After a few more hours of erratic driving, I dove in around Ithaca.

"It's paradise," the Cornell paleontologist Warren Allmon, who agreed to be my dive instructor for the day, said. Allmon killed his engine at the end of a dirt road overlooking Ithaca's Lake Cayuga.

"I mean, paleontologically, it's just paradise."

Allmon directs the Paleontological Research Institution in Ithaca, whose Museum of the Earth houses what it can of the spoils of upstate New York. It wasn't hard to convince him to spend the day cracking open local rocks, so we hiked down to a stream in the woods that was slowly excavating the ancient seafloor. A large, incongruous block of limestone diverted the stream, and I asked Allmon where it came from. He pointed to an imaginary spot somewhere in the treetops. Compared to the rocks at our feet, this errant stone was still millions of years in the future. And, in the other direction, buried 2,500 feet below us, were giant seams of sea salt from an ancient Persian Gulf millions of years in the past. But, before us, entombed in the banks of the stream, was a mucky tropical sea bottom, where thin, frangible layers of gray siltstone marked the passage of centuries.

"Geologists take it for granted that rock equals time," Allmon said. "I don't know of another experience that we all have in our daily lives where a solid substance represents time."

He took out his rock hammer and unsentimentally laid into the bluffs, splintering the rock into jagged slabs. The rocks revealed a spattering of tiny seashells, swirling burrows, and a hash of body armor and compound eyes. The seashells once housed marine worms called brachiopods, but the burrows were a mystery. As for the armor and eyes, they belonged to trilobites—vaguely technological sea bugs that thrived in the ocean for almost 300 million years. Before their eternity in the stone, these eyes caught glints of starlight dimly streaming through the murk.

"If you were snorkeling here, there would be really low visibility, not many waves, a storm or two every so often," Allmon said, gazing out at the woods. "Kind of like the Gulf Coast without a breeze, that kind of thing. I can't conceive that this was more than a hundred feet deep."

The mud in these turbid waters was delivered from an epic mountain range in the east, which was then hemorrhaging sediment under the assault of tropical storms and monsoon rains. The Yankee Himalayas were shoved into the sky by eastern New England, long part of an island chain that had rifted off an Antarctic supercontinent and traversed a southern ocean—and which was now crashing into North America. Disoriented by this jumbled geography, I was gently reassured, "We're still talking about a world in which Pangaea had yet to come together."

In the course of millions of years, as these mountains wore down, the dissolved rock flushed into the rivers and floodplains of an Appalachian Bangladesh, and on out into the ocean, where it settled on a sinking seafloor, piling up here in the middle of the state, almost two miles thick. Down the road, in Trumansburg, at Taughannock Falls, this tremendous pile of time is visible, if only in part, as a 200-foot wall of piled-up seafloor that frames a wisp of falling white water. Just downstream, at the more modest Lower Taughannock Falls, this giant stack of gray is interrupted by a startling platform of limestone. The dramatic change roughly marks the Taghanic Event—a mass extinction that razed corals, brachiopods, and squid-like creatures stuffed in elegant shells all over the world. It was one of almost 20 global mass extinctions in the history of complex life, a list that includes 5 cataclysmic outliers, when the planet nearly died, and one that might someday include us. The Taghanic Event was an ancient global-warming disaster, complete with rising seas and oxygen-starved oceans. This is how most mass extinctions unfold. It didn't quite achieve the peerless horror of the worst five Armageddons in Earth history, but, elsewhere in New York, the rocks do record one such doomsday.

As you drive farther west, and farther offshore, the shallow seabeds of Ithaca grade to outcrops of deep-water black shale that glisten with natural gas. The rocks are greasy with the organic residue of buried sea life, and are now fracked with great enthusiasm. This life sank to the bottom of putrid, anoxic seas that pulsed throughout the Devonian period, burying massive amounts of organic carbon, in wave after wave of extinction. Where West Virginians dig up forests from the Carboniferous period as coal, and Permian sea life spouts from Texas oil derricks, much of the natural gas we frack today comes from the smothering seas of the Devonian extinctions. Spying some of these ominous black shales rising from the side of I-90, and knowing that there was boundary in the rocks nearby marking one of the worst mass extinctions of all time, I pulled off at the nearest exit, walked into the geology department of SUNY Fredonia, and poked around for help. I ended up walking into the office of Gary Lash.

In 2011 Lash was named one of *Foreign Policy*'s "Top 100 Global Thinkers," earning the nod after estimating that the local Marcellus Shale was the largest unconventional natural gas reserve in the world—a discovery that toppled the board game of global energy geopolitics. The first commercial gas well in the United States, drilled into the marine shales of the Devonian in 1826, isn't far from Lash's office. It's in front of a dentist's office, marked by a plaque on a rock. A little farther down the road, these rocks crop out along Lake Erie, and roughly mark the culmination of an apocalypse 375 million years ago.

The great Devonian mass extinction remains something of a mystery. There were oxygen-starved oceans, fueled by an explosion of massive algae blooms—perhaps even driven by runoff from the land, as the emerging world of trees carried out their massive geoengineering project, greening the continents. Other research adds invasive species spread by surging seas, preposterous volcanoes, and extreme climate change to the chaos for good measure. Whatever form this destroyer took, it laid waste to 99.99 percent of the largest reef systems the world has ever known—the so-called "megareefs" of the Devonian, 10 times more extensive than our own. Trilobites, tentacled drifters, fish wrapped in heavy armor

—nothing was spared. Lash had his own grisly ideas about the disaster, one involving catastrophic methane releases from the deep and intense global warming.

"It was just a bunch of bad things all converging at once," he said.

He photocopied a map of the area and pointed me to the extinction exposure in neighboring Dunkirk, New York, on the shores of Lake Erie.

"I wish I didn't have classes, or I'd go down there with you," he said.

As I was leaving, he offered some mysterious parting words.

"Have you ever huffed some of the shale?"

He started laughing.

"When you get down there, take a piece of the black shale and break it open. Take a hit of that."

I followed Lash's directions and arrived where New York ends, at Lake Erie. It was an odd place. I parked in a field under a coal-fired power plant and strolled onto a sandy beach strewn with Russian zebra mussels and plastic. The beach was framed by hulking blocks of black shale, glimmering with a yellow, sulfurous sheen. *Here's the end,* I thought, noting the change in strata from green-gray to deep black, marking the final extinction pulse of the Devonian cataclysm. I picked off a flake of black shale, as directed, and cracked it open in my fingers. It smelled like gasoline.

On the drive up to Ithaca, something was going awry. Rolling hills bursting with green slowly dulled to an unearthly pallor, and road signs normally reserved for traffic alerts now flashed updates about outer space: SOLAR ECLIPSE TODAY. I stole a look at the waning sun through my windshield, at some ambiguous risk to my retinae. The missing piece of star was carved out by the same moon that once careened around a Devonian world, its ancient tug evident in the regular layering of seafloor in outcrops along the highway. In the Devonian the moon was 10,000 miles closer, and summoned surging tides that flung fish onto tidal flats, daring them to walk. As a result, the sun rose and fell 400 times a year, and corals—nestled in lagoons and fringing barrier reefs—registered this ancient astronomy in their skeletons. In Ithaca, Allmon showed me a head of coral from outside Rochester with yearly packages of daily growth lines numbering 400. *What was that day like?* I wondered, picking out a line. These reefs were annihilated by mass extinction, and it is believed that our reefs will also mostly die off in the second half of this century.

What came after the Devonian was everything: fish ultimately emerged onto a landscape already furnished by plants, and some of them even spread their fins and learned to fly. But what will come after us? In a hundred million years, the same cratered stone will still be tethered to our planet by space-time, hurling around it at impossible speed. Its moonlight will still shine on the creatures below, but whose eyes will gaze back?

CHRIS COLIN

This Sand Is Your Sand

FROM *Outside*

THIS IS A classic story of greed, nature, fear, change, large dogs, and violent golfing, but I didn't know that when it began. Nor did I know that I'd get sucked into the thing myself. All I knew was John Harreld walking to the water's edge on a summer morning in 2015 in Guerneville, California.

For neighbors in Harreld's quiet river community, 90 minutes northwest of San Francisco, the image of the then 43-year-old starting his day in the sand had become a regular, if provocative, sight. Several times a week, he'd plant himself in a folding chair overlooking a gentle bend in the mellow Russian River. Sometimes a buddy joined him; other times he sat alone. The job was simple: Enjoy the beach. Sip coffee. Maybe spot an otter slithering up the lazy current. Most of all, do these things in clear view of the camera hidden in the trees.

The man who'd hung the camera believed that Harreld was trespassing. Harreld believed he was reclaiming land that belonged to the people. On a literal level, that's all the two were fighting about: just 20 feet of gravelly beach.

But at the heart of the fight swirled bigger and fundamentally contradictory ideas. Remarkably basic questions about America itself were woven through their dispute. There would be no bloodshed—the same could not be said elsewhere—but the level of fury would nevertheless become shocking for the once friendly neighbors.

Harreld and his wife, Judith, live in a one-story home a short stroll from the river, with sunflowers and a tomato garden out front. He's slim and youthful, with crinkly Paul Rudd eyes and a jovial, laid-back air. When he's not working as a medical-device analyst, he tends to be diving, rock climbing, investigating shipwrecks, or showing elaborate Lego creations he made in his home's dedicated Lego studio. He's a detail guy. One of these creations is a scale model of Chichén Itzá so meticulously accurate that when you shine a flashlight on the balustrade, it mimics the equinox effect at the actual temple.

On this June day, as he walked down the dirt path to the beach for morning sentry duty, he was pursuing his newest hobby: being the Rosa Parks of obscure riparian law.

Fifty yards or so above this quiet stretch of the river is a vacation home owned by Mark O'Flynn, a lawyer from San Francisco. Nearly four years ago, O'Flynn posted his first NO TRESPASSING sign. Like many property owners, he had come to equate public access with broken glass, poop in the bushes, and bad music blaring from drunk strangers' speakers.

The problem, as Harreld saw it: the sign stood in flagrant violation of federal law. As he would explain to anyone who'd listen, the beach was subject to a public easement below a line called the ordinary high-water mark—a calculation roughly analogous to the average high tide. In 1981 the US Supreme Court confirmed in *Montana v. United States* that this easement trumps private ownership in navigable rivers. Navigability, in turn, is established by proving that the river has qualified as a "highway of commerce," used by ferries carrying tourists, say, or loggers floating felled trees downstream, both of which happened on the Russian River.

In other words, Harreld believed, that gravelly beach was everyone's to enjoy, regardless of what signs were posted.

Harreld had known O'Flynn socially. His house is a minute's walk away, and Harreld and his wife had been over for dinner. When he saw the sign, he emailed Mark to ask if it was his. Indeed, Mark replied.

There would be no more neighborly meals. Over the years that followed, a full-on cold war blossomed. Someone would run a shin-high line of wire across the path. One of Harreld's allies would remove it. O'Flynn would hang cameras in the trees. Someone would paint over the latest NO TRESPASSING sign.

And now, on this Thursday morning in 2015, Harreld and some friends approached the beach only to find the path blocked by "cut bamboo, tree branches and matts of algae," according to the June 18 entry in a detailed journal he's been keeping since the conflict started. The next day, the group began removing the makeshift barricade, and O'Flynn ran out of his house.

What ensued borders on slapstick, with O'Flynn, according to Harreld, "pulling the bamboo out of my hands, then dashing over to pull some out of my friend's hands, then dashing back to me." (I don't have O'Flynn's account; he won't discuss the dispute in detail.) Soon two sheriff's deputies arrived, but they declined to take action.

Perhaps at this point you're marveling at the amount of free time that middle-aged white men have. I marveled, too. But as I got deeper into the dispute, I came to see that this picayune squabble wasn't all that it seemed. Behind the folly of turf wars and the arcana of river law, a larger conflict was playing out, one rooted in a profound disagreement over how we think about nature and how we divide it.

All over the country, that disagreement has become a feature of the landscape. I wanted to see it up close—and to see what happens when someone tries shifting the status quo in a small way. Like many rivers around the country, the Russian is full of vacation homes and resorts claiming private beaches, ordinary high-water mark be damned. It seemed some civil disobedience was in order —or, rather, civil obedience, if John Harreld was correct. So I gathered together three friends and some boats.

"How do property owners feel about boaters landing on the beaches in front of their land?" I asked the woman at the canoe-rental place in Forestville, the next town over from Guerneville. It was June 2017, the beginning of the season for her operation.

She cocked her head kindly, the way you do when talking to an idiot.

"There's *mucho* public beaches to enjoy," she said. "I imagine canoeists wouldn't want to go up on anyone's private beach, would they?"

Au contraire. Remember "The Swimmer," that John Cheever short story where the lush decides to swim his way across all the pools in his county? That was my plan, except that instead of swimming pools, I would hit as many putatively private beaches along the Russian River as possible—first on a 10-mile stretch with my buddies, then on another 4 miles or so during a handful of smaller excursions. Instead of whiskey, we'd be fueled by a cocktail of righteousness and florid legalese.

"Public ownership of physically navigable rivers, including the land up to the ordinary high-water mark, pre-dates property deeds. What the property deed says or doesn't say about the river is irrelevant." So reads a river-law primer on the National Organization for Rivers website. And as the Supreme Court ruled, private ownership of the beds and banks of navigable rivers is "always subject to the public right of navigation."

My friends and I paddled into the gentle current to see if that felt true.

The modest little towns of the Russian River Valley—Forestville, Rio Nido, Guerneville, Monte Rio—are wonderful and complex places, tight-knit blue-collar communities that mix unevenly with Bay Area weekenders. There's an artsy scene and a gay scene and an ex-hippie scene and a homeless scene.

On the water, meanwhile, it's pure Platonic river. From the languid sway of the bay laurel trees to the disco shimmer of the quaking aspens to the Dude-like serenity of the redwoods, a vague drugginess wafts over the place. The water beckons all types. One minute you're chatting with a binocular-toting naturalist, the next with occupants of a 12-person raft built to look like a giant pizza. It's a communal vibe, which makes the idea of an exclusive riverbank all the more jarring.

My friends and I spent the first hour pulling up on every no-trespassing beach we could find, though knowing we were within our rights didn't make it less awkward. Imagine learning that punching someone in the face was actually legal—it would still be sort of tough to feel OK about it. But not a soul came running out to scowl at us. "Maybe nobody will say anything to you. It'd be totally different if you were black or brown and sidled up on someone's property," an African American friend had pointed out before the trip.

But by hour two, everything began to change. A broad gravel bar opened out into the river, cordoned off behind a length of rope and some buoys. PRIVATE BEACH NO CANOES, a sign read. Yes canoes, we decided. Three minutes later we were enjoying a fine game of bocce. Ten minutes after that came the first bellow.

A shirtless man in his 60s or 70s was standing on his deck 100 yards or so up from the river. He had a pair of binoculars trained on us and was yelling something I couldn't quite hear.

"What?" I called.

"I said read the fucking sign in front of you!"

He was deeply tanned, with a leathery chest and long, thinning hair. He proceeded to unload. He was sick of people from God-knows-where marching all over his private oasis. As he yelled —now on the lawn in front of me—I did my best to politely represent the law. It bounced off. "How would you feel if I hung out in your fucking backyard?" he asked.

"There isn't an easement on my backyard."

He waved his hand. "If someone breaks their leg on this beach, I'm liable," he said. I told him this wasn't so—the state is liable, per the public easement.

"You don't get it," he replied. "I pay taxes on this beach."

"Everyone does," I said gently.

It went on like this. I thought that if we could just keep talking, we'd find some common ground, even if we never agreed that his ground was common. He thought otherwise and told me to fuck off again before ultimately walking away.

Over the next few hours, at a dozen ostensibly private beaches, my friends and I would encounter better and worse. At one beach, a friendly young couple lolled on a pair of inner tubes a few feet from shore. They were renters in a small community that claimed this particular beach as its own, but as far as they were concerned, anyone could use it if they cleaned up their mess. Half a mile away, we ran into a group at another beach that had been claimed by a community association, the Odd Fellows Recreation Club. "Your law is wrong," one woman said when I explained my position. At another private club, a woman shook her head and told me, "It's always been this way and always will be."

Throughout my journey, I heard a common refrain from homeowners: that they weren't trying to keep neighbors away, just outsiders. The species of outsider varied. Sometimes they were oblivious city folk, sometimes trashy teens and their hip-hop. Other times they were from Santa Rosa—said, occasionally, as code for Mexican. In each case, the alleged trespassing was ruining the river, disrespecting its traditions, polluting its beauty.

There's a long history of racism and xenophobia wrapping themselves in the cloth of conservation, from Hitler to the early eugenics days of the Sierra Club; concepts like purity and cleanliness proved darkly malleable. I certainly don't purport to know the hearts of anyone I spoke with. But the level of hostility toward interlopers was striking.

Two-thirds of the way into our trip, we came upon a pirate flag flapping above a small spit of sand. I hollered up a guileless hello to the two men standing nearby. In response, one disappeared into some trees. He returned with two large dogs, which he led down to us.

Clearly we were meant to be frightened away. But I wasn't ready to leave, so I tried to de-escalate the situation with chatter: a mindless remark about dog breeds, then an explanation of what the hell we were doing there.

"I know about the goddamn high-water mark," the man spat at us.

All this time, his friend had been holding something. I saw now that it was a golf club. He stepped up to a makeshift tee and squared his shoulders. Before I could register it, he was winding up and blasting a ball in our direction.

He missed us by a good 15 feet. But now he was teeing up again, clobbering another poor Titleist at us. This one came close enough that we heard a *fffffftt* as it shot over our heads. Five minutes after pulling up on this beach, we were hauling away at top speed. Already I was thinking about the next leg of my expedition, where the locals were said to be far less friendly.

One gray August morning in San Francisco, I climbed some stairs in a building in the Cow Hollow neighborhood to meet Mark O'Flynn in his small law office.

In person, O'Flynn calls to mind a football coach—close-cropped hair, stern but courteous bearing, minimal chitchat. He wouldn't let me record our conversation, so I scribbled notes while he discussed the nuances of public-trust doctrine, the history of California statehood, and the like. In 2015 he told the *Sonoma West Times and News* that he doesn't object to the idea of a public easement. He told me he believes that his particular property is exempt, having originally been sold before the California Constitution provided for that easement in 1879—in other words, he claims he was grandfathered in.

It became clear that O'Flynn felt like the victim in the saga. He just wanted his nice river home, without having to deal with trash and skinny-dippers and other people's music.

O'Flynn framed his argument in granular legal terms, but it's not hard to imagine—or sympathize with—the underlying sentiment: the world gives you years of peaceful picnics here on this sand, and then suddenly someone wants to flip the script on you.

"In much of the country, you don't have national forests or state parks—rivers are some of the only recreational places," said Eric Leaper, executive director at the National Organization for Rivers, a 40-year-old group dedicated to defending public rights on rivers. But they're also a finite resource, he added, and people get territorial.

That's what happened one afternoon in July 2013, on Missouri's Meramec River. As with the Russian, summer brings a steady stream of canoeists, rafters, and inner tubers to the river, and tension between boaters and locals was running high. A landowner named James Crocker fell into an argument with a man named Paul Dart Jr. Dart was canoeing down the river with friends when they stopped on a gravel bar that Crocker believed was his. The two men yelled, Dart claimed it was his right to be there, and Crocker pulled out a 9mm pistol. He shot Dart in the face, killing him. Later it would come out that they were more than 100 yards from Crocker's property.

Dart's death drew support to the cause, and within the niche community of public-access enthusiasts, Crocker's ultimate murder conviction played as vindication. For while federal law may be plain, getting states to honor it—or even know it—is another matter. A parallel legal universe exists at the local level, with this sheriff tolerating a length of barbed wire across a river and that state thumbing its nose at the Supreme Court ruling. In Colorado, for instance, courts have essentially ruled that merely letting rivers flow across their land is basically all that property owners owe the public.

The public-access folks have certainly had victories. In a significant ruling last November, the Utah Supreme Court upheld the right of anglers and others to use the banks and beds along a stretch of the Weber River.

Cullen Battle, a retired private attorney in Salt Lake City, represented the Utah Stream Access Coalition in the Weber River suit. A lifelong fly fisherman, Battle has watched states cede untold acreage to the property-rights movement.

"The public attitude used to be that rivers are public highways. But in the 1950s it started to change," Battle told me. Landowners started claiming rivers and riverbanks as private property—and, particularly in the western states, they started to get away with it. Defending public access requires expensive legal efforts and a deciphering of 19th-century court rulings that, as Battle says, "is like interpreting Scripture." More often than not, communities simply defer to those NO TRESPASSING signs. "If you keep saying it over and over again," he said, "people start believing you."

A millennium and a half ago, the Roman emperor Justinian declared that the public's access to all navigable waterways was inalienable. "By the law of nature these things are common to all mankind; the air, running water, the sea, and consequently the shores of the sea," the Code of Justinian states. It paved the way for the public-trust doctrine, a guiding principle compelling governments to hold the most vital natural resources for the collective good. To this day, "navigable in fact" rivers remain the property of all. But not everyone reads their Byzantine legal texts anymore. And so it's a short leap to coating your PRIVATE PROPERTY signs with lubricating oil, as Harreld alleged he observed on one occasion, thereby making them harder to remove.

By June 2015, the beach war was boiling over. Complaints on both sides were lodged with the county's permit office, the California State Lands Commission, the sheriff, and just about anyone with a dot-gov email address. At one point, according to Harreld, O'Flynn erected a robust cinder-block and lumber bench in the middle of the beach trail, attaching to it a CERTIFIED WILDLIFE HABITAT sign—"the kind anyone can order online," Harreld noted. Harreld attached his own sign, which declared O'Flynn's sign false and unlawful. This was promptly taken down—and then replaced again by evening. The next day, Harreld received a cease and desist notice from O'Flynn.

And that was all in three days. For the better part of two years, each side hoped the California State Lands Commission would step in and issue a ruling to end the madness.

Harreld concedes that convenience had stoked his indignation as much as anything—in an era when the guys with the power are off in Washington, he said, above the law, here was one right down the street. But then a funny thing happened: his actual concern for the river caught up with his theatrical defense of it.

"It got under my skin," Harreld said. "All those hours of measuring tides, taking photos, sitting in the sand—I found that I really loved it."

According to Don McEnhill, executive director of a local non-profit called Russian Riverkeeper, that is precisely why the public should care about squabbles like this one. From access comes stewardship.

"If people don't use the river, they won't care about it, and it turns into a toilet," he said.

In the property-rights movement, McEnhill sees owners as people who "want to intimidate, harass, and bully the public into vacating their rights." He said the Russian River situation was open and shut. "Beaches are unequivocally in the public domain, and the state and federal Supreme Court have said so."

When I told Harreld about the final stop on my occupy-the-beaches experiment, he insisted on coming. We set off in kayaks on a hot morning in Guerneville, heading west toward Monte Rio. We portaged around a dam and wound past a small island, and soon the only sound was the purling beneath our paddles and starlings nipping around dementedly against a *Simpsons* blue sky.

Suddenly I felt a surge of sympathy for Team Homeowner. Beneath their anger was a simple human truth: when you've got something good, it's hard to let go. On an ever dirtying planet, these folks had the fortune to occupy a lovely sliver. Maybe McEnhill is right and sharing that sliver will invest the public in keeping it clean. But maybe you give the beaches to the people and the people toss Budweiser cans onto them.

Doesn't matter, Harreld said as we paddled—the law is the law. So we zeroed in on a spot a couple of miles downstream from O'Flynn's beach. For more than a century, the exclusive Bohemian Club has held its annual encampment here, on a 2,700-acre parcel along the river. The event is both a groovy arts mixer and a highly secretive gathering for establishment elites; every Republican president since Coolidge has reportedly shown up, alongside your Kissingers and Rockefellers and corporate executives. The exclusive nature of it has led countless journalists and activists to try sneaking past the intense security.

I didn't want to sneak in. I just wanted to throw a Frisbee on their pretty beach.

We arrived to find a stretch of rocky sand, with a dock and a bullpen-like shade structure at the end. We pulled up and sat in the baking sun. The club's intense security apparatus is legendary. I'd heard rumors of snipers in the trees, the Secret Service lurking about, and trespassers being hauled straight to jail. Be careful, the community-services officer from the sheriff's office had told me. You don't want to get on one of their lists.

I was contemplating all this when a big and vaguely Wilford Brimley–ish man emerged from the woods.

"Time to be on your way, boys," he said. He wore khakis, a gray mustache, and what appeared to be a security badge around his neck.

I replied that we intended to stay a while longer, per the public easement, but had no intention of going beyond—

"I don't care about high-water marks," he said bluntly. "What's your name?"

I told him, and he smirked. "Here to write one of your articles?"

Had the sheriff's office told them who I was? Did the establishment elites have even greater snooping technology than I had imagined possible?

Harreld and I spent the next few minutes trying to discuss the matter with him. We were calm and friendly. At one point, I invited him to throw the Frisbee with us. But inside, something had flipped in me. It didn't make sense. Nobody else was using the beach. Why not let others on it? If they made a mess, sure, fine them or whatever, but otherwise, who cares?

Evidently, the security guy. He told us he'd be calling the sheriff to arrest us if we didn't leave immediately.

Harreld said he'd happily chat with the sheriff, and for the next two hours we skipped stones, swam, and waited. The sheriff never came—the sheriff's department later told me that it had received no calls—and at last we climbed into our boats and paddled back to where we'd begun: O'Flynn's beach.

Harreld and I parted ways, and I got in my car, deflated. How far we'd drifted from old Justinian's vision. I could think of only one thing to do: go to a different beach, an emphatically public one where I could find at least a semblance of harmony.

It was late on a Sunday afternoon when I arrived, but the sand was still packed—dudes grilling, teens lounging, parents urging kids to accept one more smear of sunscreen. I wandered from one end to the other. And then: there it was.

"Hey! Private beach!"

I recognized the growl instantly. It was the shirtless man I'd encountered on the first leg of my journey, shirtless again, drinking beer with friends under a scattering of umbrellas. Apparently I'd strayed into the private section.

The friends were in their 60s and 70s. "Sweet Melissa" was playing from a radio; it was someone's birthday. The shirtless man and his friends didn't want me there, but they deigned to let me stay a moment so they could vent about all the changes to the river and the country in general. "It's just too many people," he said. "It didn't use to be. Then *they* started coming." He shot me a meaningful look. Later I asked for everyone's contact info, for fact-checking purposes.

"If you think I'm giving you my information, you're out of your mind," one of them said.

"Fake news!" shouted another. "You're all fake news!"

As I walked away, I found myself thinking about that last comment more than anything. In a weird way, it felt as true as anything I'd bumped up against on the river. O'Flynn, Harreld, the Bohemian Club, these guys, me—for everyone, the river was a realm of competing narratives. As in the fake-news paradigm, truth has become subjective and reality is shaped accordingly. For some this is the place to fight back against interlopers and bad manners. For others it's a venue for fighting greed. Legal confusion, meanwhile, hovers over it all.

"All it takes is a guy with a NO TRESPASSING sign, and if you don't know better, that's the law," said Dave Steindorf, executive director of the river-stewardship nonprofit American Whitewater. "The sheriffs often act as the local private-security force for property owners. Without the public knowing it, our access to rivers gets eroded."

Last year the California State Lands Commission issued a report declaring that the area up to O'Flynn's grass lies below the ordinary high-water mark and therefore is public-trust land. Harreld was right: nobody can block the public from enjoying the river. The bench barricade, ironically, had already been swept away by high water during winter flooding.

Triumphant after the ruling, Harreld spoke of O'Flynn with a new note of grace. He was sympathetic about finding a solution to the litter problem. And regarding the security cameras, he suggested using them to prosecute anyone who might pollute the beach, rather than to deter people from going there in the first place. "I would one hundred percent support this," Harreld said. O'Flynn still declines to speak in detail about the case.

As of this writing, a new sign, lovingly handmade, stands by the water: HELP RESTORE THE RIPARIAN COORIDOR — PLEASE LAUNCH WATERCRAFT UP RIVER.

DOUGLAS FOX

The Brain, Reimagined

FROM *Scientific American*

A YOUNG WOMAN WITH wavy brown hair and maroon nails lay
on a gurney in a hospital room in Copenhagen. Her extended left
arm was wired with electrodes. A pop pierced the air every few seconds—an electric shock. Each time, the woman's fingers twitched.
She winced. She was to receive hundreds of shocks that day.

The woman, attended to by several physicians in laboratory
coats, was renting out her arm for 1,000 Danish kroner, about
$187. Thomas Heimburg, a physicist trained in quantum mechan-
ics and biophysics, sat on a stool, safely out of the way, sketching
on his iPad the details of a harsh experiment that he hoped would
produce profound results.

The physicians had injected the woman's arm with the anesthetic lidocaine—a dose strong enough to deaden her limb for
surgery. At first, the nerves in her arm did not respond to the
shocks. But the attendants gradually dialed up the current. At this
moment, the jolts were 40 milliamperes, nearly 10 times their original strength—similar to the electricity coursing through a five-
watt lightbulb.

Pop—another shock. The woman's hand twitched like a dying
snake. Heimburg paid no notice as he stared at a computer monitor on the wall. A waveform depicting the electric signal in the arm
muscle and nerve leaped across the screen in one large spike—evidence that the ever increasing shocks had started to overcome the
anesthetic. The nerve was now firing as strongly as it did before
the woman was anesthetized. Heimburg was pleased. "The things

that are written in books," he said quietly, "they are in contradiction to this."

Heimburg, who works at the Niels Bohr Institute in Copenhagen, famous for physics research, hopes to contradict lots of things written in books. This experiment, which I witnessed in December 2011, was designed to investigate a long-standing medical mystery.

Physicians have administered general anesthetics for 170 years. They have discovered dozens of effective compounds. When given at progressively higher doses, the drugs all silence nerve functions in the body and brain in the same distinct order: first memory formation, then pain sensation, then consciousness, and eventually breathing. This same sequence happens across all animals, from humans to flies.

Yet no one knows how anesthesia actually works. The molecular structures of nitrous oxide, ether, sevoflurane, and xenon are so different that it is unlikely they exert their common effects by binding to equivalent proteins in cells, as other drugs do.

Heimburg thinks anesthetics work in a radically different way: by changing the mechanical properties of a nerve. If that is true, it means that nerve cells, or neurons, throughout the body and brain are mechanical machines, not the electric circuits scientists have believed in for decades. In Heimburg's view, the electric pulses are simply the side effects of a physical shock wave that ripples down the nerve, similar to the way sound waves travel. He thinks anesthetics silence nerves by soaking into the fatty membranes that encase nerve fibers, rendering them too soft to transmit the shock waves, like a guitar string too slack to twang.

It was tempting to dismiss Heimburg as nutty when I watched that experiment. But in the seven years since then, he and his colleagues have rolled out an array of evidence: delicate measurements of how mechanical waves move through single nerve cells and of how much and how quickly the membranes can expand and contract, as well as studies showing how anesthetics alter these properties. Other scientists are starting to take an interest. Now Heimburg is preparing for a crucial experiment that could clinch his case: measuring the heat emitted by a single nerve cell as a pulse shoots through it.

Heimburg's work continues to demonstrate that the nerve pulse is more complex than most biologists may realize. The mechanical components may have been overlooked because of an accident of history: 50 years ago off-the-shelf instruments could readily measure the tiny electric impulses in neurons but not the mechanical ones. Hardware limitations influenced which discoveries scientists made and which ideas entered mainstream scientific thought. Heimburg's experiments are now reopening a decades-old scientific schism.

The story of the mechanical neuron holds lessons for all of science about biases and accidents of history. It also could change our basic understanding of nerves, brains, and intelligence. Scientists have struggled to explain how brains achieve such daunting feats as face recognition and conversation while relying on proteins in neurons that are electrically noisy and unreliable. Heimburg is showing how the mechanical waves may compensate for this noise. If his theory proves out, he could rewrite biology. Or he might just be wrong.

Hot Nerves

The neural pulse that scientists have tried for so long to explain lasts for only an instant. Step on a thumbtack, and your brain senses the pain within a fraction of a second. The signal travels through nerve fibers at up to 30 meters per second.

The fibers resemble tiny hollow pipes, finer than a hair. The pipe wall is formed by an oily cell membrane. Charged sodium and potassium atoms, called ions, hover around the inside and outside of the membrane. By the mid-1900s researchers had learned to stick electrodes into nerve cells to monitor the voltage across the membrane wall. They discovered that as a nerve pulse travels down the membrane and passes the electrode, the voltage spikes for several thousandths of a second. In 1952 two British scientists, Alan Hodgkin and Andrew Huxley, reported that the spike happens as sodium ions stream through the membrane wall from outside to inside. The voltage then reverses to normal as potassium ions gush through the membrane from inside to outside. The Hodgkin-Huxley model became the foundation of modern neurophysiology.

Hodgkin and Huxley received a Nobel Prize in 1963. But a few scientists continued to unearth observations that undermined their model, observations that Heimburg has re-created, even though some of those scientists had been written off as misguided.

Ichiji Tasaki, a senior neurobiologist at the National Institutes of Health for many years, was one of them. In 1979 he conducted an unorthodox experiment. Gazing through a microscope, he gingerly placed a fleck of shiny platinum atop a fine white thread—a nerve fiber bundle of a crab, laid bare by dissecting the animal's leg—and trained a laser onto the platinum. By measuring the reflection of the laser light, he could detect motions that would show whether the nerve bundle briefly widened or narrowed as an electric pulse passed by. He and his then postdoctoral fellow, Kunihiko Iwasa, took hundreds of measurements. After a week, the answer was clear: every time a pulse shot through the nerve fibers, they briefly widened, then narrowed again, within a few thousandths of a second.

The ripple was minuscule: the membrane surface rose by only about seven billionths of a meter. But it coincided perfectly with the passing electric pulse, confirming a suspicion Tasaki had harbored for years: that Hodgkin and Huxley were wrong.

As far back as the 1940s, researchers had noticed that as an electric pulse passes through a nerve fiber, the translucent cell briefly becomes more opaque. By 1968 Tasaki and another team found evidence suggesting that as the pulse arrives, molecules in the membrane physically rearrange themselves, then revert to their original configuration after the pulse passes.

Then there was the heat. Researchers expected an electric pulse to release heat—common when electricity flows. But several teams discovered something strange. A nerve fiber's temperature rose several millionths of a degree Celsius as a pulse raced by, yet after it passed, the temperature quickly fell again. The heat had not dissipated; instead the nerve had reabsorbed most of it, also within a few thousandths of a second.

For Tasaki, the transient widening, the rearranging molecules, and the heating and cooling pointed to a startling conclusion: the nerve signal was not just a voltage pulse; it was every bit as much a mechanical pulse. Scientists who listened to nerves with electrodes were missing much of the action.

Tasaki would spend the rest of his life probing these effects. He came to believe that they originated not in the cell membrane but in a layer of protein and carbohydrate filaments just underneath it. According to his theory, as the voltage pulse arrives, the filaments absorb potassium ions and water—causing them to swell and warm—a process that then reverses itself after the pulse passes by.

As Tasaki pursued these ideas, he gradually fell out of step with the field. Other factors conspired against him. Having grown up in Japan, he spoke stilted English. "You [had] to know a lot of things to have a really substantive conversation with him," says Peter Basser, an NIH section head in neuroscience who knew Tasaki for 20 years. "And I think a lot of people thought he wasn't really as deep and perceptive as he was." And although Tasaki collaborated with visiting scientists, he did not produce student protégés who would carry his ideas forward.

Emblematic of the schism was the ideological rivalry that arose between Tasaki and another prominent NIH neuroscientist, Kenneth Cole, who adhered to the mainstream view. Although the two men occupied the same lab building from the 1950s to the 1970s, they barely spoke for 15 years, except at public presentations, where one would undermine the other by standing up in the audience and posing prickly questions.

Tasaki gave up his lab during an NIH reorganization in 1997 and moved into a small space in Basser's lab. He continued working seven days a week, well into his 90s. One day in December 2008, as he walked near his home, he lost his balance and banged his head on the ground. He died a week later at the age of 98.

By then, Tasaki's work had disappeared from sight. "I don't think anybody disputed that those things were being seen, because he was respected in the lab," said Adrian Parsegian, a biophysicist at the University of Massachusetts Amherst, who was at the NIH from 1967 to 2009. Rather Tasaki's findings "were explained away as not central" to nerve signaling—nothing more than side effects of the voltage pulse. The underlying scientific questions "didn't get resolved," he said. "One side got into the textbooks, and the other one didn't."

Fatty Liquid Becomes Chrystal

Heimburg came across Tasaki's work in the mid-1980s, while pursuing his PhD at the Max Planck Institute for Biophysical Chemistry in Göttingen, Germany. Soon he found himself immersed in long sessions at the library, poring over old papers. He would eventually connect the dots in a different way than Tasaki had. He believed that the mechanical wave, the optical changes, and the transient

heat must occur in the fatty cell membrane of nerves throughout
the body and brain, not in the protein and carbohydrate filaments
below the membrane, as Tasaki had thought.

By the late 1990s Heimburg had begun doing his own experiments, compressing artificial cell membranes to see how they
might respond to a mechanical shock wave. This work revealed
something crucial: the membrane's oily lipid molecules are normally fluid and randomly oriented, but they hover close to what
chemists call a phase transition. Squeeze the membrane just a little
bit, and the lipids condense into a highly aligned liquid crystal.

These experiments led Heimburg to declare that a nerve pulse
is a mechanical shock wave that travels down the nerve membrane.
As it advances, it should squeeze the membrane's lipid molecules
into a liquid crystal—a phase change that would release a small
amount of heat, just as water does when it freezes. Then, as the tail
end of the shock wave passes, a few thousandths of a second later,
the membrane would revert to a fluid state, reabsorbing the heat.
That brief transition into a liquid crystal and back would also cause
the nerve membrane to widen briefly, just as Tasaki and Iwasa had
seen when they shined a laser on that platinum fleck.

Heimburg's experiments went one crucial step further. They
showed how the shock wave and phase transition might be linked
to the voltage spike that occurs as the pulse passes by. Heimburg
found that he could push a membrane into its liquid-crystal state
simply by putting it under a voltage. "People applied voltage across
biologic membranes for 70 years or so, and none of these electro-
physiologists had ever checked" for a liquid-crystal structure, he
said.

Textbook diagrams portray cell membranes as thin, passive
sheets of insulation wrapped around pipelike nerve fibers. But
physicists are starting to realize that cell membranes have surprising properties. They belong to a class of materials known as piezo-
electrics, which can convert mechanical forces into electric forces,
and vice versa. Quartz watches run on this principle. This means
that a voltage pulse traveling down a membrane will carry with it
a mechanical wave. And conversely, a mechanical wave traveling
down a membrane will express itself as a voltage pulse.

When Heimburg and his fellow researcher Andrew D. Jackson
first published the theory in 2005, they had still never observed
one of these electromechanical pulses in motion. One of Heim-

burg's former students filled that gap. In 2009 Matthias Schneider, a biophysicist now at the Technical University of Dortmund in Germany, reported that he could trigger a mechanical wave by applying a voltage pulse to an artificial membrane. The pulse strength was similar to that found in nerve cells. The shock wave traveled at approximately 50 meters per second, similar to the speed at which thumbtack-triggered signals race from the foot to the brain. By 2012 Schneider had confirmed that the mechanical and voltage pulses were part of the same membrane wave.

Schneider's most important finding came in 2014, however. A key feature of a nerve pulse is that it is all-or-nothing. If a neuron receives a weak incoming shock, it will not fire a voltage pulse. If the shock is strong enough, it will fire. "There is a threshold," Schneider says. He found that the electromechanical waves on his artificial membranes were indeed all-or-nothing. The determining factor seemed to be whether the membrane was squished hard enough to force it into liquid-crystal form. Only then, he says, "you get a pulse."

Anesthesia Explained

Why had Heimburg first committed to this view of nerves and anesthesia? Hoping to find out, I visited him at his office at the Niels Bohr Institute during the same week I witnessed the hospital experiment.

Heimburg had the bookshelves of a physicist, not a biologist, crammed with volumes by dead German physicists. Among them was a row of clothbound books by Hermann von Helmholtz, who in the mid-1800s formulated a key premise of thermodynamics, that energy can change form but cannot be created or destroyed. Helmholtz, incidentally, also measured the speed of nerve pulses. "I find it absolutely mandatory to read these old texts," Heimburg said. They document the gradual discovery of fundamental connections among energy, temperature, pressure, voltage, and phase transitions. These principles underlie Heimburg's ideas about nerve function, the ideas of a physicist pushing his way into another field. "Thermodynamics is the most profound science that we have," he said. "If you know thermodynamics, you are wise."

He was quick to point out weaknesses in popular explanations about anesthesia. Biologists think anesthetics silence nerves by binding to and thus blocking ion channels—valves in a nerve membrane that open and close to allow sodium or potassium ions to flow through. Biologists say the flow of ions propels voltage pulses down a nerve fiber commonly portrayed as an electric signal. But because different anesthetics have vastly different molecular structures, Heimburg could not believe they could all bind to ion channels. That explanation was "completely ridiculous," he said, with a hint of frustration, as if pointing out something that should be obvious. Something "deeper, more profound," must be at work.

Heimburg's ideas were shaped in part by an old volume entitled *Studien über die Narkose,* or *Studies of Narcosis,* published by Ernest Overton in 1901. It recounts a particular experiment that caught his attention. Overton took dozens of different anesthetics and put each into a flask of water with a layer of olive oil floating on top. He shook each flask, then waited for the water and oil to separate again. He measured how much of each drug ended up in the oil versus the water. The more potent an anesthetic was in animals, the more strongly it moved into the oil, a striking result later confirmed for modern anesthetics. Olive oil and cell membranes are composed of the same oily molecules, called fatty acids. Heimburg surmised that the drugs might work by soaking into the cell membranes, altering their physical properties.

Experiments with synthetic membranes support that idea. When Heimburg infuses a membrane with an anesthetic, it prevents the membrane from becoming a liquid crystal. It does so by lowering the temperature (and raising the pressure) at which the phase transition from fluid lipid to crystalline lipid occurs—just as salt or sugar lowers the freezing point of water.

Heimburg reasoned that preventing this transition in a membrane would stop a mechanical pulse from advancing down a nerve fiber, explaining why anesthetics deaden nerves. And notably, he predicted it should be possible to overcome this effect. To create higher pressure to solidify a membrane using an electric shock, you have to crank up the current—exactly what the physicians did to the woman's arm at the hospital in Copenhagen. Stronger electric shocks did indeed overcome the anesthetic. If anesthesia can be overcome by pushing harder on a membrane with electricity, then it should also be reversible by increasing the physical pressure on a membrane.

Biologists demonstrated this way back in 1942. They used two different anesthetics, ethanol and urethane, to inebriate tadpoles to the point that they stopped swimming. Then the scientists put the animals in a hyperbaric chamber and raised the pressure to 136 times that of the atmosphere. The anesthetic effect vanished: the tadpoles resumed swimming. When the pressure was lowered, the animals again fell motionless. "It's very surprising," Heimburg said, with a smile. "How would you have the idea to put drunken tadpoles under pressure?"

No Tolerance for Debate

To this day, Heimburg is frustrated by the way biologists react to his ideas, which he calls soliton theory (a soliton is a self-sustaining wave that maintains its shape as it travels). He has faced opposition from the moment he published his theory in 2005 in the *Proceedings of the National Academy of Sciences USA,* despite that journal's high regard.

One critic, Catherine Morris, a prominent neurobiologist emeritus at the Ottawa Hospital Research Institute, told me that the whole line of work reeks of superiority from a physicist who thinks he can simply march into a different field and set people straight. She summed this up in a favorite witticism of hers: "It strikes me as this business that physicists do, saying, 'We can approximate this cow as a single point.'"

To some extent, Morris's reaction is understandable. It is one thing to say that nerves are mechanical as well as electrical. It is quite another to reject the concept that ion channels play a role in nerve conduction—which Heimburg and Schneider do, in their biggest and most problematic departure from mainstream biology. Never mind that scientists have discovered hundreds of ion channel proteins. Or that the ion flows can be selectively altered with drugs. Or that mutations scientists can create in the proteins change the way neurons fire. "They just blithely ignore vast amounts of biology," says Morris, who spent 30 years studying ion channel proteins.

Heimburg and Schneider acknowledge that these proteins must serve some function. But they point to experiments, some by Heimburg, showing that ions can flow across artificial membranes even without channel proteins. They attribute this flow to transient holes that appear as the membrane shifts between fluid and liquid-crystal phases, and they think it happens in nerves in the body and brain.

Their skepticism reflects a cultural tendency in physics: a belief that all things should be explainable through thermodynamic principles. Biologists, they say, have neglected these principles as they fixate on proteins. A similar brand of puritanism may have facilitated the eventual dismissal of Tasaki's theory. He "did not like the term 'ion channels,'" said former postdoc Iwasa when we spoke in late 2017. This iconoclastic outlook may have guided Tasaki to discover things that others could not have, Iwasa said, "but later on, it may not have helped" him.

Brian Salzberg agrees. He studies nerve physics at the University of Pennsylvania and began his neuroscience career in 1971, crossing paths occasionally with Tasaki. "He was a very clever experimenter, and I have no doubt that he measured real changes" in nerve thickness, Salzberg said earlier this year. "But he misinterpreted them." Salzberg says nerve fibers temporarily swell as a voltage pulse goes by in part because water molecules flow into the membrane through the same ion channels that let in sodium and then flow back out through the ion channels that let out potassium. If Tasaki had accepted the idea of ion channels, he might have been open to other interpretations of the mechanical wave.

But another powerful factor may have helped push Tasaki out of sight—holding an important lesson for all of science today.

Ideologues

It is intriguing that the thermal energy of a firing nerve may be twice as large as the energy in the electric signal that has dominated neuroscience. The fact that these nonelectric features fell out of favor may stem, in part, from a quirk of history.

Tasaki was a gifted instrument builder who cut his scientific teeth in Tokyo during World War II. Faced with severe equipment shortages, he assembled his own instruments from stray electric components. Years later in the United States, he used these skills to build exquisite, one-off instruments that measured the heat, or temporary expansion, of nerve cells.

Those devices, and expertise, never found their way to other scientists. Measuring the electric nerve signal was different. Scientists created easily transferable methods, such as inserting a tiny electrode into a cell membrane. As these techniques spread from one lab to another, so did the electrical view of nerve signaling. "There's a cultural bias," Parsegian admitted. "People look with a tool that they feel they understand, and they don't use one that they don't understand. It could have tilted the thinking."

Today the technical gaps are starting to disappear. As I checked in with Heimburg between 2011 and 2018, he gradually repeated one old experiment after another, using modern technologies to clarify the surprising things that Tasaki and others first saw decades ago. In 2014 Heimburg redid the drunken-tadpole experiment, using synthetic membranes instead of animals: as he cranked the pressure up to 160 atmospheres, the impacts of anesthetics were reversed—except that this time, Heimburg could link the effect directly to phase changes in the membrane. In 2016 he used microscopy to precisely measure, in a single cell, the mechanical wave that Tasaki and Iwasa first documented in 1979.

Heimburg, now 58, is seeking funding for what could be the most critical experiment of all: measuring the heat as a nerve pulse, or action potential, passes by. Tasaki had measured heat from bundles of fibers, but Heimburg plans to use a microchip that will measure the heat blip of a single neuron. This experiment could address a key criticism of his theory: that a nerve membrane's brief phase change from liquid to crystal should release, and reabsorb, more heat than Tasaki ever saw. Heimburg contends that the old experiments systematically underestimated the heat; because they measured many neurons, the heat reabsorption after early pulses canceled out the heat releases of later pulses. "The true signal is probably much higher," he told me in late 2017. If his measurements bear out, they could bolster his claim that the membrane transmits a mechanical wave.

Perhaps most significantly, other scientists are stepping in—outsiders who are not polarized by the old, calcified disputes. Nongjian Tao, a biosensor engineer at Arizona State University, is using lasers to track mechanical pulses in single nerve cells—like Tasaki and Iwasa did, except that Tao reflects his light directly off the nerve rather than a tiny platinum mirror, making the measurement more sensitive. He hopes to monitor hundreds of individual neurons in nerve networks at once, with lasers sensing mechanical waves as they ripple to and fro. Such work could answer a key question. "The existence of these [mechanical] effects is not in doubt," says Simon Laughlin, a neuroscientist at the University of Cambridge. "The question is whether neurons actually use them to do something useful."

Laughlin does not work on mechanical waves, but as someone who has studied ion channels for 45 years, he imagines that the waves could influence the little protein valves. Recent experiments show that the valves are extremely sensitive to mechanical forces in the membrane. If mechanical waves help to open and close ion channels, that could profoundly change our understanding of the brain, because firing neurons mediate all thinking. Ion channels are notoriously noisy and jittery: even tiny thermal vibrations can cause them to pop open or close randomly. Information theorists have struggled for decades to explain how the brain can achieve reliable cognition using such unreliable channels. But mechanical waves could mean the openings and closings are purposeful. "That's a definite possibility," Laughlin says.

There are hints that this could be true. Some neurons in the mammalian cortex seem to violate the Hodgkin-Huxley theory. When they fire at high rates, their ion channels open more quickly, as a group, than expected. One explanation is that the channels are responding en masse to a sudden change in the membrane —the arrival of a mechanical wave that opens them more or less in unison, allowing them to fire faster than they otherwise could. The speed might allow them to transmit information at phenomenally quick rates—a possible basis for cognition. In this view, a nerve pulse is both electrical and mechanical.

Heimburg and Schneider occupy a strange place in all of this. They could perhaps one day share a Nobel Prize. Or they could end up nowhere, transfixed by the same insistence that gripped Tasaki for so many decades. The fact that some neuroscientists such as Laughlin and other experts such as Tao are interested in mechanical waves would seem like an important opening for the physicists. But Heimburg was steadfast when we spoke in February. "What many people try to do is somehow rescue the Hodgkin-Huxley model by just combining it with the view that we have," he said. "But I personally . . . would not accept any kind of compromise between the two models."

CONOR GEARIN

Little Golden Flower-Room: On Wild Places and Intimacy

FROM *The Millions*

THIS FEBRUARY, ON the first day barely warm enough for it, I took off my shoes and set out on the cold, hard mud of a trail through the loess hills in Iowa. I was helping plan my April wedding, sometimes losing sight of the celebration and seeing only tasks to be done. And as excited as I was for the wedding, I knew that there was a good chance my wife and I would become temporarily long-distance in the fall while I finished my master's program and she began working. That afternoon, I wanted to be transported for a while, to be ensconced in a place that felt elsewhere. The park, called Hitchcock Nature Center, isn't quite out of reach of the Omaha skyline across the river. It's bordered by pastureland, hayfields that spend half the year as short as a lawn, and by cornfields. Jets headed for Eppley Airport howl at low altitude overhead. In other words, human development encroaches on Hitchcock from every side.

But once you're in the borders of the park, like passing through the gate of a besieged city out of Tolkien, you're in a sanctuary, a place set apart. The loess hills are sideways savannas, wave-crest slopes. They're a geological anomaly that occurs only in western Iowa and China: steep, whimsical ridges built of loose, wind-blown silt—*loess* means "loose" in German. Loose hills are places of constant change. The eroding, nearly vertical slopes make it hard for trees to hang on, and historically fires dealt the coup de grâce in favor of a community of grass and fire-resistant bur oak. At Hitchcock, conservationists have restarted the wildfires in controlled form, coaxing them and reining them in. The loess hills therefore shift, from spot to spot and through time, passing from small groups of trees, to open-grown bur oaks with wildflowers below, to the rugged bunchgrasses of the eastern prairie—big bluestem, Indian grass, switchgrass. You might have a favorite tree and come back next week and find it with a corkscrew lightning scar, burnt to a crisp. You might have a favorite flower—get out while it's having its week in bloom. The expanses of grass are sweet summery green one day, deer-hide tan another, and blackened from fire the next.

It was on one of its charcoal-stained days that I saw Hitchcock in February. The western edges of the park had been burned just a week or so before. The litter layer that had built up in the under-story of the grass served as fuel for a burn right down to the soil, leaving only the blackened tufts of the center of bunchgrasses, spiky underfoot. The ash covered the bare ground, and recent rain was seeping it into the soil profile, returning rich organic material back to the earth.

That makes it sound charming. Let's play it straight: I looked out from a hilltop and saw a shorn, devastated landscape. Black ground under a cold gray sky. A lava field in Iowa. Used right, fires are good for a grassland's health, a bit like the way that an emetic can be good for someone who swallowed poison. It's not pretty, but it's medicine.

Yet burned grasslands have a kind of gothic glamour for those with that kind of taste, especially on gray, misty days. I tried to avoid the sharp remains of bunchgrasses and stay on the smooth, hard-packed soil of the trail. It was cold but not too cold, a just-habitable window of late winter. I took a picture from my favorite vista that looks out over a ridge reaching into farm country, a black volcanic slope in the cornbelt, a small grove of trees on the ridge just dark arms reaching out in supplication.

This outpost of the loess hills makes me wonder—how small can a natural place be and still be an intact place with an identity of its own, make you feel like you're inside it, a space set apart from the built and altered places beside it? Not that we want nature to fit into a tiny box and not exist outside our boxes. But if we don't understand and care for the smaller manifestations of wildness close at hand, how can we ever care for the great wildernesses?

One way I try to answer that question is by visiting small nature spots more than once, tracking their identity through time. If a place isn't expansive in size, maybe it unfolds in another dimension—in the number of moods it takes on, the number of ways that light can transform it. Monet painted a series showing the same view of the River Seine across at least 15 mornings. The way I see it, it's the closest a painter has come to being an ecologist.

When I came home I compared my moonscape photo with other pictures of Hitchcock from September of last year. The difference was one of fairy-tale contrast. I had close-ups of purple gayfeather and vervain set against a fuzzed-out background of green grass. And the same view of the ridge from that last growing season: everything in Shire-green and easy midmorning sunlight.

It was a shock to my midwinter gothic mood. It bounced me right out of the wedding-planning stress, back into a place where I could be excited. I realized that small prints of a watercolor painting might make a decent wedding favor. I got out a six-by-eight-inch piece of watercolor paper and pulled up the September version of the loess hills on my laptop. On second look there was a dab of Sicily or Morocco along with the Shire: scrubby hillsides with olive-toned trees descending down to a hayfield, the paths the tractor took through the grass taking the place of breaking Mediterranean waves. I got to work roughing out the compositional lines in pencil, then early washes of pale green and brownish yellow, and the sky the softest blue I could make it. And that little triangular grove of trees clinging to the ridge, nearly being thrown off, like a rider from a horse, or a surfer from a wave.

I returned to the loess hills trail this September, now married but living long-distance from my wife as I finished the last semester of my master's degree. This time my mood was midwinter but the landscape still summery and cheering. I felt out of joint, out of sync. I tried to think less analytically, tried to just absorb the color and warmth of the place. I put my shoes back on when I saw someone coming up the trail because I felt silly about walking barefoot. I was asking a different but related question from my previous one about small wild places: How little intimacy can I survive on? It can cause imagination to expand, or inversely to fit in smaller places. Hamlet: "I could be bounded in a nutshell, and count myself a king of infinite space, were it not that I have bad dreams." I wanted the park to act as my time-traveling vessel, to connect me with a different era of my life.

Instead of continuing along the narrow ridge like I usually do, I took the descending trail toward the hayfield. It takes a wide curve and just touches the grove of trees from my painting. I realized I had only been up close to these trees once before, on that burned-up late-winter day. In my painting they looked like a solid mass, like you could hardly hope to walk through them. I now moved toward them, now seeing the grove instead as a place I could enter. I felt like I was stepping into my painting. That watercolor was the lens through which I saw this place, a two-dimensional frame of reference that was now becoming a three-dimensional diorama. It tied back, also, to the golden glow of the wedding weekend, when my wife and I were near each other, when we celebrated with nearly 200 friends and family. I put my shoes back on, stepped off the trail, and pushed through the tallgrass to get closer. It turned out that this grove wasn't so dense. Though it had a thick outer wall of small dogwood trees and more than a little poison ivy, once I crashed through that, it had an open canopy allowing for an understory of grass and flowers now glowing gold in the late-after-noon light. Goldenrod, purple thistles, white snakeroot. A place not solid but better understood as a room. Like a golden room of flowers with trees as a roof.

I was seeing Annie Dillard's "tree with the lights in it," one of those rare moments for a naturalist when the universe catches a holy flame. I insist they are rare, whatever else you might hear.

There was a slumping barbed wire fence that kept me from the deeper interior of the grove. At least it would keep visitors from trampling the flowers, and probably wasn't high enough anymore for deer. Not wanting to intrude any longer on the sanctuary, I turned back on the trail.

Now with the perspective of emerging from that refuge instead of seeing it from the outside, somehow the nearby hayfield did not look so out of place next to the grasslands and this grove. It was like the ocean lapping against a beach: always there but never overcoming the beach totally, a dynamic balance.

Amid change and human use, the park was still a kind of island, a holdout of what the Loess Hills looked like 200 years ago. And within it, it contained smaller islands: little prairies among woods, little woods among prairies. The irregular, rollicking slopes con-stantly changed one's perspective about which habitat was most abundant, and which was a precious remnant.

Hitchcock suggested an answer to my question about how small a place could be and still feel like an intact place: pretty small. To look closely at nature is to change one's scale constantly: from an appreciation of a horizon to a heavy bumblebee heavily stepping among goldenrod blooms, from an aster opening its flowers to the poised, rolling ridge on which it grew. A shelter can be as small as the creature looking for safety can allow it to be. There were days, when I was growing up in St. Louis, when a mother deer left a fawn in our backyard so she could go forage and make more milk. Just by placing the fawn behind a tree trunk, neither the five humans nor the middle-aged dog roaming the yard found the fawn until we saw them both leaving. A little fawn-sized depression in the grass marked where the infant had lain. To the fawn, it was a kind of safety.

But while animals, humans included, are clever at finding shelter in the small, we are also looking for the tremendous, the gratuitous sweep of a river valley, the ocean, the prairie without fences. There's a value in the expanse that can't be counterfeited even by the most detailed, magnified look at the small. We want not just to be comforted for today but without bounds. This dance between desire and compromise is my daily drama being in a long-distance relationship. It's why I didn't look to stay longer than a few moments at the little golden grove. I was thinking about the Atlantic, hearing a jet go overhead and thinking of my next trip to my wife.

BEN GOLDFARB

The Endling: Watching a Species Vanish in Real Time

FROM *Pacific Standard*

THERE IS A word, sad and resonant, for the last member of a dying species. The word is *endling*. Martha, who perished at the Cincinnati Zoo in 1914, was the endling for the passenger pigeon —the final representative of a bird once so prolific its flocks blackened the sky. The Tasmanian tiger's endling, Benjamin, froze to death in the Hobart Zoo one night in 1936, when his keepers accidentally locked him out of his enclosure. Lonesome George, the last Pinta Island tortoise, expired peacefully in 2012, at around 100 years old.

It is entirely possible that the endling for a bashful porpoise called the vaquita is today swimming somewhere off the Mexican coast. Vaquitas dwell exclusively in the Gulf of California, the tongue of the Pacific Ocean that laps the Baja Peninsula, in a tiny pocket of turbid sea that could fit three times within Los Angeles and its suburbs. At just five feet long, vaquitas are the world's smallest cetaceans, the order that includes whales, dolphins, and porpoises. They eat fish and squid, which they locate with high-frequency clicks. They avoid the rumble of boat engines, prefer traveling in inconspicuous duos, and refrain from jumping, splashing, or slapping their tails. They are a headache to study. For all their secrecy, they are adorable—endowed with a snub snout, fetching dark eye-patches, and black lips whose coy smile, researchers have written, recalls a marine Mona Lisa. Cross Flipper with a very shy panda and you've bred a vaquita.

Although the Mexican fishermen who began plying the upper Gulf in the early 1900s occasionally encountered it, the vaquita —Spanish for "little cow"—wasn't officially recognized as a species until 1958, after scientists deduced its existence by examining odd skulls that had washed ashore. We know less about the vaquita's habits than we do about practically any cetacean's. Among the first people to survey *Phocoena sinus* was Bob Pitman, an ecologist who has seen more whales, dolphins, and porpoises than perhaps any person on Earth. In August of 1993 Pitman sailed into the upper Gulf on a ship called the *Ocean Starr* to track common dolphins. When the *Ocean Starr* crossed paths with two vaquitas, he convinced its captain to follow. For several days, the ship motored back and forth across the Gulf, its crew scanning the surface through binoculars for the vaquitas' black, triangular dorsal fins. On August 11 Pitman saw 25. "I think there is a good possibility," he told me, "that no one will ever see that many in a single day again."

In the years since Pitman's survey, the vaquita, never abundant, has entered a tailspin that is almost certain to end in its demise. In 1997 nearly 600 vaquitas swam the waters of the Gulf. A decade later there were 250. Then there were fewer than 100. Then 60. A 2016 report warned that the vaquita was "racing toward extinction"; a 2017 follow-up lamented that the collapse had "continued unabated." Today, fewer than 30 vaquitas remain. They are the world's most endangered marine mammal. "Every time I see one," Pitman told me, "I wonder: Is this the last one I'm going to see? Is this the last one *anyone*'s going to see?" Like most people invested in the porpoise's survival, he often sighs heavily. The word *intractable* is a fixture of his vocabulary. "We talk about extinction as a glib abstraction. But it's real, it's happening, and vaquita are next in line."

The damnedest thing about the conundrum is this: if ever an endangered species should be easy to save, this is it. Vaquitas are cute and, in their own introverted way, charismatic. They lack salable tusks, pelts, or meat. We know exactly where they live: not in some distant corner of Asia or Africa, but in our backyard. You can rent a car in San Diego in the afternoon, as I did, cross the border at Calexico, and arrive in vaquita country in time to eat dinner at a restaurant called, of course, La Vaquita.

Vaquitas, unfortunately, are collateral damage. They share their habitat with a fish called the totoaba, a mammoth cousin of the sea bass whose swim bladders are a delicacy worth up to $100,000 per kilogram in mainland China and Hong Kong. Although totoaba fishing has been banned since 1975—they, too, are critically endangered—poaching is rampant. Vaquitas, roughly the same size as totoabas, are prone to getting entangled and drowning in illegal nets. Demand for totoaba bladders soared in 2008, driven by an influx of cash into the Chinese economy; the dried organs became popular investment vehicles, a commodity as fungible as gold bars. Seventy-five hundred miles away, Mexico's black market erupted. Scientists, fishermen, and tourists soon began finding beached totoaba carcasses with the bladders cut from their bellies, meat left to rot. "Prices now are higher than cocaine," Lorenzo Rojas-Bracho, the Mexican biologist who chairs the vaquita recovery team, told me in the spring of 2017. "It's madness." That year, conservationists recovered 396 nets from the upper Gulf—48 tons of illegal gear.

As poaching surged, so did vaquita bycatch. Between 2008 and 2015, 80 percent of the world's vaquitas vanished—hauled up dead by fishermen, dumped surreptitiously back into the ocean. Biologists went two years without spotting a fin. The government, frantic, suspended all gillnetting in 2015, a measure that shut down a $50 million legal fishery for shrimp and fish but failed to stem poaching. American conservationists proposed a boycott on Mexican seafood to compel tougher enforcement. Miley Cyrus and Leonardo DiCaprio passed the hat on vaquitas' behalf. Some fishermen, ruined by onerous regulations and incensed at the impositions of foreigners, believe vaquitas are already extinct. Others claim they're a hoax. "Most people here," one fisherman in the coastal town of San Felipe told me, "think they don't exist."

The Gulf of California, known also as the Sea of Cortez, seems, at first, an improbable place to search for a marine mammal. The long finger of the Baja Peninsula is desert, all blinding sand, sere scrub, and crimped purple mountains; the upper Gulf—the Alto Golfo—materializes like a mirage, a band of pale shimmering water beneath a faded-denim sky. The Gulf here, where the beleaguered Colorado River limps to the finish line after its tortuous journey through the American Southwest, looks more like a sump than a sea. In summer, it can approach bathtub temperatures. These are among the most fecund waters on Earth, churned by tides into a nourishing soup trafficked by creatures from fin whales to great white sharks. When John Steinbeck sailed the upper Gulf in 1940, he found it "almost solid with fish—swarming, hungry, frantic fish, incredible in their voraciousness." Among those frantic fish are schools of coveted totoaba, which surge into the upper Gulf to spawn in late winter and early spring: the most profitable time of year to be a poacher, and the most dangerous to be a vaquita.

One March morning during the height of totoaba season, I climbed into a black inflatable raft at the San Felipe harbor, a dusty port cluttered with rusted shrimp trawlers and patrolled by pelicans. At the craft's wheel stood Raffaella Tolicetti, an athletic Italian, dark-haired and earnest, with a seabird's silhouette tattooed on her shoulder. Emblazoned on the raft's front panel was a modified Jolly Roger, its grinning skull hovering above a crossed trident and Little Bo Peep crook. This was the piratical logo of the Sea Shepherd Conservation Society, the controversial band of nautical vigilantes who had come to Mexico to save a species beyond saving.

Tolicetti motored me out to Sea Shepherd's 183-foot-long mothership, the SSS *Sam Simon*, anchored a half-mile offshore. ANTI-POACHING was printed in towering letters on the hull. Gaping shark jaws had been painted on the bow, as though the *Sam* was preparing to engulf any lawbreaker that crossed its path. We scaled a rope ladder onto the deck, which bristled with cranes and radar towers and smelled of iron and fish. A shaggy Australian named Alistair Allan, erstwhile punk guitarist and current first mate, arrived to show me around the warren of cramped bunks, low-ceilinged hallways, and steep staircases belowdecks.

"People spend months on this ship," he said. "We're kind of a weird family." The volunteer crew—quartermasters, engineers, welders—skewed young and European, their skin tattooed with creatures from every corner of the animal kingdom: penguins and wolves, snakes and squid, green squirrels and dancing boars. Allan's own bicep was adorned with a vengeful-looking seal gripping a spiked club in its flippers, perched atop the word JUSTICE.

Sea Shepherd was founded in 1977 by an activist named Paul Watson, shortly after he was excommunicated by Greenpeace for his radical tactics. Under Watson's leadership, Sea Shepherd emerged as a sort of extrastate maritime constabulary, hounding poachers and whalers across the high seas despite conspicuously lacking official enforcement powers. Sea Shepherd scuttled fishing boats and slashed enemy ships with a giant blade dubbed the Can Opener; after one of Watson's vessels slammed into a Costa Rican boat and sprayed its crew with fire hoses, local officials charged Watson with attempted murder. (Those charges were later dropped, though Costa Rica maintains a warrant for Watson's arrest, for shipwrecking.) The group waged its most famous fight against Japanese whalers in Antarctica, a battle immortalized by the reality show *Whale Wars*. During one memorable campaign, the *Sam Simon*—named for a creator of *The Simpsons*, who'd donated $2 million—was rammed by the *Nisshin Maru*, a massive factory whaling ship. When the vessels collided, Tolicetti recalled to me, the floor tilted and steel screeched, "as though we were a can of soda." The *Sam*, I noticed, still bore the scars: dented railings, smashed deck lights.

The vaquita campaign—dubbed Operation Milagro—was different. No longer was the *Sam Simon* cruising the lawless high seas; the upper Gulf lies in Mexico's national waters. There were no collisions or can openers. Instead, when the *Sam* and its sister ship, the *Farley Mowat*, stumbled upon poachers, they filmed them and alerted the Mexican Navy. The group passed the rest of its time peacefully dragging grappling hooks through the sea, snagging illegal nets. The pirate had become the police officer. When I visited the *Sam Simon*, no sooner had I boarded than the ship hooked a gillnet, which came up loaded with five totoabas, Doberman-sized fish with metallic scales and clouded eyes. "The nets, the illegal fishing—it never stops," one crewman grumbled.

The excitement had just begun. That afternoon, the *Sam*'s radar detected a panga, a small motorboat, about two miles out, squarely within a designated vaquita refuge. Allan lifted his binoculars to divine the boat's intentions. "The giveaway is that there's a large number of people on the bow," he said. "That's where they have to haul the net from. That's telltale." He set course toward the panga, then radioed the aft deck. It was time to fly.

I scrambled sternward and found the *Sam Simon*'s resident pilot, a mohawked Irishman named Jack Hutton, hastily assembling a white drone. A burst of radio chatter authorized takeoff. The drone leapt aloft, its rotors buzzing like an angry wasp. Hutton paced the deck, guiding the machine from a joystick that hung around his neck. A Samsung tablet mounted to the controller showed us the feed from the drone's camera, the sea rolling by at 40 knots. The crew gathered around, the air pregnant with chase.

"We have visual via the drone," Hutton said. The drone swung over the panga, whose captain opened throttle. The drone pursued. I leaned over Hutton's shoulder to watch the drama unfold on the screen. The drone hovered nearer the boat, which skipped along like a well-flung stone, casting a foamy V-shaped wake as it fled. It was a beautiful scene, as bright as a Mondrian canvas: the teal water, the white wake, the yellow slickers of four fishermen standing in the stern, eerily nonchalant. The drone's eye zoomed in and, with a shock, we saw a dozen totoabas lying like firewood on the panga's floor. Sixty thousand dollars of contraband, on the low end, in a country where the per capita income averages $3,000 a year. I found that I'd forgotten to breathe.

"Dude," one crewman murmured. "This is such a good shot."

After a few minutes of cat-and-mouse, the drone neared the end of its battery life, and zoomed back to the *Sam Simon*. The crew was jubilant. "That's one of the first times we've gotten footage of fish inside a panga," Allan said. Photos were relayed to the navy. Any minute, we figured, the cavalry would arrive.

But the afternoon took a disheartening turn. The panga cruised into shallow water, where the deep-hulled *Sam Simon* couldn't pursue. There it paused to pull another gillnet, and was soon joined by more pangas—four altogether, hauling illegal nets in broad daylight. Tolicetti bent over her phone, exchanging texts with a naval officer. "The guy keeps saying: 'Please don't lose them, please don't lose them,'" she fretted. "I'm like, OK, but just come!"

They didn't come. The sun sank behind the folded peaks of the Sierra de San Pedro Mártir; the stark day faded to dusk. In one report, I'd read that in the upper Gulf the navy had stationed a surveillance team whose arsenal included "a helicopter, Persuader and Maul marine patrol airplanes, six Defender rigid inflatable boats, two jet skis, four interceptors, three small boats, five pickup trucks, and two Unimog vehicles." That evening, though, it didn't dispatch so much as a canoe. The pangas converged and broke away like dancers, fearlessly exchanging fish and nets. I wondered if the *Sam Simon*'s crew missed the good old days of vigilantism.

Tolicetti and Allan ducked into the galley and returned with bowls brimming with pasta and beans—vegan, of course. Tolicetti's phone pinged again. "Oh no," she said. "Oh no. Holy cow."

"What?" Allan demanded.

"They found vaquitas dead."

"Vaquitas, plural?"

Tolicetti could only nod, her stricken face illuminated by the screen's glow. The *Sam Simon* turned to shore.

In 2006 a young Mexican-born marine biologist named Catalina López Sagástegui arrived in the upper Gulf to help make sense of the vaquita mess. She was no stranger to clashes between fishing communities and marine mammals—she'd worked with gray whales and bottlenose dolphins further south along the Baja Peninsula—but she quickly realized that she'd stumbled into a far nastier fight. Fishermen and conservationists could barely sit at the same table without triggering a shouting match. Much of the animosity, she surmised, stemmed from the fact that the parade of nonprofits had never fully acknowledged that fishermen were members of the ecosystem, as bound to the Gulf as the vaquita itself. "One of the first things fishermen told me when I got there," López Sagástegui said, "was: 'You don't have to live here—you get to go home.'"

López Sagástegui and her colleagues made gradual inroads, convincing some fishermen to go to sea with trackers on their boats so that scientists could collect data about the industry. And then the illegal totoaba trade detonated. When López Sagástegui came to the upper Gulf, she told me, "I thought it was the worst situation in the world." She laughed sadly. "Now when I look back on it, that was *easy*."

A week before I arrived in the Gulf, years of building pressure found violent release. In a coastal town called Golfo de Santa Clara, dozens of fishermen attacked officials from Mexican environmental agencies. The fishermen were furious that they hadn't been permitted to catch corvina, a slender relative of the totoaba that also spawns in the Gulf. The corvina fishery is among the region's most lucrative industries, and, unlike totoaba gillnets, corvina nets are generally too fine to ensnare vaquitas. But the government feared that *totoaberos* could use the corvina fishery as cover, and delayed issuing permits. Denied their final reliable source of income, fishermen lashed out, beating three inspectors and burning government property. Newspapers showed boats and trucks smoldering in the sand.

Two powerful forces seemed to be feeding each other, like colliding fronts building into a thunderstorm. First, there was the justifiable anxiety afflicting the area's fishermen. Not only had the corvina fishery been postponed, but the temporary gillnet ban was threatening to become permanent—even as a compensation program, which paid fishermen a monthly wage to offset their lost income, was nearing its end. (Several months later, the ban was made permanent and the compensation program extended.) Although the government had spent years developing vaquita-safe trawl nets, the new gear didn't catch shrimp as well as the old, and scientific reports lamented that the net-research process was plagued by an "absence of coordination and oversight."

Second, Mexico's infamous drug cartels were reportedly tightening their grip on the totoaba game. Mexican investigative journalists noted that totoaba-trafficking routes closely paralleled the Sinaloa Cartel's. Rumor had it that a crime boss had been murdered over totoaba-trafficking competition. Although fish trafficking was finally criminalized in April of 2017, poachers are rarely prosecuted. One night in San Felipe, while eating dinner at La Vaquita, a few scientists and I watched two pickup trucks tear down the street, pangas dragging from their trailer hitches, totoaba-gauge gillnets piled above the boats' gunwales. Many minutes later, a police car meandered along the road, sirens off; when it reached a fork, it cruised off in the opposite direction that the pangas had taken. "No es posible," one of my companions murmured with a sad smile.

Although San Felipe was more tranquil than Santa Clara, unease hung over the town. Fishermen who'd spoken out against poaching and corruption had received death threats. Now they refused to talk to me, even anonymously. A program that would pay 40 law-abiding fishermen to sweep the sea for illegal nets had been paused, for fear the government couldn't protect its helpers from reprisal. Even Sea Shepherd seemed spooked. Oona Layolle, Operation Milagro's French-born coleader, told me that poachers had recently escaped arrest by firing off assault rifles, a frightening escalation. (Months later, on Christmas Eve, *totoaberos* would shoot down one of the group's drones as it hovered overhead.) Soon after I left, fishermen demonstrated against Sea Shepherd's interference, burning a panga in downtown San Felipe to protest the foreigners' incursion.

Through the backlash ran a current of legitimate grievance. Several nights after my tour aboard the *Sam Simon,* I met two fishermen whose lives had been rearranged by vaquita recovery: spindly, white-mustached Victor Manuel Horozco and his husky, goateed colleague, Rafael Sánchez Gastélum. We rendezvoused on the Malecon, a seaside riviera that bustled with teenage bicyclists popping wheelies and vendors selling roasted corn and men playing tubas. A mural depicting a female vaquita and her calf swimming through a sunlit ocean adorned a nearby sanitation plant.

Gastélum did most of the talking. He'd been fishing for shrimp and finfish since he was 13, he told me, though these days he mostly took tourists sportfishing. Rather than raging against vaquita conservation, he'd accepted it. That was where the money was now, and recovering the porpoise seemed like the best way to revitalize his industry. He'd adopted vaquita-safe fishing gear, helped scientists monitor the dwindling population, and planned to assist the search for illegal nets. He received 8,000 pesos a month—around $400—as compensation for not fishing commercially. Sure, he'd seen vaquitas, though he wasn't awed by them. "It's just an animal," he said, shrugging.

Although Gastélum was one of the fishing community's most vocal spokesmen for saving the vaquita, he hardly seemed ecstatic about his new career in conservation. The monitoring payments barely covered his gas. The vaquita-safe trawls didn't work nearly as well as the old gillnets. The monthly checks were peanuts, less than he'd sometimes made in a single day of fishing. The more pangas and permits you owned, he pointed out, the larger your compensation package. "The person who gets two hundred thousand pesos, he is happy at home, he doesn't have to go fishing," he said. Even well-meaning recompense simply served to enrich the powerful. "My whole family has to eat," Gastélum added. He was working with scientists because he had no other choice. "If they allowed us to fish, I would be fishing."

Horozco listened silently as his friend talked, leaning against the Malecon's railing, the black mass of the Gulf to his back. When Gastélum at last wound down, he chimed in. The older man by a couple of decades, Horozco recalled a better time—a time not only before the whole vaquita mess, but when the Colorado River still reached the sea, before American dams and canals sucked it dry to water Las Vegas casinos and Imperial Valley crops. On the way down to San Felipe, I'd driven past La Salada: the river's former delta, once a lush mesquite paradise, now a forbidding desert, bright and hard and featureless as a frozen lake. "People would fish with hand nets from the road," Horozco said wistfully. "Now there's no water and no shrimp."

Unlike some locals, Horozco didn't deny that *totoaberos* were responsible for the vaquita crisis. But he also wanted me to know that Mexico's northern neighbor had been pillaging San Felipe's water for decades. It was easy to blame totoaba poaching and conservation restrictions for destroying the town's fishing industry, but truthfully its collapse had been unfolding since engineers in my own country drew up blueprints for the Hoover Dam. If the town's marine economy crumbled once and for all, Horozco wasn't sure what he'd do.

"This is a fishing town," he said. "It has always been. I can't become a miner or a construction worker. A job on the sea—I can do that."

How do you track the demise of a species so elusive that locals debate its very existence? You don't look; you listen. In 2011 a group of scientists, led by Armando Jaramillo-Legorreta, planted the Gulf with dozens of acoustic detection pods: narrow cylinders that record the vaquita's vocalizations, rapid bursts of clicks that, slowed down, tick like the dial of a rusty combination lock. That first year, some of the pods picked up over a thousand clicks per day. By 2017, none registered more than a hundred. Every year the Gulf grows emptier, lonelier, quieter.

To a conservation biologist named Barbara Taylor, the creeping silence was eerily familiar. More than a decade earlier, in 2006, Taylor and other researchers surveyed China's Yangtze River for the baiji, a freshwater dolphin endangered by dams, pollution, fishing, and shipping. If biologists could capture the final few baiji, the thinking went, they could relocate them to protected lakes. For two months, boats cruised the Yangtze with hydrophones, listening for the dolphin's call. Factories lining the banks belched effluent into the river. The team called off the search each afternoon when the smog grew too soupy. Every day, Taylor grew more certain they'd arrived too late. Later that year, the baiji was declared extinct—the first cetacean exterminated by humans. Unlike Martha the passenger pigeon or Lonesome George the tortoise, the baiji's endling never had a name.

Today, Taylor helms the scientific team monitoring the vaquita's decline, which means that, for the second time in a dozen years, she is documenting the dire final days of a vanishing cetacean. She lives up a steep driveway in the San Diego hills, where, after my trip to Mexico, I visited her and her husband, a marine-mammal biologist named Jay Barlow. I'd last spoken to her at a conference in Wisconsin the previous summer, at which she'd won a big award. Her once-obscure porpoise was finally generating headlines, but Taylor felt more like a hospice worker than an emergency room doctor. "I've been working on this for decades, and now I'm having a whole bunch of people show up and thank me for failing," she told me over a mozzarella salad. A black vaquita decorated her white sweatshirt.

Taylor is also a painter, an avocation that serves the dual roles of public relations and therapy. After lunch, we admired watercolors depicting bowhead whales, harbor porpoises, and, of course, vaquitas. She showed me her latest work-in-progress, which featured a flock of four vaquitas flying in formation behind a California condor—a giant vulture that had narrowly avoided extinction itself. She was considering adding another vaquita, this one hatching from a condor egg. "Would that be too cheesy?" she wondered, and laughed.

The condor was a carefully chosen symbol. Beginning in the 1990s, scientists saved the bird by capturing adults, breeding them, and reintroducing their offspring to the wild. In 2016 Taylor's recovery team announced that it would try to rescue vaquitas in similar fashion. The proposal, dubbed Vaquita CPR, was undeniably desperate. Porpoises, skittish and easily stressed, are more difficult to catch and house than, say, bottlenose dolphins. Unknowns abounded: whether vaquitas could be husbanded in floating pens; whether they could be kept healthy and well-fed and libidinous. Sea Shepherd worried that precious porpoises would die in the process; other groups feared the Mexican government would use the program as an excuse for slacking off on enforcement. "It's hard to feel good about putting animals in floating sea pens," Taylor admitted to me. "But what would that vaquita's life be like if it wasn't swimming around in that sea pen? Odds are, it would be dead."

In October of 2017, Vaquita CPR, endowed with a 67-person team, a $5 million budget, and a 135-foot-long repurposed Bering Sea crabber called the *Maria Cleofas,* launched operations. The project started auspiciously: the team managed to routinely locate the reclusive porpoises, including several mother-calf pairs. On October 18, the *Maria Cleofas* came upon a trio of vaquitas and successfully herded one, a juvenile, into a net strung behind a smaller capture boat. Veterinarians eased the calf into a transport box, dampened her dorsal fin with wet cloths, and boated her shoreward to El Nido, or the Nest, the circular mesh sea pen. But the calf didn't take to her new environs, darting blindly around the pen, colliding with handlers, and floating at the surface, apparently exhausted. The crew, fearing for her health, rushed her back to the capture site and released her, hopefully to reunite with her mother. The failure concerned Vaquita CPR's staff, but it didn't defeat the project. Perhaps the group needed to try an adult.

Seventeen days later, at 4:18 p.m. on Sunday, November 4, the opportunity arose. An adult female got her flippers and fluke snared in the capture net; the crew had her untangled and onboard within two minutes. At 4:26 she received an injection of a sedative called diazepam and three minutes later was transferred to another boat, the *Defender,* in a soft-sided sling. "She was really calm throughout that whole process," Frances Gulland, the Marine Mammal Center senior scientist who helped oversee the capture, told me. "Her breathing rate and her heart rate were steady. We thought: 'Oh, great, she's going to be the older matriarch-type female. She's going to be a great animal to start out with.'"

The ride to El Nido aboard the *Defender* took an hour. The sun set; the sky turned pink and then purple. At 6:42 p.m., the group slipped the porpoise into the net pen, and her tranquility evaporated.

"She swam really fast at the sides, and then right before hitting the net would do kind of a somersault, like an Olympic swimmer doing turns," Gulland recalled. "She never slowed down. She actually accelerated. She was swimming faster and faster." Scientists surrounded the pen, their faces furrowed with concern. At 6:57 she slowed down and floated like a log at the surface. Half an hour later she went limp. The team hurried her out of the pen and into the ocean, an emergency release. The vaquita raced away and then, to their horror, turned back, hastening toward the net

pen, disoriented and frantic. She would have collided with the side
of an adjacent boat if they hadn't netted her again. This time her
heartbeat was faint. She'd stopped breathing.

For nearly three hours desperate veterinarians ministered to the
deteriorating animal. She was intubated, ventilated, and hooked
up to intravenous fluids. The team massaged the cool, rubbery skin
of her chest, felt the thump of her heart slow through her blubber layer and then speed up. Her blowhole gasped open, closed,
open. At 10:10 p.m., she went into cardiac arrest. Eleven minutes
later the team declared her dead. When Gulland necropsied her
body that night, beneath fluorescent lights in a concrete-walled
room at a facility called Camp Uno, she found the porpoise's heart
muscle had turned pale. She'd suffered capture myopathy, an extreme stress reaction. Her dorsal fin and tail, Gulland noticed,
were notched with faded net scars.

In the episode's aftermath, most media outlets portrayed it
as the nail in the vaquita's coffin: a "final bid," a "last-ditch effort." But a skeletal crew of vaquitas—a dozen, according to one
watchdog group—still hung on in the Gulf, and they appeared
to be breeding. If the tragedy had a silver lining, Taylor told me,
it was that it had focused the world's attention, more than ever,
on the Gulf of California. The Mexican government had pledged,
for the umpteenth time, to strengthen enforcement; maybe now
they meant it. Perhaps, too, the surviving vaquitas had survived
for a reason. "There's a lot of natural selection going on here,"
Taylor added. "The ones that are left, they're not random. These
animals recognized nets. They knew how to go around nets." Conservation biologists are, by definition and necessity, a hopeful lot;
concession was unthinkable, even if the game appeared over. Let
down at every turn by humans, perhaps vaquitas would yet protect
themselves.

In its 11th hour, the vaquita, like the baiji before it, has become a
cause célèbre—elevated from anonymity just in time for us to remark on its likely passing. Meanwhile, the international spotlight
still hasn't shone upon legions of other obscure cetaceans, many
dwindling by the day: Western Africa's Atlantic humpback dolphin,
Southeast Asia's Irrawaddy dolphin, India's Ganges River dolphin.
These animals share a key trait: they inhabit the rivers and coastlines
of developing countries where growing numbers of people derive

their livelihoods from the water. Saving them will require tough laws, rigorous enforcement, and scientific research. But it will also require that conservationists help coastal communities develop economic opportunities that don't jeopardize these species' lives.

Over the years, ill-conceived attempts to diversify the Gulf's economy have come and gone. Most bear the whiff of the tragicomic. In 2007, for instance, the Mexican government offered fishermen millions of dollars to turn in their gillnets and launch ecotourism businesses—which promptly collapsed when the great recession hit the following year. The most effective projects, the marine biologist Catalina López Sagástegui has come to believe, are the ones that begin from the bottom up: a single scientist penetrates a community, gains its trust, and labors for years to fit conservation solutions into its extant culture. The Gulf, she told me, had made fitful progress in that regard—evidenced, for instance, by fishermen agreeing to surrender their gillnets to try vaquita-safe gear. "We have the world looking at Mexico, and I think a lot of people will come out of this knowing a lot more." For the vaquita itself, though, that hard-won knowledge has almost certainly coalesced too late: by the time you read these words, its endling may well be dead.

For now, the vaquita's immediate future hinges on the valiant and imperfect efforts of well-intentioned vigilantes—foreigners working not with fishermen but against them. Two days after Sea Shepherd's high-speed drone chase, the SSS *Sam Simon* deviated from its routine to search for a corpse. Raffaella Tolicetti had shown us photos, taken by locals, of two recent vaquita victims: an adult male, his black skin ragged and torn, and an infant, wrinkled from the womb. The calf had washed ashore, but the adult remained adrift. The carcass, if Sea Shepherd could recover it, had a morbid value, ironclad proof that the species persisted.

It was a glorious day, the sea flat as a billiard table. Marine life rioted around us: dolphins frolicked in our wake, frigate birds flirted with the radar tower, fin whales crested to starboard. If vaquitas were this playful, I thought, they'd already be saved. The wariness that served them well in nature was also their downfall. Invisibility, in the realm of conservation, is death, no matter how cute you are. Whale-watching had saved the whales, but no ecotourism industry would bail out a creature so rare and retiring that its very existence was a matter of dispute.

We didn't find the vaquita's body that afternoon, though it would turn up days later, gnawed by fish and baked by the sun. Our search was curtailed by another crisis: the *Farley Mowat* had pulled a net containing a record 66 totoabas, and Sea Shepherd's media gurus had to document the haul. The *Farley* carried the rotting fish ashore, where officials lined them up and knifed out their prized bladders. The dissection took hours. The smell was horrific. I wasn't allowed to attend, but when I spoke with Sea Shepherd's media director all he could talk about were the flies, the swarms that emerged to blacken the boats, the trucks, the nets, the dead fish baking on the concrete dock. The vroom of millions of tiny wings. "It was like a nightmare," he said. They buried the whole foul mess in the desert.

A few days later I drove north up Highway 5, back to Los Estados. Uncountable legions of sphinx-moth caterpillars had hatched in the night, and their bodies, green and corpulent and unavoidable, inched across the pocked pavement, turning to jelly beneath my tires. As I passed through the sunbaked wastes of La Salada, the radio turned to static. Without auditory stimulus, my mind wandered, and I thought of something Oona Layolle told me. When Paul Watson, Sea Shepherd's legendary Ahab, had first proposed a vaquita campaign, other captains had rejected it as hopeless. Only Layolle volunteered, her motives less messianic than journalistic. "Even if they disappear we can be there to document, record, and show the world that this cannot happen again," she told me. If the vaquita blinked out, it would die as a media martyr, its expiration blogged and tweeted and filmed by drones. An endling for the Digital Age. Whether we'll learn anything from its demise is up to us. Our track record does not inspire confidence.

GARY GREENBERG

What If the Placebo Effect Is Not a Trick?

FROM *The New York Times Magazine*

THE CHAIN OF Office of the Dutch city of Leiden is a broad and colorful ceremonial necklace that, draped around the shoulders of Mayor Henri Lenferink, lends a magisterial air to official proceedings in this ancient university town. But whatever gravitas it provided Lenferink as he welcomed a group of researchers to his city, he was quick to undercut it. "I am just a humble historian," he told the 300 members of the Society for Interdisciplinary Placebo Studies who had gathered in Leiden's ornate municipal concert hall, "so I don't know anything about your topic." He was being a little disingenuous. He knew enough about the topic that these psychologists and neuroscientists and physicians and anthropologists and philosophers had come to his city to talk about—the placebo effect, the phenomenon whereby suffering people get better from treatments that have no discernible reason to work—to call it "fake medicine," and to add that it probably works because "people like to be cheated." He took a beat. "But in the end, I believe that honesty will prevail."

Lenferink might not have been so glib had he attended the previous day's meeting on the other side of town, at which two dozen of the leading lights of placebo science spent a preconference day agonizing over their reputation—as purveyors of sham medicine who prey on the desperate and, if they are lucky, fool people into feeling better—and strategizing about how to improve it. It's an urgent subject for them, and only in part because, like all apostate professionals, they crave mainstream acceptance. More important, they are motivated by a conviction that the placebo is a powerful medical treatment that is ignored by doctors only at their patients' expense.

And after a quarter century of hard work, they have abundant evidence to prove it. Give people a sugar pill, they have shown, and those patients—especially if they have one of the chronic, stress-related conditions that register the strongest placebo effects and if the treatment is delivered by someone in whom they have confidence—will improve. Tell someone a normal milkshake is a diet beverage, and his gut will respond as if the drink were low fat. Take athletes to the top of the Alps, put them on exercise machines, and hook them to an oxygen tank, and they will perform better than when they are breathing room air—even if room air is all that's in the tank. Wake a patient from surgery and tell him you've done an arthroscopic repair, and his knee gets better even if all you did was knock him out and put a couple of incisions in his skin. Give a drug a fancy name, and it works better than if you don't.

You don't even have to deceive the patients. You can hand a patient with irritable bowel syndrome a sugar pill, identify it as such and tell her that sugar pills are known to be effective when used as placebos, and she will get better, especially if you take the time to deliver that message with warmth and close attention. Depression, back pain, chemotherapy-related malaise, migraine, post-traumatic stress disorder: the list of conditions that respond to placebos—as well as they do to drugs, with some patients—is long and growing.

But as ubiquitous as the phenomenon is, and as plentiful the studies that demonstrate it, the placebo effect has yet to become part of the doctor's standard armamentarium—and not only because it has a reputation as "fake medicine" doled out by the unscrupulous to the credulous. It also has, so far, resisted a full understanding, its mechanisms shrouded in mystery. Without a clear knowledge of how it works, doctors can't know when to deploy it, or how.

Not that the researchers are without explanations. But most of these have traditionally been psychological in nature, focusing on mechanisms like expectancy—the set of beliefs that a person brings into treatment—and the kind of conditioning that Ivan Pavlov first described more than a century ago. These theories, which posit that the mind acts upon the body to bring about physical responses, tend to strike doctors and researchers steeped in the scientific tradition as insufficiently scientific to lend credibility to the placebo effect. "What makes our research believable to doctors?" asks Ted Kaptchuk, head of Harvard Medical School's Program in Placebo Studies and the Therapeutic Encounter. "It's the molecules. They love that stuff." As of now, there are no molecules for conditioning or expectancy—or, indeed, for Kaptchuk's own pet theory, which holds that the placebo effect is a result of the complex conscious and nonconscious processes embedded in the practitioner-patient relationship—and without them, placebo researchers are hard-pressed to gain purchase in mainstream medicine.

But as many of the talks at the conference indicated, this might be about to change. Aided by functional magnetic resonance imaging (fMRI) and other precise surveillance techniques, Kaptchuk and his colleagues have begun to elucidate an ensemble of biochemical processes that may finally account for how placebos work and why they are more effective for some people, and some disorders, than others. The molecules, in other words, appear to be emerging. And their emergence may reveal fundamental flaws in the way we understand the body's healing mechanisms, and the way we evaluate whether more standard medical interventions in those processes work, or don't. Long a useful foil for medical science, the placebo effect might soon represent a more fundamental challenge to it.

In a way, the placebo effect owes its poor reputation to the same man who cast aspersions on going to bed late and sleeping in. Benjamin Franklin was, in 1784, the ambassador of the fledgling United States to King Louis XVI's court. Also in Paris at the time was a Viennese physician named Franz Anton Mesmer. Mesmer fled Vienna a few years earlier when the local medical establishment determined that his claim to have cured a young woman's blindness by putting her into a trance was false, and that, even worse, there was something unseemly about his relationship with her. By the time he arrived in Paris and hung out his shingle, Mesmer had acquired what he lacked in Vienna: a theory to account for his ability to use trance states to heal people. There was, he claimed, a force pervading the universe called animal magnetism that could cause illness when perturbed. Conveniently enough for Mesmer, the magnetism could be perceived and de-perturbed only by him and people he had trained.

Mesmer's method was strange, even in a day when doctors

routinely prescribed bloodletting and poison to cure the common cold. A group of people complaining of maladies like fatigue, numbness, paralysis, and chronic pain would gather in his office, take seats around an oak cask filled with water, and grab on to metal rods immersed in the water. Mesmer would alternately chant, play a glass harmonium, and wave his hands at the afflicted patients, who would twitch and cry out and sometimes even lose consciousness, whereupon they would be carried to a recovery room. Enough people reported good results that patients were continually lined up at Mesmer's door waiting for the next session.

It was the kind of success likely to arouse envy among doctors, but more was at stake than professional turf. Mesmer's claim that a force existed that could only be perceived and manipulated by the elect few was a direct challenge to an idea central to the Enlightenment: that the truth could be determined by anyone with senses informed by skepticism, that Scripture could be supplanted by facts and priests by a democracy of people who possessed them. So when the complaints about Mesmer came to Louis, it was to the scientists that the king—at pains to show himself an enlightened man—turned. He appointed, among others, Lavoisier the chemist, Bailly the astronomer, and Guillotin the physician to investigate Mesmer's claims, and he installed Franklin at the head of their commission.

To the Franklin commission, the question wasn't whether Mesmer was a fraud and his patients were dupes. Everyone could be acting in good faith, but belief alone did not prove that the magnetism was at work. To settle this question, they designed a series of trials that ruled out possible causes of the observed effects other than animal magnetism. The most likely confounding variable, they thought, was some faculty of mind that made people behave as they did under Mesmer's ministrations. To rule this out, the panel settled upon a simple method: a blindfold. Over a period of a few months, they ran a series of experiments that tested whether people experienced the effects of animal magnetism even when they couldn't see.

One of Mesmer's disciples, Charles d'Eslon, conducted the tests. The panel instructed him to wave his hands at a part of a patient's body, and then asked the patient where the effect was felt. They took him to a copse to magnetize a tree—Mesmer claimed that a patient could be treated by touching one—and then asked the patient to find it. They told patients d'Eslon was in the room when he was not, and vice versa, or that he was doing something that he was not. In trial after trial, the patients responded as if the doctor were doing what they thought he was doing, not what he was actually doing.

It was possibly the first-ever blinded experiment, and it soundly proved what scientists today call the null hypothesis: there was no causal connection between the behavior of the doctor and the response of the patients, which meant, as Franklin's panel put it in their report, that "this agent, this fluid, has no existence." That didn't imply that people were *pretending* to twitch or cry out, or lying when they said they felt better; only that their behavior wasn't a result of this nonexistent force. Rather, the panel wrote, "the imagination singly produces all the effects attributed to the magnetism."

When the panel gave d'Eslon a preview of its findings, he took it with equanimity. Given the results of the treatment (as opposed to the experiment), he opined, the imagination, "directed to the relief of suffering humanity, would be a most valuable means in the hands of the medical profession"—a subject to which these august scientists might wish to apply their methods. But events intervened. Franklin was called back to America in 1785; Louis XVI had bigger trouble on his hands and, along with Lavoisier and Bailly, eventually met with the short, sharp shock of the device named for Guillotin.

The panel's report was soon translated into English by William Godwin, the father of Mary Shelley. The story spread fast—not because of the healing potential that d'Eslon had suggested, but because of the implications for science as a whole. The panel had demonstrated that by putting imagination out of play, science could find the truth about our suffering bodies, in the same way it had found the truth about heavenly bodies. Hiving off subjectivity from the rest of medical practice, the Franklin commission had laid the conceptual foundation for the brilliant discoveries of modern medicine, the antibiotics and vaccines and other drugs that can be dispensed by whoever happens to possess the prescription pad, and to whoever happens to have the disease. Without meaning to, they had created an epistemology for the healing arts —and in the process, inadvertently conjured the placebo effect, and established it as that to which doctors must remain blind.

It wouldn't be the last time science would turn its focus to

the placebo effect only to quarantine it. At a 1955 meeting of
the American Medical Association, the Harvard surgeon Henry
Beecher pointed out to his colleagues that while they might have
thought that placebos were fake medicine—even the name, which
means "I shall please" in Latin, carries more than a hint of con-
tempt—they couldn't deny that the results were real. Beecher had
been looking at the subject systematically, and he determined that
placebos could relieve anxiety and postoperative pain, change the
blood chemistry of patients in a way similar to drugs, and even
cause side effects. In general, he told them, more than one-third
of patients would get better when given a treatment that was, pharmacologically speaking, inert.

If the placebo was as powerful as Beecher said, and if doctors
wanted to know whether their drugs actually worked, it was not
sufficient simply to give patients the drugs and see whether they
did better than patients who didn't interact with the doctor at all.
Instead, researchers needed to assume that the placebo effect was
part of every drug effect, and that drugs could be said to work only
to the extent that they worked better than placebos. An accurate
measure of drug efficacy would require comparing the response of
patients taking it with that of patients taking placebos; the drug effect could then be calculated by subtracting the placebo response
from the overall response, much as a deli-counter worker subtracts
the weight of the container to determine how much lobster salad
you're getting.

In the last half of the 1950s, this calculus gave rise to a new way
to evaluate drugs: the double-blind, placebo-controlled clinical
trial, in which neither patient nor clinician knew who was getting
the active drug and who the placebo. In 1962, when the Food and
Drug Administration began to require pharmaceutical companies
to prove their new drugs were effective before they came to market, they increasingly turned to the new method; today, virtually
every prospective new drug has to outperform placebos on two
independent studies in order to gain FDA approval.

Like Franklin's commission, the FDA had determined that the
only way to sort out the real from the fake in medicine was to iso-
late the imagination. It also echoed the royal panel by taking note
of the placebo effect only long enough to dismiss it, giving it a
strange dual nature: it's included in clinical trials because it is recognized as an important part of every treatment, but it is treated

as if it were not important in itself. As a result, although virtually every clinical trial is a study of the placebo effect, it remains underexplored—an outcome that reflects the fact that there is no money in sugar pills and thus no industry interest in the topic as anything other than a hurdle it needs to overcome.

When Ted Kaptchuk was asked to give the opening keynote address at the conference in Leiden, he contemplated committing the gravest heresy imaginable: kicking off the inaugural gathering of the Society for Interdisciplinary Placebo Studies by declaring that there was no such thing as the placebo effect. When he broached this provocation in conversation with me not long before the conference, it became clear that his point harked directly back to Franklin: that the topic he and his colleagues studied was created by the scientific establishment, and only in order to exclude it—which means that they are always playing on hostile terrain. Science is "designed to get rid of the husks and find the kernels," he told me. Much can be lost in the threshing—in particular, Kaptchuk sometimes worries, the rituals embedded in the doctor-patient encounter that he thinks are fundamental to the placebo effect, and that he believes embody an aspect of medicine that has disappeared as scientists and doctors pursue the course laid by Franklin's commission. "Medical care is a moral act," he says, in which a suffering person puts his or her fate in the hands of a trusted healer.

"I don't love science," Kaptchuk told me. "I want to know what heals people." Science may not be the only way to understand illness and healing, but it is the established way. "That's where the power is," Kaptchuk says. That instinct is why he left his position as director of a pain clinic in 1990 to join Harvard—and it's why he was delighted when, in 2010, he was contacted by Kathryn Hall, a molecular biologist. Here was someone with an interest in his topic who was also an expert in molecules, and who might serve as an emissary to help usher the placebo into the medical establishment.

Hall's own journey into placebo studies began 15 years before her meeting with Kaptchuk, when she developed a bad case of carpal tunnel syndrome. Wearing a wrist brace didn't help, and neither did over-the-counter drugs or the codeine her doctor prescribed. When a friend suggested she visit an acupuncturist, Hall balked at the idea of such an unscientific approach. But faced with the alternative, surgery, she decided to make an appointment. "I was there for maybe ten minutes," she recalls, "when she stuck a needle here"—Hall points to a spot on her forearm—"and this awful pain just shot through my arm." But then the pain receded and her symptoms disappeared, as if they had been carried away on the tide. She received a few more treatments, during which the acupuncturist taught her how to manipulate a spot near her elbow if the pain recurred. Hall needed the fix from time to time, but the problem mostly just went away.

"I couldn't believe it," she told me. "Two years of gross drugs, and then just one treatment." All these years later, she's still won-der-struck. "What was that?" she asks. "Rub the spot, and the pain just goes away?"

Hall was working for a drug company at the time, but she soon left to get a master's degree in visual arts, after which she started a documentary-production company. She was telling her carpal-tun-nel story to a friend one day and recounted how the acupuncturist had climbed up on the table with her. ("I was like, 'Oh, my God, what is this woman doing?'" she told me. "It was very dramatic.") She'd never been able to understand how the treatment worked, and this memory led her to wonder out loud if maybe the drama itself had something to do with the outcome.

Her friend suggested she might find some answers in Ted Kaptchuk's work. She picked up his book about Chinese medicine, *The Web That Has No Weaver*, in which he mentioned the possibility that placebo effects figure strongly in acupuncture, and then she read a study he had conducted that put that question to the test.

Kaptchuk had divided people with irritable bowel syndrome into three groups. In one, acupuncturists went through all the mo-tions of treatment, but used a device that only appeared to insert a needle. Subjects in a second group also got sham acupuncture, but delivered with more elaborate doctor-patient interaction than the first group received. A third group was given no treatment at all. At the end of the trial, both treatment groups improved more than the no-treatment group, and the "high interaction" group did best of all.

Kaptchuk, who before joining Harvard had been an acupunc-turist in private practice, wasn't particularly disturbed by the find-ing that his own profession worked even when needles were not actually inserted; he'd never thought that placebo treatments were fake medicine. He was more interested in how the strength of the treatment varied with the quality and quantity of interaction between the healer and the patient—the drama, in other words. Hall reached out to him shortly after she read the paper.

The findings of the IBS study were in keeping with a hypothesis Kaptchuk had formed over the years: that the placebo effect is a biological response to an act of caring; that somehow the encounter itself calls forth healing and that the more intense and focused it is, the more healing it evokes. He elaborated on this idea in a comparative study of conventional medicine, acupuncture, and Navajo "chantway rituals," in which healers lead storytelling ceremonies for the sick. He argued that all three approaches unfold in a space set aside for the purpose and proceed as if according to a script, with prescribed roles for every participant. Each modality, in other words, is its own kind of ritual, and Kaptchuk suggested that the ritual itself is part of what makes the procedure effective, as if the combined experiences of the healer and the patient, reinforced by the special but familiar surroundings, evoke a healing response that operates independently of the treatment's specifics. "Rituals trigger specific neurobiological pathways that specifically modulate bodily sensations, symptoms and emotions," he wrote. "It seems that if the mind can be persuaded, the body can sometimes act accordingly." He ended that paper with a call for further scientific study of the nexus between ritual and healing.

When Hall contacted him, she seemed like a perfect addition to the team he was assembling to do just that. He even had an idea of exactly how she could help. In the course of conducting the study, Kaptchuk had taken DNA samples from subjects in hopes of finding some molecular pattern among the responses. This was an investigation tailor-made to Hall's expertise, and she agreed to take it on. Of course, the genome is vast, and it was hard to know where to begin—until, she says, she and Kaptchuk attended a talk in which a colleague presented evidence that an enzyme called COMT affected people's response to pain and painkillers. Levels of that enzyme, Hall already knew, were also correlated with Parkinson's disease, depression, and schizophrenia, and in clinical trials people with those conditions had shown a strong placebo response. When they heard that COMT was also correlated with pain response—another area with significant placebo effects—

Hall recalls, "Ted and I looked at each other and were like: 'That's it! That's it!'"

It is not possible to assay levels of COMT directly in a living brain, but there is a snippet of the genome called rs4680 that governs the production of the enzyme, and that varies from one person to another: one variant predicts low levels of COMT, while another predicts high levels. When Hall analyzed the IBS patients' DNA, she found a distinct trend. Those with the high-COMT variant had the weakest placebo responses, and those with the opposite variant had the strongest. These effects were compounded by the amount of interaction each patient got: for instance, low-COMT, high-interaction patients fared best of all, but the low-COMT subjects who were placed in the no-treatment group did *worse* than the other genotypes in that group. They were, in other words, more sensitive to the impact of the relationship with the healer.

The discovery of this genetic correlation to placebo response set Hall off on a continuing effort to identify the biochemical ensemble she calls the *placebome*—the term reflecting her belief that it will one day take its place among the other important "-omes" of medical science, from the genome to the microbiome. The rs4680 gene snippet is one of a group that governs the production of COMT, and COMT is one of a number of enzymes that determine levels of catecholamines, a group of brain chemicals that includes dopamine and epinephrine. (Low COMT tends to mean higher levels of dopamine, and vice versa.) Hall points out that the catecholamines are associated with stress, as well as with reward and good feeling, which bolsters the possibility that the placebome plays an important role in illness and health, especially in the chronic, stress-related conditions that are most susceptible to placebo effects.

Her findings take their place among other results from neuroscientists that strengthen the placebo's claim to a place at the medical table, in particular studies using fMRI machines that have found consistent patterns of brain activation in placebo responders. "For years, we thought of the placebo effect as the work of imagination," Hall says. "Now through imaging you can literally see the brain lighting up when you give someone a sugar pill."

One group with a particularly keen interest in those brain images, as Hall well knows, is her former employers in the pharmaceutical industry. The placebo effect has been plaguing their business for more than a half century—since the placebo-controlled study became the clinical-trial gold standard, requiring a new drug to demonstrate a significant therapeutic benefit over placebo to gain FDA approval.

That's a bar that is becoming ever more difficult to surmount, because the placebo effect seems to be becoming stronger as time goes on. A 2015 study published in the journal *Pain* analyzed 84 clinical trials of pain medication conducted between 1990 and 2013 and found that in some cases the efficacy of placebo had grown sharply, narrowing the gap with the drugs' effect from 27 percent on average to just 9 percent. The only studies in which this increase was detected were conducted in the United States, which has spawned a variety of theories to explain the phenomenon: that patients in the United States, one of only two countries where medications are allowed to be marketed directly to consumers, have been conditioned to expect greater benefit from drugs; or that the larger and longer-duration trials more common in America have led to their often being farmed out to contract organizations whose nurses' only job is to conduct the trial, perhaps fostering a more placebo-triggering therapeutic interaction.

Whatever the reason, a result is that drugs that pass the first couple of stages of the FDA approval process founder more and more frequently in the larger late-stage trials; more than 90 percent of pain medications now fail at this stage. The industry would be delighted if it were able to identify placebo responders—say, by their genome—and exclude them from clinical trials.

That may seem like putting a thumb on the scale for drugs, but under the logic of the drug-approval regime, to eliminate placebo effects is not to cheat; it merely reduces the noise in order for the drug's signal to be heard more clearly. That simple logic, however, may not hold up as Hall continues her research into the genetic basis of the placebo. Indeed, that research may have deeper implications for clinical drug trials, and for the drugs themselves, than pharma companies might expect.

Since 2013, Hall has been involved with the Women's Health Study, which has tracked the cardiovascular health of nearly 40,000 women over more than 20 years. The subjects were randomly divided into four groups, following standard clinical-trial protocol, and received a daily dose of either vitamin E, aspirin, vitamin E with aspirin, or a placebo. A subset also had their DNA sampled

—which, Hall realized, offered her a vastly larger genetic database to plumb for markers correlated to placebo response. Analyzing the data amassed during the first 10 years of the study, Hall found that the women with the low-COMT gene variant had significantly higher rates of heart disease than women with the high-COMT variant, and that the risk was reduced for those low-COMT women who received the active treatments but not in those given placebos. Among high-COMT people, the results were the inverse: women taking placebos had the lowest rates of disease; people in the treatment arms had an increased risk.

These findings in some ways seem to confound the results of the IBS study, in which it was the low-COMT patients who benefited most from the placebo. But, Hall argues, what's important isn't the direction of the effect, but rather that there *is* an effect, one that varies depending on genotype—and that the same gene variant also seems to determine the relative effectiveness of the drug. This outcome contradicts the logic underlying clinical trials. It suggests that placebo and drug do not involve separate processes, one psychological and the other physical, that add up to the overall effectiveness of the treatment; rather, they may both operate on the same biochemical pathway—the one governed in part by the COMT gene.

Hall has begun to think that the placebome will wind up essentially being a chemical pathway along which healing signals travel —and not only to the mind, as an experience of feeling better, but also to the body. This pathway may be where the brain translates the act of caring into physical healing, turning on the biological processes that relieve pain, reduce inflammation, and promote health, especially in chronic and stress-related illnesses—like irritable bowel syndrome and some heart diseases. If the brain employs this same pathway in response to drugs and placebos, then of course it is possible that they might work together, like convoys of drafting trucks, to traverse the territory. But it is also possible that they will encroach on one another, that there will be traffic jams in the pathway.

What if, Hall wonders, a treatment fails to work not because the drug and the individual are biochemically incompatible, but rather because in some people the drug interferes with the placebo response, which if properly used might reduce disease? Or conversely, what if the placebo response is, in people with a different variant, working against drug treatments, which would mean that a change in the psychosocial context could make the drug more effective? Everyone may respond to the clinical setting, but there is no reason to think that the response is always positive. According to Hall's new way of thinking, the placebo effect is not just some constant to be subtracted from the drug effect but an intrinsic part of a complex interaction among genes, drugs, and mind. And if she's right, then one of the cornerstones of modern medicine—the placebo-controlled clinical trial—is deeply flawed.

When Kathryn Hall told Ted Kaptchuk what she was finding as she explored the relationship of COMT to the placebo response, he was galvanized. "Get this molecule on the map!" he urged her. It's not hard to understand his excitement. More than two centuries after d'Eslon suggested that scientists turn their attention directly to the placebo effect, she did exactly that and came up with a finding that might have persuaded even Ben Franklin.

But Kaptchuk also has a deeper unease about Hall's discovery. The placebo effect can't be totally reduced to its molecules, he feels certain—and while research like Hall's will surely enhance its credibility, he also sees a risk in playing his game on scientific turf. "Once you start measuring the placebo effect in a quantitative way," he says, "you're transforming it to be something other than what it is. You suck out what was previously there and turn it into science." Reduced to its molecules, he fears, the placebo effect may become "yet another thing on the conveyor belt of routinized care."

"We're dancing with the devil here," Kaptchuk once told me, by way of demonstrating that he was aware of the risks he's taking in using science to investigate a phenomenon it defined only to exclude. Kaptchuk, an observant Jew who is a student of both the Torah and the Talmud, later modified his comment. It's more like Jacob wrestling with the angel, he said—a battle that Jacob won, but only at the expense of a hip injury that left him lame for the rest of his life.

Indeed, Kaptchuk seems wounded when he complains about the pervasiveness of research that uses healthy volunteers in academic settings, as if the response to mild pain inflicted on an undergraduate participating in an on-campus experiment is somehow comparable to the despair often suffered by people with chronic, intractable pain. He becomes annoyed when he talks about how quickly some of his colleagues want to move from these studies to clinical recommendations. And he can even be disparaging of his own work, wondering, for instance, whether the study in which placebos were openly given to irritable bowel syndrome patients succeeded only because it convinced the subjects that the sugar was really a drug. But it's the prospect of what will become of his findings, and of the placebo, as they make their way into clinical practice, that really seems to torment him.

Kaptchuk may wish "to help reconfigure biomedicine by rejecting the idea that healing is only the application of mechanical tools." He may believe that healing is a moral act in which "caring in the context of hope qualitatively changes clinical outcomes." He may be convinced that the relationship kindled by the encounter between a suffering person and a healer is a central, and almost entirely overlooked, component of medical treatment. And he may have dedicated the last 20 years of his life to persuading the medical establishment to listen to him. But he may also come to regret the outcome.

After all, if Hall is right that clinician warmth is especially effective with a certain genotype, then, as she wrote in the paper presenting her findings from the IBS/sham-acupuncture study, it is also true that a different group will "derive minimum benefit" from "empathic attentions." Should medical rituals be doled out according to genotype, with warmth and caring withheld in order to clear the way for the drugs? And if she is correct that a certain ensemble of neurochemical events underlies the placebo effect, then what is to stop a drug company from manufacturing a drug —a real drug, that is—that activates the same process pharmacologically? Welcomed back into the medical fold, the placebo effect may raise enough mischief to make Kaptchuk rue its return, and bewilder patients when they discover that their doctor's bedside manner is tailored to their genes.

For the most part, most days, Kaptchuk manages to keep his qualms to himself, to carry on as if he were fully confident that scientific inquiry can restore the moral dimension to medicine. But the precariousness of his endeavors is never far from his mind. "Will this work destroy the stuff that actually has to do with wisdom, preciousness, imagination, the things that are actually critical to who we are as human beings?" he asks. His answer: "I don't know, but I have to believe there is an infinite reserve of wisdom and imagination that will resist being reduced to simple materialistic explanations."

The ability to hold two contradictory thoughts in mind at the same time seems to come naturally to Kaptchuk, but he may overestimate its prevalence in the rest of us. Even if his optimism is well placed, however, there's nothing like being sick to make a person toss that kind of intelligence aside in favor of the certainties offered by modern medicine. Indeed, it's exactly that yearning that sickness seems to awaken and that our healers, imbued with the power of science, purport to provide, no imagination required. Armed with our confidence in them, we're pleased to give ourselves over to their ministrations, and pleased to believe that it's the molecules, and the molecules alone, that are healing us. People do like to be cheated, after all.

JEREMY HANCE

The Great Rhino U-Turn

FROM *Mongabay*

AS WE WALK out into the zoo enclosure, Cossatot comes over to greet me. Cossatot is a capybara, the size of a very big dog; his species is the world's largest rodent. He quickly determines from smelling my hands that I've neglected to bring him a treat. Looking a bit put out, he goes back to lounging in his one-man kingdom. But where Cossatot reigns was once the domain of an even larger, far more endangered animal. Little does Cossatot know, but his kingdom has made history. I'm visiting the old Sumatran rhino enclosures of Cincinnati Zoo with Terri Roth, head of the zoo's Center for Conservation and Research of Endangered Wildlife (CREW), and Paul Reinhart, leader of the team that cared daily for the rhinos.

Roth jokes that the enclosures have fallen far from their glory days when they housed arguably the rarest large terrestrial mammal on Earth. The two enclosures—one for the male rhino, Ipuh, and the other for the female, Emi, and her calves—are now the domain of Cossatot and a pair of nervous emus. Above them are half-million-dollar metal structures that look like giant rectangular umbrellas, built to shade the rhinos' eyes from the sun, just as the canopy does in the rain forest, and prevent severe eye damage.

"Every day we walk in here and I look at those pictures," Reinhart says, pointing to photos of all the rhinos—Ipuh and Emi, Andalas and Suci, and most beloved of all, Harapan—that once called Cincinnati home.

"I miss all of them," he says.

Last Chance for the US

In February of 1995, one year before Terri Roth would take the job as director of CREW, two Sumatran rhinos died within five days of each other at San Diego Zoo. This left just three Sumatran rhinos in the whole of the United States: Rapunzel, Emi, and Ipuh, the sole male.

Over a decade before, in 1984, conservationists had kick-started a grand plan to capture Sumatran rhinos in the wild and breed them in facilities in Indonesia, Malaysia, the United Kingdom, and the United States. The bill for this large-scale undertaking was paid by the US and UK zoos. Although conservationists were able to capture 40 rhinos over 11 years, the program had turned into a catastrophe. By 1995, nearly half of the 40 rhinos were already dead due to poor feeding practices, disease, accidents, and simple ignorance. Moreover, not a single rhino had been bred in captivity. Now, there would be no more rhinos coming to the United States. Due to a lack of success, the catching had ground to a halt, with the last rhino caught in Sabah, a Malaysian state on the island of Borneo, in 1995.

By this time, the UK had only one rhino, a male named Torgamba. Peninsular Malaysia had eight, but no luck breeding. Sabah had five, but only one female. Indonesia had two in captivity, both females.

The US zoo community, down to its last three rhinos, had one final shot at doing what it promised it could back in 1984: making a baby rhino. But the US rhinos were scattered: Ipuh was at Cincinnati, Rapunzel at the Bronx Zoo, and Emi at LA Zoo.

Roth says it was the Cincinnati Zoo director, Ed Maruska, who convinced the other zoos to send their females. "They thought, 'Well, if anyone will do it, then Maruska will do it.'"

By August 1995, just months after San Diego lost its two rhinos, the three survivors were all brought together at Cincinnati.

"I'd become very much smitten by the beast," Maruska remembers of seeing his first Sumatran rhino. "It was a hairy animal. It was very unusual, very primitive looking. I thought in every shape or form, Cincinnati's going to be a part of this program."

Maruska then made his second big move: he hired Roth in 1996.

"Ed said, 'We have got to breed these rhinos. It's the last chance,'" Roth remembers.

No pressure.

Cracking It

Terri Roth's office is full of rhinos: sculptured metal rhinos, stuffed toy rhinos, plastic rhinos, wood rhinos, and cinema-sized rhino posters. She doesn't have an unnatural obsession with rhino replicas: nearly all of these are gifts, she says. Roth, who owns a small cattle farm over the border in Kentucky, has become something of a celebrity in the small circle of rhino people, because she accomplished something that many had begun to despair would never happen.

It's not that no one had been trying to breed the rhinos from the time they were brought into captivity—they had. But the animals would fight, sometimes viciously, whenever they were brought together. And even when mating happened, something was off: the females weren't getting pregnant.

The first thing Roth wanted to figure out was what was going on with the reproductive cycle of the females. They trained the two females, Emi and Rapunzel, to undergo ultrasounds without anesthesia, which would have been too risky. They quickly discovered that Rapunzel would never have children: she had a large mass in her uterus. So now the US was down to an Adam and Eve scenario: Emi and Ipuh.

They focused all their energy on Emi.

"We were working on her, three times a week, ultrasounds, conditioning her for blood collection, monitoring her for about eight months, and I still couldn't figure out her reproductive cycle," Roth says. "You're beating your head against the wall."

Basically, Emi wasn't ovulating.

It was in the summer of 1997 that Roth made a risky, but fateful, decision. She decided to put Emi and Ipuh together, even though they didn't know Emi's ovulating schedule. Solitary in the wild, males and females will frequently fight like hell when brought together. Sometimes the fights result in injury to one of the animals —not something you generally want to risk with a species on the edge of extinction.

To mitigate the risk, Roth and her team decided to do all they could to make sure Ipuh and Emi weren't too ornery.

"We thought, 'Let's do it when it's hot. Let's do it after the male has had its breakfast.' We put them out. Ipuh would eat his browse. He'd go into the pool, sink down, and then we put Emi in.

"The keepers were all on alert. They were ready to jump in if they're needed to. It actually worked pretty well because [Ipuh] was not really interested in picking a fight," Roth says. "He was just in the water . . . Sometimes, Emi would go over to the pool and blow at him." There were "no big skirmishes," Roth recalls. "For the most part, they ignored each other."

Reinhart adds that "it could have gone wrong every single day," but they kept going.

And they did, for 42 days. Forty-two days of keepers having to put two near-one-ton, critically endangered animals into a potentially perilous situation.

"Then one day, there was just this total difference," Roth says. "We put [Ipuh] out there and [Emi] went towards the pool and he started coming out of the pool.

"It was the most agreeable situation, shockingly. There was no chasing. There was [no] sparring, he just came out of the pool, started following her around, and after a while started mounting her. We were ecstatic."

But Ipuh was not exactly a seasoned lover.

"He tried . . . we even got lights out and left them together as late at night as we possibly could," Roth says. "He mounted her and mounted her and mounted her and finally got exhausted but never was able to breed her."

But that first encounter did lead to something historical.

"Two days after that . . . I did the ultrasound where I saw she ovulated for the first time," Roth says.

A light bulb went off. It turns out the Sumatran rhino is an induced ovulator, which means the female needs something to kick-start her reproductive cycle. In the case of this species, Roth believes it's the interaction with a male that allows a female to ovulate. Roth says they don't even have to copulate to ovulate —they just need to spend time with a male at the right time in their cycle.

"We've even seen situations where the males run her around

but not even [mount] her, and she's ovulated. I think the ovulation is partly a response to the excitation of being with the male," Roth says.

Still, the team didn't know how long Emi's cycle would be, so they started up again with the daily introductions, and 21 days later Ipuh and Emi tried again.

"She conceived on that one. That was the first pregnancy, which was shocking because it's pretty quick," Roth says. "We saw the little fetus developing. We saw a heartbeat. We sent out the press release. A week later, the embryo was gone. We thought, 'At least we knew he was fertile. We knew things were working.'"

Indeed, the team had the information they needed: they knew Emi needed interaction with a male in order to ovulate, they knew her cycle was around 21 days, and they knew how long the follicle would grow during the cycle. Breeding had started to go well; pregnancy, not so much.

"Then she lost the second one, then she lost the third one and it actually became more challenging for me because people started saying, 'It's because of the ultrasound exams that she's losing the pregnancy,'" Roth says. "I was forced to reduce the amount of work I was doing with her, instead of increasing it. Then we were learning less."

At this point, Roth started to run lots of blood tests to see if she could find anything amiss, comparing them with blood samples from other captive rhinos in Indonesia and Malaysia.

Then Emi lost her fourth pregnancy, and her fifth.

"Finally, I just said, 'Let's just put her on the progesterone supplement, because we don't think it will harm anything, and it seems like it could only help and not hurt,'" she says. Progesterone is a hormone produced in the ovaries that becomes elevated during pregnancy. This was in 2000, four years after Roth was hired.

It worked—the sixth pregnancy finally stuck. But no one had any idea how long it would last.

"The only thing I could find was somebody had at some point said it was a seven-month gestation, which we didn't believe because no rhino is that short," Roth says.

In fact, 16 months later, Emi gave birth to a baby boy: Andalas. He was not only the first Sumatran rhino born in captivity in 112 years, but the first tangible success of that tragedy-filled program launched in 1984.

The Cincinnati Effect

Cincinnati Zoo, the second-oldest in the country, sits smack-dab in the city among the rolling hills surrounding the Ohio River. Generally considered one of the world's top zoos, it has a long history of breakthrough captive-breeding successes, from giraffes to trumpeter swans to bison.

But perhaps none of the zoo's past glories could compete with the birth of Andalas.

In many ways, Tom Foose, conservation coordinator for the Association of Zoos and Aquariums (AZA) and the driving force behind the 1984 meeting to launch the captive-breeding program, had a point about *why* US and UK zoos should have a crack at breeding Sumatran rhinos: the world's best zoos had both the expertise and the technology to have the best chance of success.

"That's what I love about the Sumatran rhino story, because it's a perfect example of how zoos can contribute," Roth says. And Cincinnati was even more distinct than many zoos. Not only did it have a long history of captive breeding and expertise, but it also had an entire research facility, CREW, devoted to this kind of work.

"We often have a discussion here at CREW about the disconnect between the reproductive sciences and conservation. There is so much power in that kind of technology, but it's used so little in real conservation efforts," Roth says.

At the same time, Cincinnati Zoo, and the zoo community in general, suffered relentless criticism over the program.

"We felt it constantly. Partly because Westerners, partly because zoo, probably partly because female," Roth says.

Maruska says they took "a lot of fire" even from the wider zoo community. They were accused of taking wild animals out of their habitat just to exhibit them; they were told they'd never succeed.

"We faced the same with the California condor," Maruska says. "We had people from the Audubon Society saying, 'Let the birds die in dignity.' Well, there is no dignity in extinction. Come on."

Roth remembers that the zoo was even accused of making up pregnancies during the period when Emi was losing one after another.

"And then the negative stuff about, 'They're losing pregnancies, they must be doing something wrong there. Cincinnati is a bad environment,'" she says. "But we just kept at it. I just kept our eyes on the goal, and this is what we need to accomplish."

Roth and the Cincinnati team may be the single most important reason for the eventual success. Roth was able to make astoundingly difficult decisions and then, perhaps even more importantly, stay the course when the criticism became overwhelming.

"Terri was the person that really did the job," Maruska says.

It just took them—and everyone, in fact—much longer to produce calves than anyone could have expected at the 1984 meeting.

"Hell, I think we did a yeoman's job with a handful of animals," Maruska says. "I believe that if we had our full complement of animals, we'd [have] been a lot farther than we are today. I really do."

The next step for Roth, however, was proving that Andalas wasn't a fluke, and that Emi and Ipuh could replicate their little miracle.

Fast-Forwarding

In 2004 Emi gave birth to her second calf, Suci. Then in 2007 she gave birth to her third, Harapan. She successfully carried both these calves without the use of synthetic hormones.

"People thought it was really risky, but I really wanted to prove that they could do this themselves in a managed breeding program," Roth says. Still, she believes the progesterone was vital for that first pregnancy in getting Emi over the "hump."

"Once they're producing, just keep them producing, because everything is healthy, and everything is working right, you don't want to stop that," she says.

Unfortunately, Emi died in 2009 of iron storage disease, though at the time the team had no idea what was wrong. It's an "insidious" disease, according to Roth, that can only be diagnosed after death.

In 2013 the zoo decided to euthanize Ipuh. Suffering from cancer, Ipuh had stopped eating and was barely able to walk.

"It's hard to describe when they were born; it's even harder to describe when an animal passes away," says Reinhart, who spent 22 years caring for Ipuh. "[He] contributed so much to the species and the knowledge and the propagation of these animals and he stayed with us to the very end." Today, his preserved body rests at the University of Cincinnati.

An even bigger heartbreak came a little over a year later when Suci, Emi and Ipuh's daughter, died from iron storage disease, the same sickness that took her mother.

"With Suci, we suspected it when she started showing the same symptoms that Emi did," Roth says. For a while, Suci, just nine years old, improved with aggressive treatment, but a few months later her health worsened. "Her liver was just too damaged," Roth says.

She believes iron storage disease was an issue at Cincinnati because the rain forest rhinos have evolved to live with multitudes of parasites and biting insects that constantly drain them of blood.

"They're trying to absorb as much iron as they can from what little iron they get on their diets because they have this constant load of parasites. They're bleeding, and they're having to build up tissues that parasites have chewed down, so they need it all the time," she says. "We bring them into our zoos or our facilities and we get rid of all the parasites, and they don't have that outlet anymore, so they're not losing iron anymore."

By the time of Suci's death, the Sumatran rhino program had shifted significantly. During the period when Cincinnati Zoo was struggling to produce just one calf, many experts began to feel the best thing for the species would be to bring them into managed sanctuaries in their local environment. This way, the rhinos would have direct access to their wild, natural foods and, many experts believed, this might help induce mating and decrease the chance of disease.

In 1998 Indonesia opened the Sumatran Rhino Sanctuary (SRS) deep in Way Kambas National Park, a park also home to some of the last wild Sumatran rhinos on Earth. Two females were brought from zoos in Indonesia that year, as was Torgamba, all the way from the UK. Unfortunately, breeding between these pairs was never successful.

Still, by the late 1990s the SRS and the Sungai Dusun rhino center in Malaysia—where six rhinos would die in 2003—were beginning to be seen as the future of the program.

In 2007 the United States sent Andalas, the first calf born in captivity, thousands of miles to the SRS in the hope that he could find an unrelated mate. It was time for the Cincinnati staff to transfer what they learned overseas.

"We really work hard here, that whatever we develop here it's not about 'mine, mine, mine,'" Roth says. "That's why I was just so pleased that they were able to do it in Indonesia."

Andalas mated successfully with Ratu, a wild rhino found roaming near a village in 2005 and brought to the SRS for her safety. Their union produced Andatu, a male, in 2012, and Delilah, a female, in 2016.

"It is a really good template; the hardest thing is to get people to follow it," Roth says of the subsequent breeding successes. One of the most challenging bits is simply allowing the animals to spar, which Roth believes is a natural part of their breeding process.

"You have to have confidence that you knew what you're looking at," she says. "And to hold your ground and say, 'No, keep them together, keep them together, keep them together,' because after an hour or two, they're going to settle down and they'll breed."

Cincinnati then made one of its toughest decisions yet: to send Harapan, its last rhino, and the public's favorite, to Indonesia.

"Many of us really wished we could just get more rhinos. We wished we could have gotten a female from Indonesia and bred her with Harapan, and kept the program going. That was hard," Roth says. But it was clear that the best thing for the species was to let Harapan go.

In 2015 Harapan took the same journey from Cincinnati to Sumatra as his elder brother, in the hope that he, too, would breed successfully with one of the females at the SRS. Harapan was the last rhino in Cincinnati—and the last Sumatran rhino in the US.

"We miss him here," Reinhart says. "He's in a better place, but he was our last born and we really loved him here. I do miss him still."

Harapan's arrival in Sumatra marked not only the last Sumatran rhino leaving the western hemisphere, but also 20 years since the close of the original captive-breeding program. The 1984 program was still a long way away from achieving a sustainable population in captivity that could, if nothing else, ensure the species wouldn't go the way of the dodo, the Tasmanian tiger, and the woolly rhino. However, conservationists had ensured that by 2015 there was still a chance to do so.

HOLLY HAWORTH

The Fading Stars: A Constellation

FROM *Lapham's Quarterly*

MORE THAN 400 years ago, around 13.8 billion years after the beginning of the universe, in September 1608, a Dutchman from Zeeland came to The Hague announcing he'd invented an instrument "by means of which all things at a very great distance can be seen as if they were nearby." The prospect was thrilling. European explorers had spent centuries crossing wide oceans in hopes of finding distant lands to claim for their countries, and they were doing so with increased fervor—bringing the resources of faraway places closer to home by establishing trade routes; collapsing cultural distance by eradicating languages, religious practices, ceremonies, myths, and stories; forcing distant others to charade as Europeans with imposed cosmologies and worldviews—since Christopher Columbus had pulled a new world closer, 116 years before the seeing instrument was introduced.

The soldiers who were convened at The Hague that day seized on it as a way to spot enemies on the horizon. They were gathered to talk about independence of the Dutch Republic from Spain's control, and they paused to consider the arrival of the new contraption. The commander of the Spanish forces said to the commander of the Dutch Republic, "I could no longer be safe, for you will see me from afar." Ship captains of every European nation would soon have an instrument to spy on one another, to bring distant lands into focus.

News of a "spyglass" reached Galileo Galilei, a professor of mathematics at the University of Padua, in the spring of 1609, when a friend shared with him a description of the invention.

Galileo worked out how to make one for himself from a tube and two lenses. But instead of pointing it horizontally to look for enemies, he canted it upward to spy on the faraway night sky. At once, it gave up some deep, dark secrets about the moon, that nearest object to us which had in many cultures been an orbed goddess or the home of one. According to church doctrine and Aristotelian philosophy, it was an immaculate heavenly body—perfectly smooth, perfectly spherical. But what Galileo saw instead were shadows signaling rifts, craters, and mountains. It was not goddess-like but rather a body of rock with topography, an object of scientific study that might be mapped and surveyed.

This moon had already lived 10,000 lives. Hawaiians—not yet "discovered" by Captain James Cook, who would arrive with a telescope in hand in 1778—had 30 different names for it, in accordance with the phases of its cycle, such names as *Hilo*, "faint thread"; *Hoaka*, "arch over the door"; *Mohalu*, "unfold like a flower"; *Akua*, "god," for the full moon; and *Mauli*, meaning "ghost" or "last breath," for the waning crescent. This same moon was for the Japanese what held the waters of life (*silimizi*) and the waters of rejuvenation (*bakamizi*). The same moon that a woman in the Congo 37,000 years ago kept track of by etching notches into a baboon fibula. Moon that for the Yakuts had taken in a young orphan girl as she fetched water with her yoke and pails. Moon that for the Maya a rabbit lived in. Moon that an old man lived in. Moon that had through it all ruled the tides and thus the ocean and all waters of the planet. Moon the cycles by which all cultures always had planted, fished, traveled, harvested, and hunted. Moon that was alive to the faraway eye until men would come to walk on it, imprint its fine untouched dust with their boot soles, plant the American flag to proclaim it—360 years after Galileo first spied it, 4.5 billion years into its existence—desolate, dead, a once mysterious place finally conquered by knowledge. One giant leap for mankind.

In the cloudy spaces of the Milky Way, Galileo saw clusters of stars. In the places that had before looked empty, the dark gulfs of space between stars, he saw even more stars. He saw more stars than had ever been seen.

He rushed to publish his findings in *Starry Messenger* the following year. The mathematician Giovanni Faber, a member of the Italian scientific society Accademia dei Lincei (Academy of the Lynx-Eyed), waxed rhapsodic in response:

Yield Vespucci, and let Columbus yield. Each of them
Holds his way through the unknown sea, it is true.
But you, Galileo, alone gave to the human race the sequence of stars,
New constellations in heaven.
O bold deed, to have penetrated the adamantine ramparts of heaven.

The fired-up language of penetration accompanied the use of the telescope from the beginning, the telescope as a kind of phallus probing the night sky, which had so long been associated with the feminine as a kind of womb, ruled by the moon that was like an egg when full, women's menstrual cycles waxing and waning with it every 28 days. For Faber it was as if Galileo's telescope had fertilized the sky and *made* the new stars in formerly vacuous space.

Encouraged, Galileo kept penetrating the darkness night after night. In the winter of 1610, he wrote a cryptic message to the astronomer Johannes Kepler: "The mother of love emulates the figures of Cynthia." It was common practice for a scientist to create coded documentation, by way of a dated letter, of an important finding he intended to reveal later to make sure he could prove when the time came that he was due the credit of discovery. The mother of love in Galileo's code was the planet Venus, and the figures of Cynthia that she emulated were the phases of the moon (Cynthia was another name for the moon-associated Roman goddess Diana, who was shown in many depictions wearing a veil). Galileo had peered into the darkness of the sky chamber and seen the lady Venus undressing. Over the course of the 584 days of her cycle, she slowly slipped off her gown of light and put it back on again.

The implication was that Venus orbited the sun, supporting Copernicus's theory of a sun-centered universe; this would finally bring Galileo before the Roman Inquisition on charges of heresy and land him in the prison of his own house in 1633, where he would remain for the last decade of his life, blind.

But in 1611, the year after publishing *Starry Messenger,* Galileo was inducted into the Accademia dei Lincei and adopted the signature Galileo Galilei Linceo, "Galileo Galilei the Lynx-Eyed." The human race had always had the stars, but Galileo the Lynx-Eyed had given them new stars through the instrument that was, at the ceremony of his induction into the academy, named the telescope, from the Greek *telescopos,* meaning "far looker." The all-male acad-emy, on its founding in the early 17th century, had taken the name of the lynx, the animal thought to have the keenest eyesight at night; its members were expected to "penetrate into the interior of things in order to know the causes and operations of nature, as it is said the lynx does, which sees not only what is outside but what is hidden within."

With a few adjustments, Galileo had also pointed his telescope down, turning it into a compound microscope. The invention of the telescope and the microscope at nearly the same time allowed scientists to probe the mysteries of the infinitely vast and the in-finitely small. Galileo posited that the universe and everything in it was mechanistic, ordered, abiding by rules of law, and that with the new seeing instruments, its parts could be studied, its workings discovered, flies and bees no less well-oiled machines than solar systems and galaxies.

New lands, too: explorers found hidden worlds—continents and islands—within the black oceans, staked their flags into the soil, then set to work mapping the terra incognita. The lands were not unknown to the dark-skinned inhabitants but were for the light-skinned powers until they could gain a carnal knowledge of the place, often raping the women of the American wilderness, a wilderness they called virgin, felling the forests full of shadow so that light could come into the clearings.

These rapacious advances into new territories had their own limits. The cultural historian Marjorie Hope Nicolson wrote that by the early 17th century, "the whole world seemed known and mapped," and a kind of melancholy set in. As the English scholar Robert Burton worked on his *Anatomy of Melancholy,* finally pub-lished in 1621, in which he attempted to dissect the feeling that stemmed from this too-much-knowing, to see it inside and out, Galileo's seeing machines opened realms to further exploration.

Like Faber, Giambattista Marino, the most famous poet in Italy at the time, compared Galileo to Columbus in his 1623 epic poem, *L'Adone.* While the latter "cleav[ed] the breast of the ocean, vast and deep," Galileo explored the heavens and found "new stars and lights once hidden to all men." Nature, so often spoken of in femi-nine terms, hid herself from men; that she was vast and deep had become irksome. Knowledge came with light, mystery with dark-ness. And the darkness goaded men on relentlessly.

With Galileo's findings, further darkness beckoned to be illuminated. "From Mars and the moon, from Venus and Mercury," Nicolson wrote, bright lights came shining. There now "appeared a new expanse of space, a vast illimitable ocean without bound."

Even as Galileo sat under house arrest, unable to travel, he kept his far lookers set up at the windows. In 1638 the English poet John Milton visited him there, in Florence, where he was growing old, his eyes dim.

He showed Milton the sights of heaven, and it was those "thousand thousand stars" that blazed behind Milton's eyelids when he began to write his epic poem *Paradise Lost* 20 years later, himself blind by then. What Milton had seen through Galileo's telescope remained his enduring source of imagination while darkness imbued his greatest period of creativity. He said that the Muse visited him at night and gave him his verses; then in the morning, he would ask to be "milk'd," dictating his lines to a scribe.

Milton's drama of the fall of man and humanity's place in the cosmos plays out against a vast backdrop of interstellar space, sparkling with "innumerable stars." The epic poem seems at times to settle in with the mystery: "The great Architect / Did wisely to conceal and not divulge / His secrets to be scanned by them who ought / Rather admire." But more often Milton associates the hidden and the dark places of the cosmos with Satan, the darkness with the unknown, Eve with Satan and thus with darkness. In the darkness, chaos lurks. In the darkness, things are "swallowed up and lost / In the wide womb of uncreated night." God is light, and goodness dwells in illumination. Milton might well have been writing an epic lament of blindness, while blind to the epic irony that his greatest work was conceived in the dark, that the mesmerizing shimmer of the thousand thousand stars can be seen only in the dark, that the Milky Way is only milky in the dark, that it was only in the darkness he received his Muse's milk.

The worrisome thing about Galileo's observations of the phases of Venus, what didn't sit right with the Church, was that if Venus orbited the sun, then the Earth and our moon did, too, which meant once and for all that we were not at the center of the universe. If we revolved around a star that was just one of millions of stars in a galaxy, itself one of millions in an illimitable universe, each star a possible sun around which planets might spin, then the Earth was, as Adam says in *Paradise Lost*, "a spot, a grain," and we were not, perhaps, the apple of God's eye.

For Europeans who were claiming new Edens all over the globe, it was as if they had been cast out of the garden yet again, into a lonely, darkened cosmos. As if humankind had been banished to this Earth that got dark at night, left alone with its fears by God, who dwelled with the angels in eternal light. The "darkness visible," Milton wrote, "served only to discover sights of woe, / Regions of sorrow, doleful shades, where peace / And rest can never dwell."

The telescope was credited with lending the light of knowledge to the darkness, gifting sight to the sky that was once obscure. "Much dost thou owe to heaven that grants to thee / the invention of the wondrous instrument," Marino wrote in *L'Adone*, "but far more heaven owes to thy superb / device, which makes its beauties manifest." Making telescopes was a tricky art, a matter of grinding lenses of sufficient quality. Though Galileo the Lynx-Eyed didn't invent the instrument, he became known for it due to his fine craft—which placed heaven, Marino believed, in the astronomer's debt, even though heaven had granted Galileo his lenses of glass: glass made of sand that was itself created by the ocean's pounding, the ocean made to pound by the moon, the tides' ebb and flow deepened and heightened by the moon's gravity. Still, for Marino, the moon remained less wondrous than the instrument used to view it.

The fascination spread and expanded; astronomers built longer and longer telescopes. By the time *Paradise Lost* was published, the far lookers of light had reached heights of 30 feet. By 1670 they were 50 feet tall, and by the end of the century, some were more than three times bigger still. These were used to make detailed maps of the moon, to observe Saturn's newly discovered rings and satellites, to discover polar ice caps on Mars, to watch the eclipses of Jupiter's moon Io.

Soon astronomical telescopes were becoming "too large to be handled by a single person and too expensive for private individuals," writes the historian Richard Dunn in *The Telescope: A Short History*, so state-funded observatories were set up in all the major cities of Europe. Nations competed to build the largest. In 1675, under the aegis of Charles II, England built its Royal Greenwich Observatory, and soon the empire's first astronomer royal, John Flamsteed, began a 40-year labor to catalogue 3,000 points of light in the sky, assigning them numbered designations, applying systematic rigor to the age-old stars. Stars whose cycles were recorded in cuneiform, the first script, on clay tablets in Babylonia. Stars that were hunters who had chased animals into the sky that were also now stars. Stars that to the Greeks were nymphs, nursemaids, and teachers. Stars whose rising, grouped in certain constellations, heralded sailing season for ocean cultures. Stars by which navigators made paths across the sea. Stars that to the Cherokee, who lived in the land I am from, were of the same nature as pine trees. Stars that were young ball-playing boys who didn't help their mothers gather food for dinner. Stars that danced in a circle. Stars that were sisters. Stars that were brothers. Stars that were ancestors.

The stars of night when night meant darkness. You could still see them then.

In a poem written in 1637, the year before he visited Galileo, Milton wrote of "the stars, / That nature hung in heaven, and filled their lamps / With everlasting oil, to give due light." This may be the first comparison of the stars to oil lamps, and a hint of what was to come: if humans were stranded in a lonely cosmos, separated from God, why not burn some of the Earth's own oil to light the way through the dark, to make our own earthbound stars.

Having for many years kept curfews that restricted movement through nighttime streets, major cities of Europe and the Americas began enacting laws—at the same time the telescope began its popular rise—requiring public street lighting. In London a 1599 act required every householder to hang a lantern outside at dusk each night from the first of October until the first of March while providing an exception for the shorter nights of summer as well as between the seventh day after the new moon and the second day after the full moon as, according to encyclopedist Johann Georg Krünitz, the "grand light that rules the night lights up the streets sufficiently, and lanterns are superfluous." The act became more stringent, though, in Milton's England of 1646, carrying a shilling fine for those who disobeyed. A few decades later, in 1684, an Englishman named Edward Wyndus applied for a patent on the "discovery of a new experiment for the great and durable increase of light by extraordinary glasses and lamps." With a convex lens, these lights "threw out long rays of light," the scholar E. S. de Beer later wrote, and "dispelled the unknown terrors of the night."

Around the same time, public lighting was introduced in Paris by the city's first police chief, who was appointed in 1667 by Louis XIV, the Sun King. The lights, called "artificial suns," became a symbol of the king's power and of the authority of the police. Under the Sun King—whose personal astronomer tutored Danish astronomer Ole Rømer, who determined the speed of light at the Paris Observatory in 1676—darkness was punishable. Soon there were a thousand lanterns in the city, spaced 20 yards apart.

By 1700 there were more than 5,000 lights in Paris. The Sun King died in 1715, but the lights burned on. By 1750 there were 8,000. Refractors were used to strengthen their power: in 1763 new *réverbères* were invented to make lanterns shine many times more brightly. In the first volume of his *Tableau de Paris,* Louis-Sébastien Mercier wrote, "Now the city is extremely brightly lit. The combined force of 1,200 réverbères creates an even, lively, and lasting light."

The city of London began mandating longer times that lamps should burn, for more of the year, and at shorter distances from one another. It set higher penalties for those who failed to comply. Finally, in 1735 the Court of Common Council decided that lighting should last from sunset to sunrise throughout the year. Prior to the act, it was thought that there had been a thousand lamps in the city lit about 750 hours a year. Afterward, 4,679 lamps were set up, lit more than 5,000 hours each year.

From the Hôtel de Cluny in Paris, the astronomer Charles Messier scanned for diffuse objects in the sky with a telescope. These he would compile in a catalogue of 110 galaxies, nebulae, and star clusters. But city residents were more dazzled by *réverbères,* hailed for how they "turned night into day." Paris, center of the Enlightenment, became the City of Light, *ville lumière.* Poems at the time celebrated the *globes brillantes* and *astres nouveaux,* the new stars. These new stars would outshine the old stars. This globe would be brilliant in the dark universe.

In 1820, two years before the Catholic Church began permitting the teaching of heliocentrism, effectively unbanning Galileo's *Starry Messenger,* the British Empire established an observatory in its South African colony, at the Cape of Good Hope. The Cape and its environs were home to the Damara, Xhosa, Zulu, Ndebele, Namaqua, Tlôkwa, Khoikhoi, Tswana, and /Xam, among dozens of other subgroups. Each had its own complex cosmology and relationship with the sun, stars, moon, and planets. Their attendant knowledge, stories, and rituals were soon to fade, but some were recorded by colonists. The nomadic Khoikhoi used the heavenly bodies to reckon their position on the land and guide their nighttime navigation. The Nguni, a grouping that included Xhosa, Zulu, and Ndebele peoples, kept close track of the heavens for the timing of their agricultural practices. For the Namaqua, the appearance of the planet Jupiter in the sky signaled the start of the fruitful season. For the /Xam, the stars gave some of their heart to the people to help them stave off hunger. Many of the Cape cultures divided the year by its moons, each moon with a name that marked seasonal natural phenomena. A record was made of a grass bushman who said the moon is "a thing which knows things; for it sees things which will come to pass. It is the one who knows them, things which we do not know." In one /Xam story condensed by a colonist, "the moon angers the sun so much that the sun stabs the moon with its knife (rays) until the moon is reduced to a backbone. At this point, the sun spares the little piece, which goes home to grow back to its full size, only to begin the cycle anew." The telling of such stories created an intense intimacy with the night sky.

British astronomers sent to track the heavens of the southern hemisphere at the Cape observatory loathed and feared these dark-skinned storytelling natives. They sent letters of bitter complaint home about both the "dirty, black, flea-y" inhabitants and the snakes that lurked in the grass. They promoted the observatory as a beacon of light and knowledge in a land that they would come to call the Dark Continent.

John Herschel—son of astronomer William Herschel, who had first spotted Uranus and built a telescope in England so enormous that it served as an emblem of the Royal Astronomical Society—came to the Cape and set up a private observatory in 1834. His goal was to catalogue the stars and constellations and give them "rational" names. In a letter to the royal astronomer, Thomas Maclear, who resided at the official Cape observatory just a few miles away, he insisted that none of the natives' names for the stars be preserved. Maclear wrote back, "All the practical astronomers will back you in expunging the heterogeneous jumble of birds, beasts, men, and things . . . that appear ridiculous in astronomical research."

This expungement was part of a global shift in the way the night sky would come to be viewed, not as particular to the individual cultures in whose lives they played both practical and mythical parts but rather, as the historian Elizabeth Green Musselman writes, a "celestial blanket [that] covered the globe in one fabric." The blanket of stars heralded "a one-world empire," and those with the telescopes named each point of light and the constellations they formed when strung together, overlaying the vast kaleidoscope of names and stories that the heavenly bodies held for peoples in every tiny portion of the globe. The International Astronomical Union would eventually standardize the constellations with official names in order to avoid what one modern astronomer called "a chaotic situation."

The British made a striking observation about the bush people: their eyes, wrote one colonist, were "little inferior in optical power to small telescopes." Another said of them that "the eye operates with a precision and force, which a person who has never witnessed the like would scarcely be disposed to credit . . . They will often discern with distinctness what others require a telescope to distinguish." For these native inhabitants with their telescopic vision, the stars were already close; in their eyes, the astronomers who used long tubes and lenses to view the stars but did not pattern their lives around them or tell stories of them must have seemed very far from the twinkling lights of the heavens.

The colonial stargazers had come far from home to get closer to them, and they beamed with enthusiasm in their letters about the skies of southern Africa, which were crystal clear. Back home the frenzy for lighting up the night had taken its toll on visibility. In 1805 a cotton mill in Manchester became the first factory to be illuminated with gaslights at night; two years later came gaslights through London streets, followed by the first electric lamps. Industry now continued into the night, and the skies back in Britain were smogged by coal and washed with lights. The heavens in Europe had already receded.

But at the Cape, they seemed very near. One colonist wrote in a letter home:

Many of the most beautiful and interesting phenomena of the heavens are here more beautiful and interesting, from the pure and gentle transparency of the atmosphere through which we gaze at them. A spectator accustomed to the hazy skies of the north turns oft with untired gladness to the distinct and vivid objects of our horizon, the hues and changes of the twilight, or the brilliancy of the azure noon. The nearer to us of the heavenly bodies partake of this distinctness; the setting planets blaze like masses of flame on the sea; the moon . . . [is] very dissimilar to its vapory and languid disk in the English sky.

The John Herschel who wanted to rid the stars of the local inhabitants' stories of them wrote home to a fellow astronomer, "I cannot help believing that we are nearer the stars here." He could not help believing it, although as a rational scientist who had taken measurements of the whole Earth, he knew that couldn't be true. And yet.

In Egyptian myth the night sky is a woman's body. She arches her back, stretches out over the land. In some ancient tellings, she swallows the sun god Ra each night and births him in the morning. She is the vault of the sky, the mistress of all, the holder of souls, the producer of all heavenly bodies. It was she whom astronomers around the world spied on, mapped, catalogued, claimed as their own discovery.

More than a decade before the British observatory was built on the Cape, a Khoikhoi woman was kidnapped after her father and husband were killed in a Dutch extermination raid. She became a slave in Cape Town, then ended up in the hands of an English doctor, who took her to the bright Piccadilly Circus stage in London, where she was exhibited as a dark specimen of the hidden interior of an unmapped continent. The Khoikhoi were called Hottentots by colonists who could not understand their language, a subtle system of clicks and clacks, and the woman would become known as the Hottentot Venus. Her black skin and her anatomy —particularly her "oversized" buttocks and enlarged labia—were a spectacle, material for study by scientists who wanted to classify the inferiority of blackness. They were given special invitation to come see her. In her first performance, she was poked and prodded by fingers and the pointed tips of parasols and canes.

Later she was sold to an animal trainer, who took her to Paris, City of Light. Onstage, she was illuminated so that her audience could gawk in wonderment—as a saying went in her culture, "like a lion looking at the moon." When she died young in 1815 of an inflammatory disease, leading European scientist Georges Cuvier, then a professor of comparative anatomy at the Museum of Natural History, cut apart her genitalia, that most mysterious, darkest part of her dark body, examined them under a microscope, then put them in a jar for public display at the Museum of Man.

By the time Herschel examined the night's black body with his telescope at the Cape observatory, the Khoikhoi culture had already disintegrated. A school friend wrote Herschel a poem the year he arrived: "If Herschel should find a new star at the Cape," it read, "He will salt the star's tail to prevent its escape / And call it 'The Hottentot Venus.'" For in 1834 she had already burned out as a star attraction in Europe, and the folds of her labia had been examined in full—feminine darkness dissected and exposed under science's laboratory lights.

If this story that begins with the invention of the telescope is an origin story, of the birth of modern astronomy—the science that has uncovered more secrets of the universe than we'd ever known were possible—it is also the story of the death of the night sky and whatever enveloping mystery it once might have held us within, the exit from the dark womb of our own origins.

Perhaps it's some grand cosmic prank that the story of our advancing knowledge of the universe's darkness is also the story of our retreat from that darkness, our enshrouding ourselves in a haze of lights of our own making. Or perhaps these are only moments in history that collided and will, in the end—though when will it ever end, the universe and all this birthing and dying, all these collisions?—be revealed as only so many plot points pulled together into a constellation in what is otherwise a chaos of stardust assembling and disassembling and reassembling into strange clusters of matter. Like lynx, for instance, who can see in the dark, or like what's known as the twilight zone of the ocean, to which marine biologist Sylvia Earle traveled in 1979 and saw in that remote darkness "the flash and sparkle and glow of bioluminescent creatures," such as corals that emitted "little rings of blue fire" when she touched them.

It matters, though, surely, that only 1 percent of Europeans and Americans can still clearly see the Milky Way at night, that only 20 percent of terrestrial Earth now has true darkness. Italy, home to Galileo, who gave us the stars, is now considered to be among the most light-polluted—which is to say starless—countries. We know this because of satellites we have launched, our astronomical technology now being used to scrutinize our own bright globe in this self-obsessed age, the satellites' orbits like the universe's longest selfie sticks.

Because of the telescope, we now know that the universe is expanding, and the black gulfs of space between celestial objects are growing ever wider—the astronomer Edwin Hubble published this observation in 1929. The story of the far looker, then, is in part the story of how it was discovered that everything is getting farther away from us.

Most astrophysicists now think that the bulk of the matter in the universe—85 percent—is dark matter. This is a hypothesized type of matter, only implied by gravitational effects but invisible to the electromagnetic spectrum, not interacting with light at all. Because of the telescope, we now know that our galaxy's center is a supermassive black hole from which no light can escape. In other words, this whole thing we call the "cosmos"—a short simple name to contain the vastest, deepest, most ineffable of mysteries —is mostly darkness: invisible dark matter and dark energy, unmappable and uncrossable black holes, an unknown. No light can fully penetrate it. And yet the darkness goads men on. Lynx have largely disappeared worldwide, and the vertical migrations of bioluminescent phytoplankton—the largest migration in the world —are disrupted by ships beaming lights as they cleave the breasts of the black oceans.

EVA HOLLAND

Saving Baby Boy Green

FROM *Wired*

JESSICA GREEN WAS getting impatient. She was 19 weeks pregnant and waiting for her ultrasound images at Whitehorse General Hospital, but it was taking forever. She'd never had to wait this long before. Her fiancé, Kris Schneider, had already headed back to work for the day, and Green wanted to do the same. She told the receptionist that she would pick up the images later and headed out. It was late October in Whitehorse, the capital city of Canada's northern Yukon Territory, and winter was beginning to set in.

The ultrasound technician caught up to her in the parking lot. Green couldn't leave, the tech said. She needed to be admitted, right away. Green remembers responding with some sort of instinctive, mulish refusal: "I can't."

But she knew her pregnancy was considered high-risk: she was 37, she'd conceived via IVF, and she was carrying twins. She followed the tech inside and headed up to the maternity ward, where she learned that her cervix was shortening precipitously, a precursor to labor—it was already down to 1.1 centimeters, less than half of what it should have been. A baby's lungs and guts take a long time to fully develop in the womb, and her tiny babies still lacked the abilities to breathe or digest food on their own. But the barrier between them and the outside world was fading away.

Within a few days, a doctor performed an emergency cervical cerclage—effectively, he sewed her cervix shut—to protect the twins. That procedure came with serious risks: both twins might die. But doing nothing might also mean losing them, so Green and Schneider had opted for action. After the surgery, Green gritted her teeth through a week of strict bed rest at home, but then pain and heavy bleeding chased her back to the hospital, where she was admitted and given morphine, fentanyl, and laughing gas while the staff waited to see if her labor would hold off. When she began to dilate again, the doctors removed the cerclage sutures before they could tear through her cervix. She and Schneider now lived in her hospital room. Contractions, irregular but powerful, came and went for days.

All hope of the twins reaching full term was gone. The couple simply hoped to reach what neonatologists call the threshold of viability: the point at which medical science has the ability to keep a premature baby alive outside the womb.

A full-term human baby can seem helpless at birth, but in comparison to a preemie that baby has an impressive toolkit of skills. Aside from their underdeveloped lungs and guts, babies born too early don't yet have the reflexes or muscular control to suck and swallow simultaneously. They are prone to cranial hemorrhage, and sometimes a heart duct remains open. Their skin is thin and fragile; the veins glow eerily. They are sensitive to sound, to light, to touch. Their eyelids may still be fused shut, and the tiniest preemies may not yet even have the ability to close a fist around your finger—that essential early act, the moment when they take possession of you.

Over several decades, doctors and nurses have become better at grappling with all of these obstacles. The threshold still varies widely depending on a baby's circumstances and on the care available immediately at birth. But advances in drugs, technology, and methods of care have pushed that line earlier and earlier, and today there are preemies growing up, healthy and whole, whose survival would have been unimaginable a generation ago. These days, the line between birth and death generally lies somewhere between 22 and 25 weeks' gestation. Green and Schneider could only pray that they would get there.

Whitehorse is a small city, home to roughly 25,000 people, that sits along the only highway to Alaska. Schneider works for the post office, and Green is self-employed as a massage therapist, acupuncturist, and osteopath-in-training. The hospital where she lay bearing through jagged contractions was not equipped to deal with preemies younger than 35 weeks. So as they waited and hoped for her labor to subside, they made plans to get to Vancouver, to the neonatal intensive care unit where the very tiniest and sickest babies in British Columbia and Yukon wind up.

On November 10 one of the amniotic sacs began to leak—the one containing Baby A, who lay on the bottom of the uterus. (These were fraternal twins, so each had their own placenta and sac.) Green and Schneider were loaded onto a small plane and flown more than 1,000 miles south to Vancouver, and in the early hours of November 11, Green was admitted to BC Women's Hospital. Viability was in sight. They were at roughly 22 weeks—and after a hard conversation with their physicians, they had agreed that the doctors would attempt to resuscitate the twins if they made it to 23 weeks. The babies' heartbeats were still strong. Green went to sleep; Schneider crashed out on the floor beside her.

A few hours later, Green woke up feeling that something was wrong. A nurse came in, took a look, and rushed her to labor and delivery. The umbilical cord attached to Twin A, the girl they'd named Maia, had slipped out of the uterus and into the birth canal. Maia had no heartbeat. Now doctors had to deliver her as fast as possible before her movement through the birth canal triggered labor in Baby B, the boy they called Owen.

This meant Green had to push, even though she knew Maia wouldn't survive. She asked the doctors to put her under, to let it happen without her participation, but they couldn't—a C-section would risk Baby B too. Do it for Owen, someone said to her.

Maia came out weighing just 12.3 ounces, minuscule and bruised. The nurses handed her to Green and she held the little body against her chest. "I think she's still alive," Green said. But Maia was gone. Hospital staff dressed her tiny body in tiny baby clothes, sewn by volunteers. They took her photo, took casts of her feet—collecting mementos that her parents might spurn now but want to have later. Green was anesthetized and her cervix was sewn shut once more.

For 12 more days she remained in the hospital, enduring regular inspections of her cervix by a pack of doctors who were watching for signs of infection. Every extra day in utero could give Owen a better chance at life, but if the amniotic sac became infected, it could take him. How soon should they induce? How long could they safely wait? It was another seemingly impossible life-or-death decision.

On November 22, at about 24 weeks' gestation, Green spiked a fever. The next day Owen was delivered by emergency C-section. Schneider held Green's shoulders while the delivery team worked on the other side of a raised curtain. They caught a glimpse of their tiny son, wrapped in plastic to trap his body heat, before he was wheeled away in an incubator. At 1.4 pounds, Baby Boy Green was admitted to the neonatal intensive care unit at BC Women's Hospital. He had a 60 percent chance of survival. The NICU would be his home, and the center of Green and Schneider's world, for nearly five months.

Neonatology is a relatively young field. The first incubators for babies were invented in the 19th century, adapted from poultry incubators to create a stable and warm environment intended to simulate the womb. These early incubators were cumbersome creations of glass and metal. To fund them, they were put on public display—with living preterm babies inside them—at exhibitions across Europe and North America. Incubator babies were regular attractions at Coney Island and occasionally on the Atlantic City boardwalk throughout the early decades of the 20th century. A total of 96 preterm babies in incubators were shown to visitors at the 1939–40 New York World's Fair. (Eighty-six of them survived.)

By the 1960s and '70s, neonatology had graduated from carnival sideshow to accepted medical discipline. But the basic nature of the NICU hadn't changed that much from the Coney Island days: a typical nursery held rows of incubators, a tiny baby lying behind the plastic in each, with parents mere spectators of their day-to-day care.

Doctors' abilities to keep preemies alive a half century ago was limited. Patrick Bouvier Kennedy, the third child of John and Jacqueline Kennedy, was born five and a half weeks premature and died just 39 hours after his birth—today he would be considered only a moderate preterm baby, not in the danger zone at all. Several treatments, often working in combination, have driven that dramatic improvement. Among the most important are the invention, in the 1980s, of an artificial version of a natural lung lubricant that preemies initially fail to produce enough of on their own; antenatal steroids, widely adopted in the 1990s, to jump-start a likely preemie's lung development even before birth; continual tweaking of the mechanical ventilator and the incubator, which is now far more complex, offering controlled levels of moisture and ambient oxygen in addition to providing heat; and the ability to deliver nutritional solutions intravenously to babies who can't yet eat.

When Owen was born, neonatal units around North America were increasingly adopting a family-focused model of care: parents of preemies and other infants receiving treatment in the NICU were encouraged to join on rounds with the medical staff, to touch and hold their babies more, to change diapers and help with feedings, and to be more involved in decisions—especially life-or-death ones. Ten or fifteen years ago, many hospitals had firm rules: they would not agree to resuscitate babies born at or before 23 weeks, say, and they would not recommend the practice before 25 weeks. Now the American College of Obstetricians and Gynecologists recommends that physicians, with the parents' input, at least begin to consider resuscitation as an option at 22 weeks.

Back when Green and Schneider were waiting in Whitehorse, they'd had a tough conversation by phone with Sandesh Shivananda, a senior neonatologist and the medical director of the NICU at BC Women's. He'd told them that, at 22 weeks, the twins would have less than a 5 percent chance of survival. At 23 weeks, they would have a better chance at life, but high odds of living with severe neurological complications. Even at 24 weeks, they would likely spend several months in intensive care. He'd talked to them about the difference between "active care"—working to save a preemie's life, and "compassionate" care—easing its way from birth into death. Discussions around extreme preterm births —generally defined as 28 weeks or earlier—are similar to the ones around end-of-life care: What kinds of extraordinary measures will we deploy? For how long? To what end—saving a life, or just prolonging it?

The policy at BC Women's is to lay out the potential outcomes for parents and to work with them to form a plan, aiming for realism without being overly discouraging. It's a delicate dance, and Shivananda's goal is to give parents as much information and as much control as possible: to give them some ownership, some sliver of power, over their nightmare.

The NICU is both paradise and inferno. It's a place of modern miracles, where babies whose lungs are too small to draw breath are made to breathe, their tissues forcibly inflated and deflated by tubes connected to machines; where parents burn quietly while they watch each new heartbeat register on the glowing screen above their baby's incubator, unable to look away, in a slow immolation that can last for days or weeks or months.

"Off the bat," Schneider says, "they tell you, 'He's going to be a champion for two or three days,' and then he falls off a cliff."

"And then," Green says, "you fall off a cliff."

Owen was immediately diagnosed with extreme prematurity and respiratory distress. He was also vulnerable to sepsis. In other words: he couldn't breathe and was at risk of a severe infection. He was intubated in the delivery room, and his issues piled up from there. In the first week of his life, he was given drugs to help a valve in his heart close properly, and more drugs for his blood pressure. When he was a few days old, he had what appeared to be a seizure—more drugs. His kidneys were too small and new to function fully—more drugs. He received antibiotics for the possible infection he was born with, and then more for a suspected case of pneumonia, thought to be caused by his ventilator. He had a breathing tube down his throat for 45 days and a feeding tube threaded through his nose for four months. He received a steady supply of morphine to numb the pain of the treatments keeping him alive.

If someone so much as spoke too loudly near his incubator, his oxygen levels could drop, setting off alarms from the monitors. He received seven blood transfusions in his first two months. "It was just so tenuous," Schneider says. Green wondered, in those early days, if they had made the right decision for their son. It was an agonizing 22 days before they were allowed to hold him.

The couple moved into Ronald McDonald House, a charity-run residence on the hospital campus reserved for out-of-towners whose children faced life-threatening illnesses. Schneider took leave from his job; Green canceled months of scheduled appointments with her clients. Back home, friends took in their two dogs and raised more than $12,000 to help them make up their lost income. Green was as sleep-deprived as the mother of any other newborn: waking up repeatedly in the night to pump her milk and freeze it for when Owen was strong enough to digest it. She spent her days sitting beside his incubator, reading children's books to him in a whisper, refusing to allow herself to dwell on anything except his survival. "I remember walking into the NICU and making a choice—my feelings of anger, my feelings of grief, I really tried to keep them out of the NICU because he was so sensitive," she says. "I swear to God that he could sense the energy you brought in."

The nursery was kept as quiet as possible, but Green and Schneider were uncomfortably, intimately aware of the other parents hovering over other incubators nearby. Their feelings about those other parents were complicated. They've formed lasting connections with some, but in the NICU, envy and sadness and anger mingled with their solidarity. When another parent's baby was having a bad day, its monitors beeping out constant alarms as it struggled to grow and live, Green and Schneider felt relief that today was not their bad day—and the awful certainty that their turn would come soon enough. On one of the first days, Green glimpsed twins in side-by-side incubators, and suddenly anger and jealousy—and the pain of her loss—shot through her. One day in late January, a new mom arrived with a daughter, Bronwyn, born at 28 weeks. To Green, the baby seemed so much more stable than Owen. But after nearly 200 days of treatment in the NICU, Bronwyn died.

Technology is essential to neonatology, but there's a critical human side to the science of saving preemies too. In the late 1970s, something happened in Bogotá, Colombia, that would begin to bridge the divide between the incubator babies and their parents. A lack of equipment and concern about the risk of hospital infection led doctors at San Juan de Dios Hospital to send stable preemies home with their mothers instead of incubating them. The doctors instructed the mothers to hold the babies continuously, bare skin on bare skin, vertically against their chests, and to feed them only breast milk whenever possible. When mothers started doing this, the area's low survival rates for larger preterm babies tripled. The close contact seemed, in some ways, to replicate the womb better than an incubator—at least one in an underfunded hospital. This practice is now well known as kangaroo mother care and was written up in the *Lancet* in 1985. The paper's authors didn't endorse the home-care option for babies with access to modern NICUs. "Nevertheless," they wrote, preemies in a hospital setting "could benefit from similar emphasis on education and motivation of mothers and early skin-to-skin contact."

Three decades later, while Green and Schneider adapted to life in the open NICU, an experiment built in part on the Bogotá breakthrough was unfolding in two rooms down the hall. For the first time in North America, some new mothers could receive their postpartum care in the same private room where their infants received their neonatal care. The same nurse who checked a baby's oxygen levels and drew blood from his tiny arteries would also be checking his mother's cesarean incision site or monitoring her for excessive bleeding.

The program was part of a reimagining of the entire NICU at BC Women's. Around 2010, hospital administrators had invited past patients to consult on the design for a new building. They gave the former patients a cardboard model of the hospital and a handful of Lego figures. One woman kept moving the mother Lego character next to the baby. Why, she asked, couldn't she just get her care with her baby nearby? The answer was rote and unsatisfying. It's just not done that way. Postpartum is postpartum, and the NICU is the NICU.

But the idea of private rooms where parents could spend more time with their babies had been on the administrators' minds. "Mothers tell us, and it's in the literature, that the most stressful event of having a baby in the NICU is being separated from baby," says Julie de Salaberry, the director of neonatal programs at the hospital. This was about more than just alleviating parental distress too. One research paper, from Sweden in 2010, found that private NICU rooms reduced babies' hospital stays by an average of five days. In fact, plenty of medical literature now shows that restoring parent-child connections helps improve the lives of the tiniest preemies as surely as the drugs and the tubes and the machines do.

BC Women's opened the doors of its new building in late October last year. The new NICU, made up entirely of private rooms (including a dozen built for integrated mom-and-baby care), is intended to safely facilitate breastfeeding and skin-to-skin contact, the most basic human interactions that were once off-limits to sick babies.

Even though Owen was at BC Women's before the new building opened, skin-to-skin contact was a part of his life as soon as he was stable enough. In between the rounds of drugs and tests, he'd spend hours curled up on Green's or Schneider's bare chest, listening to their heartbeats and their breaths, so much stronger than his own. After about two months, Green and Schneider began to believe that he would make it. Finally on April 7, 2017, after four and a half months of blood tests, of tubes and wires, of constant monitoring of his oxygen levels, Baby Boy Green was discharged. Schneider had flown back to the Yukon a week earlier to get their small townhouse ready; he retrieved the dogs from their long stay with friends; he set up a bassinet in his and Green's bedroom. He met Green at the airport—his initial amazement of how other people were living their lives free of the hypervigilance and fear of the NICU finally subsiding. Owen slept the whole way home.

Owen is now 16 months old, and happy almost all the time, smiling and content to roll around on the townhouse floor. He's pale, blond, and blue-eyed; he makes eye contact and grins at strangers. He can breathe on his own now, but his lungs are fragile; a chest cold could put him back in the hospital. For months after they brought him home, Green and Schneider kept a sign fixed to his carrier that read: I'M A PREEMIE! NO TOUCHING! YOUR GERMS ARE TOO BIG FOR ME! They keep hand sanitizer with them at all times, and bottles of it sit on tables and shelves around the house. Early on, they wiped down everything they brought into their home that Owen would come in contact with—bottles, toys, new furniture—with disinfectant. On Christmas Eve, they called ahead to check if anyone at their intended dinner party had a cold; some of Green's clients will cancel, penalty-free, if they feel a bug setting in. "You want to be normal," Green says of their protocols, but you have to resist the urge to let things slide.

So far, Owen has met every developmental benchmark for his corrected age—he's within the expected height and weight, and has the motor skills you would expect in a baby who was born on his mid-March due date, instead of late in the previous November. His only limitation so far is his unwillingness to swallow solid foods—possibly an aversion from the weeks he spent with a tube forced down his throat. Eventually he's expected to catch up to his chronological age, but a medical team will be monitoring his neurological and motor development (among other things) until he's four and a half years old, to see if any hidden legacies of his early birth and his time in the NICU emerge.

In November, Green and Schneider marked the first anniversary of Maia's birth and death—and then, less than two weeks later, they celebrated Owen's first birthday with friends at a snow-covered cabin outside town. It will always be that way: every milestone for Owen will be paired, for his parents, with a reminder of what they've lost. But Green strives to appreciate her daughter's short life. She likes to think about what Maia might have experienced or perceived in utero. She would have heard her parents arguing, Green figures. She would have heard the family's dogs barking. She would have heard laughter. She also wants to find the right way for her son to know that he had a sister, and that they were born on opposite sides of a flexible, shifting line that we are gradually pushing back but whose exact location we might never be able to pin down.

In the old NICU at BC Women's, there was a bulletin board with notes and pictures from parents who'd already done their time. Green saw one from a mother who promised the current crop of parents that the fear and anxiety of the NICU would fade with months and years. "I thought, there's no way," she says. "How am I ever going to relax again?"

But it turned out to be true. She has begun to forget the language of hemoglobin and oxygen desaturation and outcomes and odds. She's forgetting what it felt like to be afraid all the time. She's forgetting the sound of the monitors beeping, the alarms going off, the glow of the screen as it announces each new heartbeat.

APRICOT IRVING

The Fire at Eagle Creek

FROM *Topic*

THE FIRE THAT tore through the Columbia River Gorge National Scenic Area began on a cliff's edge above Eagle Creek, just up the trail from the oldest national forest campground in America. In 1916 it took most of a day to drive a Model T from Portland, Oregon, to experience the beauty of this wilderness. A hundred years later, it takes less than an hour to cover that same distance. The hike along Eagle Creek to Punch Bowl Falls is a bucket-list destination. On a busy summer day, 600 people, possibly more, climb the trail to swim at the falls, but it's impossible to know for sure; there are no permits to keep track of how many people are on the trail at any given time. Websites list the hike to Punch Bowl Falls as "easy" because it is only two miles long and doesn't cover a significant elevation gain, but at points the trail is little more than a foot wide, blasted out of the basalt, and can be traversed only by gripping a cable screwed to the cliff's face. The views are breathtaking: in the spring, bright-purple larkspur bloom from the dripping moss on one side of the trail; on the other, the drop-off tumbles a hundred feet down to the creek. Unleashed dogs fall from the cliff every summer.

Julie Prentice, the summer host for the Eagle Creek Recreation Area, who wore a ball cap pulled down over her fiery red hair and agreed to park her RV on national forest land in exchange for keeping the bathrooms clean and answering a hundred questions a day, worried about the teenagers who hiked to Punch Bowl Falls in flip-flops, or the tourists who headed up the trail in heels. One even tried to push a stroller. She'd once talked a mother out of carrying her eight-day-old baby up the cliff, and she was amazed that more people didn't get hurt. The Hood River County Sheriff's Office gets called in regularly for search and rescue efforts in the area; 25 percent of all their calls have traditionally come from Eagle Creek.

The trails were always crowded on Labor Day weekend, a last summer fling, and local businesses in nearby Cascade Locks, a town near the Eagle Creek trailhead, filled their walk-in refrigerators with perishable food. The three-day weekend provided the equivalent of three months' worth of winter income—the buffer they'd need to survive the cold, dark months. It was a time to keep an eye out for rowdy tourists: just the week before, cops had pulled over a group of teenagers for throwing firecrackers into the Eagle Creek trailhead parking lot.

The summer of 2017 had been an unusually hot one for a part of Oregon considered to be a temperate rain forest, and by early September, a fire that had been smoldering since the Fourth of July weekend had closed trails at Indian Creek, seven miles up the mountain. The steep slopes were covered by loose rock and deemed unsafe for firefighters, but the fire had progressed slowly, creeping through the dense, moist undergrowth of a mature forest, singeing the woody shrubs but leaving the canopy undisturbed as helicopters and planes dropped up to 100,000 gallons of water a day. An infrared scan taken early in the morning on September 2 showed that after two months, only 373 acres had burned. Sharon Steriti, a US Forest Service officer on a two-week detail to redirect Pacific Crest Trail hikers out of the Indian Creek burn area, was on her way to help 16 hikers navigate their way out of the area when her walkie-talkie squawked an update: a new fire had just started on the Eagle Creek Trail.

Just after 4:00 p.m. on September 2, a teenager tossed lit fireworks down the cliff along the Eagle Creek Trail. The tinder-dry vegetation exploded. Those on the trail who could make it through the flames raced a mile down the steep terrain as the late-afternoon wind whipped erratic gusts around the canyon and up the high ridges, igniting the tree canopy.

At 4:31 p.m. Misty Brigham, across the Columbia River on her family's traditional fishing platform, snapped a photo from her phone of the billowing smoke. Brigham knew that the wooden

platform where her father fished could easily catch fire if the
flames jumped the river. The tribal nations of Umatilla, Warm
Springs, Yakama, and Nez Perce had dipnetted salmon from the
Columbia for 10,000 years, long before the first European settlers arrived to fell the forests and construct dams. Brigham knew
that nothing could convince her father to leave the river during
salmon season. The thought made her chest tighten. At 5:00 p.m.
she snapped another photo: a crackling wall of orange had just
crested the mountains.

As Brigham took photos, I was 30 miles downriver on Interstate
84, on my way home from Portland after helping my sister move
into a new apartment. My husband lifted his hand from the steering wheel to point out a gray pillar billowing above the horizon.
Cop cars raced past with sirens on, headed east on I-84. Officers
tore up the Wood Village exit ramp, then jumped out and stared
at the smoke through binoculars. I began searching my phone for
updates, but nothing had yet been posted on news sites. Our sons,
aged 9 and 11, were in the back seat with a stack of library books.
We never imagined that in just a few days the entire Oregon side
of the Columbia River Gorge would be under evacuation orders
and that we'd be the ones piling boxes into the back of the car.

By that evening, law enforcement had made the decision to
close the Eagle Creek campground. Julie Prentice kept kids and
parents to one side of the road so emergency vehicles could get
through. She even cleaned the toilets one last time for the news
cameras that were setting up in the parking lot, and for the worried family members who had gathered at the closed trailhead: an
estimated 150 people were still trapped up at Punch Bowl Falls on
the wrong side of the fire. Helicopters could get close enough to
drop messages, but there was no flat place to land, and the smoke
was closing in, grounding aircraft. The only way out was to hike
around the smoldering perimeter of the Indian Creek fire.

Sharon Steriti was unflustered by crises. She was eight miles away
from the stranded hikers when she volunteered to hike through
the Indian Creek burn area to join them. Night had fallen by the
time Steriti found the hikers, who were inching along the trail by
the light of their cell phones. The forest around them was lit by
an orange glow. She told the group that it was OK to be scared,
and assured them that she was in radio contact with the incident
response team. At a wide spot in the trail near Tunnel Falls, she

instructed everyone to hunker down, make some new friends. It was going to be an uncomfortable night's sleep, she told them, but they were in this together. A search and rescue team was on its way and would lead them to safety when the sun rose.

That first night, the Eagle Creek fire grew to 3,000 acres. It kept mostly to the high ridges, areas of loose rock inaccessible to ground crews and susceptible to landslides, but should it descend from the mountains toward the town of Cascade Locks, firefighters would be forced to prioritize defensible space. Houses that were surrounded by low-hanging tree branches or with firewood stacked on their decks could be secured only if time allowed. And if the fire made it all the way down to the river, and the wind kicked up, the fire's spread could accelerate beyond imagining.

The Cascade Mountains rise in a snowcapped wall from British Columbia to California. Their only sea-level passage is where they meet the Columbia River Gorge. Winds roar down the narrow passage, just as the Missoula floods did at the end of the Ice Age, when gigantic walls of water carved and then scoured the canyon, leaving behind the exposed cones of extinct volcanoes like rough-hewn pillars. In winter, cold winds from the east bring freezing rain and gusts of over 100 miles an hour. In summer, warm, dry winds transform the gorge into a kiteboarding and windsurfing paradise.

Ali High, nine months pregnant, had been folding baby clothes in Cascade Locks when the plume of smoke erupted three miles behind her house. She and her husband had left Southern California because the Columbia Gorge seemed like a safe place to start a family—no earthquakes, no tornadoes. For High, a graphic designer, the town of Cascade Locks held a rustic appeal. There was one grocery store, an elementary school with 64 students, two pubs, a soft-serve drive-in, a fish market; the population was 1,166. Hiking trails began just behind their house. High's hospital bag was already packed for her scheduled induction, but she kept glancing out the window as the smoke thickened. Her first inkling that Oregon was not quite as serene as she'd imagined had been the unexpectedly bitter winter nine months earlier. Their son had been conceived during an ice storm. Would he be born during a fire? She pushed the thought aside. The knock on the door from the sheriff's office came at 3:00 a.m. It was time to go.

As Sunday dawned over Cascade Locks, bleak and smoky,

evacuees filed across the Bridge of the Gods into Washington State. A thousand years earlier, landslides had created a temporary natural bridge across the Columbia River, almost a mile across, and this 1926 steel cantilever bridge had been named after that geological marvel. Few believed the fire would jump the river, even if the winds rose. Vacationers at the Beacon Rock Golf Course, a few miles downriver in North Bonneville, Washington, played through on the green while across the river the Oregon mountains burned. The town, with a population of around 1,000, had very nearly sold off its fire truck and closed its fire station's doors less than a year earlier. Temporary shelters had been set up for evacuees at the Skamania County Fairgrounds in the Washington town of Stevenson. Jean Foster, a self-proclaimed animal lover, asked John Carlson, the area's emergency-services coordinator, how she could help. She was ushered into an empty barn next to Rock Creek that would soon be full to overflowing with close to 300 evacuated animals: rabbits, distressed cats, dozens of chickens, three unruly goats that needed to be walked, homesick dogs, three pigs, a horse abandoned by its owner (though it was later claimed), a cockatiel, ducks, and a tarantula.

By noon on Sunday, the trapped hikers had made it safely off the trail and were reunited with worried family members at the Wahtum Lake trailhead, 14 miles and 20 hours after the fire had upended their plans. The smoke was too thick to allow aircraft to drop water on the flames, so fire crews were on the ground in Cascade Locks, and displaced business owners pooled their resources to keep the firefighters supplied with coffee, salmon chowder, and sandwiches. The smoke created a grim haze over the river. Ryan Walker, a construction worker and volunteer firefighter, studied the conditions from a roof in Skamania Landing, on the Washington side of the river. He had been trained to recognize if an inversion system was developing, and knew the damage it could unleash: if warm air pressed down from above, it could trap the smoke close to the ground, creating the perfect conditions for the fire to build up pressure, like a damper pulled out on a woodstove. And it was only a matter of time before the wind started blowing.

Early Monday morning, the incident response team determined that the Indian Creek fire, further up the mountains, was only 10 percent contained and had grown to 1,000 acres. The Eagle Creek fire, a quarter mile from the town of Cascade Locks, was 0 percent contained and covered 3,200 acres. The winds were expected to start that afternoon, with gusts of up to 25 miles per hour.

Interstate 84, on the Oregon side of the river, remained open but on standby, ready to be closed at a moment's notice. I-84 is a corridor for commerce, as well as for those in a hurry to get from Portland to the vineyards, orchards, ranches, trails, and kiteboarding spots at the dry eastern end of the gorge. Many visitors preferred the quieter scenic Historic Columbia River Highway, dedicated in 1916, with its stone barriers and bridges inspired by the scenic highways of Europe. The towns along this road are scattered and independent, clusters of homes tucked among trees and waterfalls: Warrendale, Dodson, the iconic stone-and-timber Multnomah Falls Lodge, Bridal Veil, the Angel's Rest Trail, Latourell, and the art nouveau Vista House, 733 feet above the Columbia River, from which one could see, on clear days, the full panoramic sweep of the gorge.

With 1,250 households, the town of Corbett, 20 miles downriver from Eagle Creek, is by far the largest of the communities along the highway, and the Corbett volunteer fire department had already responded to the call for aid that came from Cascade Locks around noon on Saturday. On Monday, a truck and a five-person volunteer crew drove down to defend the Bonneville Fish Hatchery, three miles from the Eagle Creek trailhead, before the wind developed a mind of its own and made evacuating human residents the far more urgent priority.

By 8:45 p.m., the Multnomah County Sheriff's Office issued a "Level 3" mandatory evacuation order for Dodson and Warrendale, seven miles west of Eagle Creek, and I-84 was closed. Bridal Veil, an area 15 miles west along the historic highway, was declared a "Level 2" evacuation zone, which meant that residents should start packing and be ready to go at a moment's notice. Residents herded agitated horses into trailers in the dark as a warm wind tore down the gorge. Bridal Veil had been established in 1886, but all that remained of the once-bustling logging town were the Pioneer Cemetery and the post office, whose doors remained open only because of a yearly influx of wedding invitations awaiting a novelty postmark. Only 45 post office boxes were still in use. The 1902 Palmer Mill fire, followed by a second mill fire in 1936, had reduced Bridal Veil to a few houses and farms, serene Bridal Veil Lodge, the Bridal Veil Lakes wedding venue, and the convent.

The Franciscan sisters, in their plain brown habits, packed quickly. They worried over the cats and the goats, until they were assured that temporary homes would be found for them. The sisters had purchased a decrepit 1915 Italianate mansion in Bridal Veil in the 1970s, after the mansion had been abandoned for a decade. The floor had been buckled by water damage, and the grounds were littered with bags of moldering trash, five wrecked cars, and two refrigerators, but over the years their restoration work had transformed the convent into a leafy retreat. Two of the nuns were in fragile health; even if their home did not survive the fire, they hoped one day to be buried beside their sisters in the small cemetery at the foot of Angel's Rest.

All that night and into the early morning hours of Tuesday, September 5, the fire leapt across the mountains like a living creature. Resin-packed conifer crowns exploded and snapped off in the intense heat, flinging lit branches ahead of the fire's edge. One ridge, then another, gave way to flames. The Indian Creek fire and the Eagle Creek fire merged into an incendiary crown that spiked above the gorge. Debris flung by the fiery tornado-like wind splashed into the Columbia River, and live embers whirled two miles through the air to land on Archer Mountain, a mile inland from the river on the Washington side, touching off an unexpected round of emergency evacuations across the gorge. Ryan Walker, the firefighter who'd worried about the inversion system from the moment he recognized it, joined a team of volunteers to scout the shoreline near North Bonneville and put out spot fires as soon as they started. A two-foot-long charcoal spear landed, still smoking, on the golf course, and the North Bonneville Fire Department salvaged it in memory of that sleepless night.

In 38 years of fighting fires, Dave Flood had never seen anything equal to the Eagle Creek conflagration. As the fire chief of Corbett, he had trained in wildland firefighting, but this was the kind of event that happened in drier states like Arizona and Colorado. Brush fires, which stayed low on the forest floor, could be put out without any real danger. But once a fire got up into the canopy, becoming a crown fire, all the water in the world couldn't put it out. It was the kind of fire he'd seen on videos, never in real life, with a plume of smoke so full of particulate matter that it collapsed under its own weight and fell onto the firefighters like a nuclear blast.

Stones ricocheted down the cliffs from a thousand feet up as the Corbett firefighters evacuated the last remaining residents of Bridal Veil and Latourell. Chief Flood knew that a misstep could put his crew in jeopardy. His training, for the first time, felt urgent. If the truck got trapped behind a landslide, his crew might be forced to run for their lives. It was time to get out.

The wild animals, too, sensed the danger. Late Monday night, a herd of elk clustered in the meadow by our house, restless and wary, huddled in the open, away from the trees. We were west of the fire, still in a Level 2 evacuation zone, but even then it was 90 degrees at midnight, and charred leaves that the fire had carried for miles floated down as we loaded our car with our six new baby chicks and a sick kitten, while our dog circled our ankles, anxious not to be left behind. A mile down the road, a neighbor snapped a photo of a cougar with what looked like a coyote dangling from its jaws. We woke our boys at 2:00 a.m. to explain that the fire was closer than we'd thought. As we waited for the knock on the door, we could hear the thunder of horse trailers evacuating.

Kim Mosiman, who had founded a horse rescue in Corbett, set up a table in the elementary school parking lot and coordinated with the Regional County Disaster Organization. Strangers across Oregon and Washington drove in with trailers for transporting livestock and offered safe pasture. Kim knew which of her volunteers at Sound Equine Solutions could stay calm in a crisis and who would be able to handle an animal in distress.

At 4:00 a.m. on Tuesday morning, the fire seemed unstoppable. Smoke and ash rained down on the Corbett fire crew as they pulled into the empty parking lot of the Vista House, the historic 1918 art nouveau rotunda with views from five sides, to try to scout the path of the fire. The glowing red edge was just a few miles down the Historic Columbia River Highway when the smoke began to clear. The winds had shifted. The fire appeared to have halted at Palmer Mill Road—the same place it had stopped in the 1902 fire. Burning branches and tree trunks rolled down into Bridal Veil Creek, sputtering to a simmer. The valley marked the beginning of a shift in vegetation: from resinous firs and hemlocks to deciduous alders and big-leaf maple, with green leaves that slowed down the fire's progress. The wind-fueled fire had apparently jumped 13 miles in 16 hours, but for now, as the wind died down, it looked like we'd been given a reprieve.

It took two weeks before I-84 reopened, on September 15, and residents in Level 3 evacuation zones were allowed to return home. Even then, with 54 engines, 12 helicopters, and 967 firefighters assigned to the Eagle Creek Fire, the blaze was only 28 percent contained. Because the burn had been so erratic, there remained dry, untouched areas of forest next to smoldering trees, and the wind was still shifting directions, moving the fire first toward the Bull Run watershed, which supplied drinking water to the entire city of Portland, then toward the orchards in Hood River. The first spattering of rain in mid-September was greeted with intense relief, but it took until the last day of November before cooling temperatures and steady rains allowed the incident command team to consider the conflagration contained.

As the smoke cleared, stories emerged. Firefighters had watched the fire jump overhead across the narrow canyon behind the Multnomah Falls Lodge, three miles east of Bridal Veil. The lodge, one of the crown jewels of the historic highway, had been built in 1925 at the base of famously photogenic falls that plunge 600 feet behind a graceful bridge. It had escaped unscathed during the 1991 fire that swept along the top of the falls—a photograph of that blaze is framed above the bar, although it now seemed tame in comparison to what had just unfolded in the gorge. Firefighters described how fire had poured down either side of Multnomah Falls, creating an amphitheater of flames. Burning trees had tumbled 600 feet over the cliff's edge and destroyed the Shady Creek Bridge. Flames had licked up to the edges of a damp circle of water pumped from the creek, the last line of defense around the stately lodge.

Chris Shaw, the lodge's executive chef, was one of the first civilians allowed back onto the closed freeway, more than a week before it officially reopened. Obligated by his contract never to leave the lodge unattended, Shaw and his wife had camped out in sleeping bags during ice storms in previous years. He had never before been forced to abandon his post.

In gratitude, Shaw put together meals at the lodge for the fire crew over an entire week, beginning on September 6, explaining that it was one of the absolute honors of his life to feed the people who had saved this building. For long, quiet months that fall, before crews arrived to repair the smoke damage, Shaw and his assistant general manager sat at a table in a small room near the front of the lodge to oversee the restoration. He was instructed to keep FEMA's mobile app open on his phone and watch for alerts about possible landslides. Multnomah Falls was usually busy with visitors who drove up the gorge to enjoy the yellow and gold leaves of autumn; on a beautiful September day, the lodge could easily serve 300 people. Instead, this year, with no tourists in sight, herons and deer wandered undisturbed through the parking lot. A bear was spotted along the closed historic highway. Eventually, chipmunks took over the empty dining room.

On a quiet Tuesday morning this summer, one week after the Fourth of July, the orange safety cones that had blocked the Angel's Rest parking area have been temporarily pulled aside, and the dirt road is full of a dozen cars and a van full of shovels, pickaxes, and hard hats provided by Trailkeepers of Oregon, a trail-restoration organization. Elaine Keavney, a board member and crew leader, began guiding crews up the ruined trail as soon as land managers deemed that it was safe to begin restoration work. Today, she asks how many of those who have driven out for the seven-hour shift are first-timers. Of the 18 people here, 3 raise their hands. There are others here who have signed up for trail work more than 30 times.

Almost a year after the Eagle Creek fire torched 48,000 acres, the Historic Columbia River Highway remains closed past Bridal Veil because of landslides, but the impact of the fire is still visible even from a car traveling 65 miles per hour on I-84. Blackened tree trunks loom over patches of bright-green undergrowth: thimbleberry, nitrogen-fixing sword ferns, and big-leaf maple stems sprouting from stumps.

Elaine double-checks paperwork and glances at everyone's shoes to make sure they are rugged enough for the work. The hazards on the still-closed trail are no joke. Hard hats must be kept on at all times, even when searching for a "facilitree" for a bathroom break. Before the fire, the loose scree of rock was held together by a thick web of moss, but the moss went up fast, glowing red for days. Without that natural glue to hold the mountain in place, a rock kicked down the talus slope could set off a landslide, dislodging fallen trees and turning them into potentially lethal torpedoes. Elaine asks how many of the volunteers have had first-aid training.

Pete Reagan, a retired family doctor in Portland, is among the three who raise their hands. He had been hiking in Switzerland when the Eagle Creek fire broke out and had followed the updates from a distance, with increasing dismay. The iconic Angel's Rest Trail, with its 270-degree views of the gorge, had been one of his yearly pilgrimage sites since 1971, just after he'd moved to Oregon. Watching it burn had transformed Pete from a trail user to a trail steward, in the language of the Trailkeepers, an organization that started small in 2007 but was flooded with new volunteers after the 2017 fire.

Guy Hamblen, another Trailkeepers crew leader, is in charge of rock work for the day, and he explains that a select team will have the job of heaving a 600-pound rock into place with levers and a sling. The extended closure of this trail has allowed forest managers to undertake erosion-control measures and incorporate switchback reinforcements that had been a growing concern for decades. The fire added new challenges—slow-smoldering roots of trees collapsed below the ground and the spring rains scoured away whole sections of the trail, creating "hell holes"—but much of the underlying damage was cumulative. With an estimated 1 million year-round hikers on the Angel's Rest Trail, it had been a challenge to stay ahead of the necessary repairs.

A few miles up the historic highway from the Angel's Rest trailhead, perched on a lip of basalt overlooking the grand sweep of the Columbia River, tourists swarm the information table in the marble-floored Vista House. Ed Murphy and Carol Addleman, retired volunteers who banter with the ease of a comedy duo, are happy to explain to visitors from as far away as Ethiopia, Hong Kong, Europe, and Australia that the elaborate rotunda was originally intended as a restroom stop, and had been nicknamed "the $100,000 Outhouse" when it went wildly over budget in 1918, but most people just want to know why so many waterfalls are still closed. Many have never heard of the Eagle Creek fire.

A man who has driven across the river from Washington State to show off the gorge to a friend visiting from New York is irritated that he couldn't find a parking spot at Multnomah Falls and had to circle eight miles out of his way to turn around. The volunteers offer sympathy and point out which trails and waterfalls are still open, then search for the schedule for the shuttle bus to Multnomah Falls from Rooster Rock State Park, less than a 15-minute drive away.

When they can find a parking spot, tourists still flock to the reopened Multnomah Falls. The blackened remains of a hollow tree greet visitors who climb the footpath to the waterfall, and the trail is closed beyond the bridge, but the lush green undergrowth is returning to the singed slopes. Lovers take selfies. Kids cling to the guardrail or lift up their faces to feel the sting of cold droplets. As soon as the historic highway reopens, it will be just as crowded as before.

Fireworks are illegal on public lands, and bottle rockets, Roman candles, and firecrackers are illegal across the state of Oregon, but permitted sellers in both Washington and Oregon are allowed to sell approved fireworks. Over the Fourth of July weekend, three separate fires broke out along the Columbia River Gorge. All three were the result of careless human actions, but being within the reach of fire hoses were quickly put out. The possibility of yet another wind-fueled conflagration is never far from the minds of residents of the gorge.

"We talk about the fire twenty-four seven," Addleman, the Vista House volunteer, admits between inundations of tourists. "When we go home, we dream about the fire."

The question "Are we loving the gorge to death?" is one that surfaces frequently in discussions about the future of the Columbia River Gorge. The legal provisions of the act that designated the gorge a National Scenic Area, in 1986, created a Gorge Commission to work alongside the Forest Service to protect and enhance the area's resources: scenic, cultural, natural, and recreational. The resulting coalition requires deft coordination with six different counties across two state lines and four tribal nations, and represents a broad spectrum of stakeholders.

To protect the Columbia River Gorge against the cumulative adverse effects of human impact requires a partnership between windsurfers, wineries, Pacific Crest Trail hikers, farmers, First Peoples (who have sovereign fishing rights), and the rural communities that staff the volunteer fire departments and visitor centers and call in fireworks infractions.

It isn't always an easy conversation. On July 10, the Oregon Department of Transportation, seizing the opportunity provided by the Eagle Creek fire closures, announced a controversial—albeit temporary—plan to convert the section of the historic highway that winds past Multnomah Falls to a one-way bike- and pedestrian-friendly road system starting in September 2018. Local residents and business owners worried that insufficient forethought had been given to the impact this would have on already crowded parking lots and that it would do little to ease congestion. Moreover, Multnomah Falls Lodge would, under the proposed plan, be significantly more difficult to access from the Cascade Locks fire station—and, once again, September is on track to be the month with the greatest fire danger. According to the Oregon Department of Forestry, 33 percent of the state was considered extremely dry in August of 2017; a year later, 95 percent of the state is now deemed extremely dry. (The land itself had the last word: landslides delayed the one-way proposal until spring.)

Meanwhile, tourism increases every year. At least 30,000 people move to Portland each year, and the gorge is an easy day trip. Permits are not required on gorge trails, and the "trail ambassador" program is an attempt to keep tabs on the people who move through popular trailheads on weekends and help redistribute the hikers. Kevin Gorman, the executive director of Friends of the Columbia Gorge, estimates that 80 percent of the area's tourists visit only 20 percent of the gorge.

Julie Prentice, who in one day lost both her job as the Eagle Creek Recreation Area host and her place to live, signed up this summer for a shift at Dry Creek Falls as a volunteer trail ambassador. Between handing out Smokey Bear stickers to kids, she picks up 65 cigarette butts from the parking area and fields questions from disappointed hikers about why the trails are still closed. She explains to them that a big fire hadn't happened in the gorge for a hundred years, and that it has cleansed the earth. It's important for humans to take a break and let Mother Nature do her work. Eagle Creek needed to rest. People get that.

Early assessment of the Eagle Creek fire showed that the damage, though it covered a broad area, was not as severe as initially feared. Only 15 percent of the forest suffered severe burns. The damage to 30 percent of the burn area was considered moderate, and the remaining 55 percent showed low-to-no burn damage. Because the winds had whipped along the ridgelines, the resulting mosaic pattern left mature, resilient trees alongside exposed burn areas, which would scatter seeds to fill in the bare patches, first with low ground cover, ideal for foraging wildlife, then deciduous trees and eventually conifers—that is, until the next fire redrew the ecological map of the forest.

Fire ecologists consider the event a dark gift to the forest. Blackened tree trunks provide habitat for woodpeckers, who feast on the wood-boring beetles that otherwise might have threatened the trees. Charcoal enriches the soil. Raptors nest in the snags. Fallen trees roll into creeks and rivers, creating shadowed spawning grounds for salmon.

But it is one thing to recognize the benefits of fire, another to have lived through an evacuation. These days, the Franciscan sisters are back in residence at Bridal Veil, although the road to their home remains closed by the fire. The sisters draft lesson plans for their students at their Franciscan Montessori Earth School and gather daily in their small chapel to pray for those around the world made vulnerable by trauma, as well as for first responders and the 15-year-old boy whose life has been forever altered by fireworks. In February, the teenager pleaded guilty to the reckless burning of public and private property and was sentenced to five years of probation. He had received death threats, and the sisters pray that he will be safe from harm, and that he will use his regret for good. On the quiet path behind the convent, pink stems of bleeding hearts nod over patches of bare charcoal and blackened trees.

Ken Smith, a Wasco elder, had watched the flames race along the ridge toward his home in Corbett from across the river, trapped in traffic while friends and family rescued his horses for him. He understands that fire, treated with respect, keeps the meadows clean, allows the huckleberries and wild strawberries to flourish, and creates homes for the animals to raise their young. Life began with fire, with the sun giving us its light, but fire is not to be treated recklessly. The salmon who swim upriver when the big-leaf maple leaves beat against the ground like a drum calling them home give themselves to us as food, but in that gift is a warning not to live wastefully. We are to be caretakers of the land, just as the animals and all of creation have cared for us. It is our responsibility to think ahead and care for those who will follow us, generations after we are gone.

Ali High, the mother in Cascade Locks who had been evacuated five days before she gave birth, is not yet ready to take her son on his first hike through the charred woods, but she tells me she will do it one day. His middle name is Phoenix.

Sharon Steriti, the Forest Service ranger who led the trapped hikers to safety, is one of the few who has been able to return to Punch Bowl Falls after Eagle Creek was closed. Rockfall has slumped into the blue depths of the pool, creating a newly precarious landscape, and the cliffside trail is every bit as sobering as before—a reminder that wilderness is indeed wild—but young oaks and maples are already springing up along a dry stretch of trail razed by the fire. Sharon smiles. "Life is tenacious. The Earth is incredibly resilient and will do an amazing job of healing if allowed to heal. Mother Nature wins. She gets the last word."

ROWAN JACOBSEN

Deleting a Species

FROM *Pacific Standard*

IN A WINDOWLESS London basement, behind three sets of
locked steel doors and a wall of glass, thousands of *Anopheles gam-
biae* mosquitoes cling like Marvel supervillains to the sides of white
mesh cubes. The room is negatively pressurized, so air is constantly
sucked inward to ensure that the mosquitoes, which have been
subjected to a new and astonishingly powerful kind of genetic engineering, never escape.

If the modifications to these whining mosquitoes were per-
fected, and they were somehow able to make their way to sub-Saharan Africa, they would have an effect on their kin unlike any
animal that has ever existed. The *Anopheles* are equipped with a
genetic tool that ensures that they are either sterile—they can't
produce viable eggs—or, if fertile, that they will pass that sterility
gene on to nearly every offspring. And the same would be true
for their descendants, which would continue to spread the genetic
sabotage into future generations.

If some future version of the mosquitoes were released, these
deadly modifications could spread through the African tropics,
crashing the population as they went. And because *Anopheles* is the
primary African vector for the parasite that causes malaria, its collapse would likely take down malaria with it. Within a few years,
the last great scourge of humanity, which kills upwards of half a
million people per year, would be vanquished on the African continent. It would be one of the greatest health achievements of all
time. And yet the intentional eradication of a species is not something we should pursue without a lot of foresight, and the release

of highly invasive genetically modified organisms (GMOs) into the wild is itself deeply disturbing.

Known as a gene drive, the ability to force particular genes into future generations of an entire species only became available to humans with the development of CRISPR, the gene-editing tool that has enabled us to make precise changes to an organism's DNA. Kevin Esvelt was a fellow at Harvard University's Wyss Institute for Biologically Inspired Engineering in 2013 when he figured out how to build a gene drive. In a 2014 paper, he proposed several applications for his invention, including hobbling weeds that had become resistant to herbicides, reducing malaria-carrying mosquitoes, and eliminating invasive rodents on islands, where they wreak havoc on indigenous birds and plants.

Many traditional conservationists were horrified by the prospect, yet other groups embraced it. The Gates Foundation made gene drive a centerpiece of its antimalaria efforts, and the eco-warriors at Island Conservation, who have long used poison to combat invasive mice and rats, seized on gene drive as a more precise weapon in their war to save native species. New Zealand is considering using a gene drive in its push to eliminate invasive rodents, weasels, and possums by 2050. Kevin Esvelt wants to engineer mice that are immune to the bacterium that causes Lyme disease, whose cycle of transmission goes from mice to ticks to people. Dengue, Zika, and several other mosquito-borne diseases are promising gene-drive targets. A lab in California is working to limit the damage caused by an invasive species of fruit fly, and labs in Australia and Texas are developing "daughterless mice" (capable of conceiving only male offspring). The first gene-drive field trials are anticipated within the next decade.

With earlier-generation GMOs, such as Monsanto's Roundup Ready crops, arguments often hinged on the potential for those genes to escape into the environment. Conservationists believed escape was inevitable, while corporations downplayed the risk, but nobody was suggesting that GMOs be let loose in nature—until now.

When I first heard about gene drive, I thought of "ice-nine," the form of water in Kurt Vonnegut's 1963 novel *Cat's Cradle* that is solid at room temperature and acts as a seed crystal for adjacent water molecules, turning them solid. At the end of *Cat's Cradle*, the frozen body of a man who has committed suicide by drinking ice-nine falls into the sea, and all the world's oceans and rivers are forever frozen, extinguishing most life on Earth. Gene drives have similar dystopian potential. In theory, a single lab could alter the entire planet. And the technology has arrived far quicker than our ability to grapple with its staggering implications.

Gene drives work by gaming inheritance, forcing their way into the genetic makeup of future generations. Sexually reproducing species usually have two versions of each of their genes, one inherited from each parent, and they randomly pass one of those to each offspring. Individuals that inherit more useful genes thrive, and are therefore more likely to reproduce and pass on those good genes, while individuals that inherit disadvantageous genes are less likely to get the chance to reproduce. In this way, evolution causes detrimental genes to disappear from the gene pool.

Conventional genetic engineering is limited by the rules of reproduction. Most engineered traits have a 50–50 chance of being passed down, and unless a trait confers some advantage to the organism, it should eventually disappear. Since most genetic engineering to date has bred traits that benefit people, not the organisms themselves, so far no GMOs have made significant inroads into nature. But a gene drive can practically guarantee inheritance. And since beneficial genes are favored by natural selection anyway, the unique value of engineering a gene drive lies in propagating a detrimental trait, possibly even all the way up to extinction.

To make a gene drive, you start with the gene-editing tool CRISPR, which consists of two parts: a gene-slicing enzyme and a string of genetic code that tells the enzyme where to cut. CRISPR is shockingly easy to use. You don't need a world-class lab, and you don't have to be a genius. I've created antibiotic-resistant bacteria in a friend's kitchen. You just order your CRISPR from a DNA-synthesis company (the going rate is $65 plus shipping), specifying the exact 20-letter sequence of DNA you want it to target. It arrives as a few drops of liquid in a test tube. You add that liquid to another test tube containing cells of the organism you want to modify, along with any new DNA you want inserted, then heat it up. The CRISPR finds the spot, makes the cut, and the new DNA gets stitched in place.

Kevin Esvelt was part of the team at Harvard that helped develop

CRISPR, and he was the first to realize that the CRISPR mechanism itself could be inserted directly into an organism's genome to create a gene drive. Once there, the CRISPR would eliminate the natural counterpart of the gene it is attached to, and the cell would copy the functioning, genetically engineered version of the gene (containing the CRISPR) in its place. The organism would then have two working copies of the CRISPR gene, one of which would be guaranteed to be passed down to each of its offspring, where the process would repeat, until virtually every individual in a population carried the engineered trait.

It was a brilliant insight, with enormous implications. According to the unwritten rules of science, Esvelt's next move should have been to quietly create a gene drive in his lab and then publish a paper announcing the achievement to the world and staking his claim to it. Instead, he paused to consider the consequences.

When I first met Esvelt in 2017 at Editing Nature, a summit convened at Yale University's Institute for Biospheric Studies to weigh the ramifications of engineering the wild, I was struck by his demeanor. He seemed haunted and tightly wound, as if he'd just come from a dark future he was hoping to save us from. His boyish smile and wispy blond hair reminded me of Tintin, but his gravelly, leading-man's voice vibrated in an unusual timbre. Like the long *dungchen* horns of Tibetan monks, it seemed to resonate with both awe for the world and sorrow for its eventual passing.

As soon as Esvelt realized how easy it would be to build a gene drive, he knew he had a potential ice-nine on his hands. "This thing self-scales," he told the biologists, conservationists, and ethicists gathered at Yale that day. "You can't run a field trial. You can't introduce it anywhere in the endemic environment without having it spread probably to every population."

After his 2013 discovery, Esvelt knew others would soon hit upon the same insight, and he felt that the runaway nature of gene drive was not something that could be trusted to biotech specialists working in isolation. "Your decision to go ahead and build it in the lab means that you are performing an experiment that could affect other people," he said. "And if you don't tell them that you're doing it in advance, you're actively denying them a voice in the decision. And frankly, that's wrong."

Esvelt pictured the headline sure to follow an accidental gene-drive release: "Scientists Convert Entire Species to GMOs.

Is CRISPR to Blame?" He feared that a botched trial could turn the public against the technology and destroy its vast potential. So shortly after their breakthrough, he and his colleagues at the Wyss Institute called a meeting of top ecologists, biologists, ethicists, and national security experts. They explained the technology to the group, and discussed the best plan of action. And their remarkable conclusion was that the only way to ethically explore the potential of gene drive was to change the culture of science. "We need to at least tell other people what we are thinking of doing before we even begin experiments," he explained. "This is difficult, because every incentive in science points against it. If you share your brilliant idea, you're inviting some larger, better-funded lab with spare hands to steal it, get it working first, publish, and get the credit."

Esvelt decided to make an example of himself. He published his paper before doing any experiments, with the hope that all gene-drive research would follow the precautions and protocols he laid out, the most important of which was preregistration of all experiments so they could be vetted by all potential stakeholders. Since then, he has spent as much time lobbying against the unwise use of gene drives as he has advocating for them, sometimes using language that distresses his fellow scientists. "We are walking forwards blind," he said in a 2016 interview that is frequently cited by gene-drive opponents. "We are opening boxes without thinking about consequences. We are going to fall off the tightrope and lose the trust of [the] public."

Not since Robert Oppenheimer has a scientist worked so hard against the proliferation of his own creation. "When you see something that is technically sweet, you go ahead and do it and you argue about what to do about it only after you have had your technical success," Oppenheimer said in 1954. "That is the way it was with the atomic bomb."

And that is how it has been with gene drive. Esvelt now runs something called the Sculpting Evolution group at the Massachusetts Institute of Technology. When I sat down in his office and asked him if he had convinced many scientists to forgo the technical sweets, he shrugged. "It will never happen unless we change the incentives," he said. "Most scientists, however supportive in theory, say they just can't take the risk." The allure of scientific immortality—or at least tenured professorship—is simply too strong, and while those working with gene drives claim to follow rigorous safety protocols, few are willing to openly share what they are inventing behind the closed doors of their labs.

We are on the cusp of a gene-drive explosion. Many agricultural pests are potential targets, as are weeds that have evolved resistance to Roundup. California's cherry growers are funding gene-drive research to eliminate the spotted-wing fruit fly, which lays its eggs in soft fruits. Tata Trusts of India recently gave the University of California at San Diego $70 million to train a new generation of Indian scientists to use gene drives for agriculture and disease control. And in the fall of 2017, the biotech firm Oxitec released genetically engineered diamondback moths (which infest broccoli, cabbage, and other brassicas) in a field trial in upstate New York. The moths carry a gene that kills females in the larval stage, and though there is no gene drive involved at present, it would be a logical next move.

The most vocal critics of gene drives have been two conservation organizations, Friends of the Earth and the ETC Group. Jim Thomas, co–executive director of ETC, told me that, for all the emphasis on curing disease and saving endangered species, he sees Big Ag lurking in the background. "Ultimately, I think that's where this technology lands," he said. "It becomes a kind of insecticide. If there's money to be made here, that's what's going to drive it." Thomas sees potential for abuse in the developing world. "How does a powerful technology shift power relations? And what does that mean for those that are marginalized and vulnerable?"

In September of 2016, 30 environmental luminaries, including Jane Goodall, David Suzuki, and Vandana Shiva, joined with ETC to publish an open letter calling for a moratorium. "We believe that a powerful and potentially dangerous technology such as gene drives, which has not been tested for unintended consequences nor fully evaluated for its ethical and social impacts, should not be promoted as a conservation tool," they wrote. "Given the obvious dangers of irretrievably releasing genocidal genes into the natural world, and the moral implications of taking such action, we call for a halt to all proposals for the use of gene drive technologies."

Friends of the Earth joined ETC in bringing the call for a moratorium to the December 2016 meeting of the United Nations' Convention on Biological Diversity, which covers the equitable use and regulation of biological resources, including genetically modified organisms. The convention has previously halted controversial technologies such as ocean fertilization and sterile-seed crops by establishing moratoria, but with gene drive it merely called for better risk-assessment. Friends of the Earth and ETC vowed to continue to rally support for a moratorium, which will be debated in Montreal this July and then voted on at the next meeting in Egypt this December.

Most gene-drive scientists accuse these groups of exaggerating the risks of genetic engineering and playing to the public's fears, but Natalie Kofler, who founded Yale's Editing Nature initiative to facilitate public deliberation around gene editing, thinks it's vital to take their point of view seriously. "The followers of those groups share a worldview with many people that I discuss this with on a daily basis," she told me. "They feel deeply that it is wrong to tamper with the DNA of wild things. There's a sacredness to it that we shouldn't mess with. And that is a worldview that is very quickly dismissed by scientists and technologists. And because it's not being acknowledged as something valid for discussion, I think it's creating a huge polarization."

Still, Kofler finds the idea of a ban on research "totally ridiculous." This is a brand-new technology, she said. "Right now we don't know nearly enough about how it works, how the public perceives it, or how it will impact the environment to take stances of opposition or support. Right now, we need to be comfortable to stay in the gray zone—to comprehensively explore this issue with the degree of openness and transparency that it deserves. So, if anything, more research—scientific and sociological alike—needs to take place."

Jim Thomas points out that there's a difference between a moratorium and a perpetual ban: "There's a feeling that taking a judicious pause and taking the time to think carefully means nothing is ever going to move forward. But that's not what a moratorium is."

When the stakes are as high as they are with gene drive, who could argue with a judicious pause? People in Africa, Esvelt says. Every year you delay work on gene drives, another half-million people die. "Who am I to tell somebody who's lost children to malaria, and has more children at risk, that they can't do it because somebody else doesn't agree? Why should some people get veto power over a technology that could save the lives of other people's children?"

*

And yet, despite that sentiment, Esvelt keeps making things more difficult for his colleagues. Last November, I along with several other journalists received an unusual email from him. "I'm writing because we have a couple of papers coming out next week that are personally embarrassing for me, but are likely consequential enough for gene drive, conservation, and science policy that you might find them interesting," he wrote. What followed was a surprising statement: "My decision to list invasive species control as a potential application of gene drive in our original 2014 *eLife* paper was an embarrassing mistake . . . It was profoundly wrong of me to even suggest it." Additional modeling, he explained, showed that gene drives were even riskier than he'd thought. For that reason, one of his new papers concluded, with the possible exception of malaria, "we should not even consider building drive systems likely to spread indefinitely beyond the target area."

The new papers triggered a wave of fresh panic in the media. "'Gene Drives' Are Too Risky for Field Trials, Scientists Say" reported the *New York Times*. Most of the coverage focused on Esvelt's mea culpa, and when we met in his MIT office shortly after, I asked him if that was the reaction he'd expected.

"Of course!" he responded. "I'm not totally naive. 'Inventor tries to stuff genie back in bottle'—that's a story. It doesn't happen very often that a scientist says, 'I was wrong.' Maybe it should happen more often."

Esvelt believed that other researchers were underestimating the risk of engineered organisms escaping a field test, even on an isolated site, in part because of a wild card beyond the scope of any mathematical model—human nature. "You build it, you try it anywhere, and someone who has an interest is going to move it illegally to take advantage. It would be totally cost-effective for someone to hire mercenaries to fly in, capture mice, and fly out again. But that's not the sort of thing most scientists think about."

I was reminded of Jeff Goldblum's chaos mathematician in *Jurassic Park*. "If there's one thing the history of evolution has taught us," he warns the park's designers, "it's that life will not be contained. Life breaks free. It expands to new territories and crashes through barriers painfully, maybe even dangerously . . . Life finds a way."

Some of Esvelt's colleagues saw the move as a publicity stunt: instead of drives "likely to spread indefinitely," Esvelt was recommending a new, self-limiting type called Daisy Drive that he had recently designed. In Daisy Drive, multiple drives are linked in an organism's genome in a kind of daisy chain. Drive A drives Drive B, and B drives C, and C drives D, and so on. But because nothing drives A, it follows normal inheritance patterns and gets quickly diluted in the gene pool. Those individuals who don't inherit A have nothing to drive B, which then gets diluted in subsequent generations. Like the stages of a rocket, the drives continue to fall away until the whole system stops working after a set number of generations. In theory, Daisy Drive allows you to affect a local population for a set amount of time.

Esvelt now hopes to use a self-limiting drive such as Daisy to combat Lyme disease in the northeastern United States, where it has become so prevalent that many people no longer risk walking in the woods and fields. Almost 40 percent of Nantucket residents have reportedly contracted Lyme disease, and that is where Esvelt has proposed to begin his "Mice Against Ticks" experiment, as well as on neighboring Martha's Vineyard. To make sure the local stakeholders understand the implications, Esvelt has been holding community forums on the islands since 2016, and most residents seem open to the idea. After an initial field test on an isolated and uninhabited island, he would release thousands of Lyme-resistant mice on Nantucket and Martha's Vineyard. If all went well, the eventual goal would be to release Daisy Drive mice on the mainland. The Lyme infection cycle would then be broken, and eventually the Daisy Drive would disappear as well. After a few generations, the mice would revert to normal.

A number of self-limiting drives have now been proposed by Esvelt and other researchers, but so far they exist only on paper, which makes Jim Thomas skeptical. "Precision in biology and ecosystems is a bit of a pipe dream," he told me. Ecosystems are remarkably complex, and viruses and parasites have tremendous capacities to evolve.

When I mentioned this critique to Esvelt, he gave me a knowing nod. "The thing everyone is overlooking is, how do you know your gene drive is going to behave over time the way you intend? We've never before engineered something that we anticipate to evolve out of our control. Perfect prediction is impossible." But unlike the skeptics, he believes you can get close enough to proceed with confidence. "You need to model very large populations over multiple generations. We can't do that in mice or mosquitoes, but we can in worms."

And they are. This winter, on the sixth floor of a nondescript MIT office building, behind a locked door with a black-and-orange BIOSAFETY LEVEL 2 warning sign, I held up dozens of petri dishes filled with what looked like twitching, emaciated commas. These were roundworms, *C. elegans,* also known as nematodes, and there were 5,000 to 10,000 of them per dish, reproducing every three days. "We can do a hundred generations in a year with a population of a hundred million," Esvelt told me. "If we really wanted to push it, we could probably do a population of a billion. I can't think of another organism that would let us do that."

One of Esvelt's postdocs placed a dish of worms under a microscope and turned on a black light. Through the lens, I could see the silvery squiggles snaking through the agar, eating bacteria. Each had a glowing red esophagus thanks to a fluorescent gene (originally from a jellyfish) that made it easier to track which ones had received the genetic modifications.

These worms will be the first organisms on Earth to harbor a Daisy Drive. Their lives will be confined to thousands of test tubes managed by a liquid-handling robot that can be programmed to move precise amounts of liquid between tubes. Each test tube will harbor an isolated population of worms, so the Sculpting Evolution team can test what happens when Daisy Drive worms invade a new, unmodified population. They can also test whether an engineered drive evolves into something unexpected, given enough time and population growth, and whether an "immunizing reversal drive" can be built that will target such a runaway drive and reset it.

Eventually, the worms could have enough genetic diversity to serve as a decent stand-in for any wild population, and all experiments on them will be preregistered for feedback from the scientific community. To keep life from finding a way, Esvelt told me the project has five layers of safety containment: physical (the roundworms are kept in a locked lab, and they aren't nearly as mobile as mice, mosquitoes, or fruit flies), ecological (there are no wild *C. elegans* to breed with on the mean streets of Cambridge), reproductive (most wild *C. elegans* are hermaphrodites and aren't interested in sex anyway), molecular (the self-limiting Daisy system), and more molecular (the gene drive targets a unique DNA sequence that has been engineered into the Sculpting Evolution worms but isn't found in wild worms).

If all gene-drive research hewed to these standards, I'd sleep better at night. But despite the recommendations from Esvelt, as well as the National Academy of Sciences, there are currently no binding rules in place. And even if everyone currently working on gene drives behaves responsibly—and they seem to be—it's easy to see how, eventually, as the technology spreads, someone, somewhere along the way, will get sloppy.

Public alarm grew louder in December of 2017, with the release of a cache of 1,200 emails between scientists and other gene-drive proponents that had been obtained through the Freedom of Information Act. "Gene Drive Files Expose Leading Role of US Military in Gene Drive Development" announced a press release, which noted that most gene-drive projects—including the London mosquitoes, Texas mice, and MIT roundworms—were being funded by the Department of Defense's Advanced Research Projects Agency as part of its Safe Genes program. Although DARPA had publicly announced it was funding the projects months earlier, this was not well known to the general public, and a number of news outlets ran with the story. The *Guardian*'s headline read, "US Military Agency Invests $100m in Genetic Extinction Technologies."

In its response, DARPA pointed out that its goals were defensive: "Our feeling is that the science of gene editing, including gene drive technology, has been advancing at a rapid pace in the laboratory," wrote the agency's chief of communications. "These leaps forward in potential capability, however, have not been matched by advances in the biosafety and biosecurity tools needed to protect against potential harm if such technologies were accidentally or intentionally misused."

The Safe Genes projects focus on learning to limit the reach of gene drives and on ways to detect and disable them, but none of that comforts Jim Thomas. "This has been the history of bioweapons research," he told me. "It's always presented as supposedly defensive: 'We have to develop these tools so we can respond in case someone else develops them.'" Thomas fears the agency's agenda may be much broader. "They're putting a finger in every single major gene-drive project so they can be close to them. So they can understand how these things work." Thomas worries that Daisy Drive is the equivalent of small-scale, tactical nukes. "Once you have this illusion that you can locally control a gene drive, then that opens the door for using it in agriculture or as a weapon." But few experts believe gene drives could make an effective weapon against other people—they are just too slow and obvious. There are easier ways to wage war.

During my most recent visit with Esvelt, I asked if he could imagine some situations where the technologies were too risky to pursue, even in a confined environment. Easily, he said. "There are areas where I would say, no research. And have!" It was after hours on a cold winter night in Cambridge, and Esvelt was looking even more pale and ragged than usual. Still, I pressed him for details. What kind of technology would be too dangerous for research? He shook his head and said, "There are some things I've thought of that I'm never going to tell another living soul."

When any new technology arrives, the debate veers toward the best- and worst-case scenarios, the big dreams and the big fears. Gene drives are going to cure malaria. Gene drives are going to become bioweapons. That's our nature. But it's easy to forget how rarely the extremes come to be.

The real test will be after we have a few minor successes controlling diseases or agricultural pests with gene drive. Suddenly we will have one of the greatest hammers ever invented, and we will go looking for nails. Every fast-reproducing plant or animal whose behavior we don't like will be a candidate for redesign. Cockroaches that hate the scent of garbage. Poison ivy that doesn't cause a rash. Fire ants with no fire. There are loose nails everywhere that just need a few whacks to make our lives more comfortable.

"Why not?" goes the counterargument. We've been hammering nature for years. Pollution, habitat destruction, pesticides, insecticides, greenhouse gases. Yale doesn't convene an ethics panel every time somebody clear-cuts a forest or dynamites a reef to harvest the fish. Why is it different once genes are involved?

And yet it is.

Anyone who's ever taken the time to hike to the pristine valley or paddle to the uninhabited island knows the sublimity of finding oneself in a place where the agenda is nonhuman. It's a reminder that there are ways of being in the world that have little or nothing to do with human ways, patterns of existence that get us out of our own heads and expand the conversation of what it means to be a quivering coil of DNA on the third planet from the sun. It's a form of diversity, and every species is a kind of culture, a cohesive and elegant web of quirks, predilections, and traditions.

We've dammed Glen Canyon. We've littered Everest with ropes and oxygen tanks. Our paw marks are all over even the wildest places. But we have yet to conquer the DNA of wild things. For the time being, that frontier has been visited by only a handful of early explorers.

In deciding if we have the right to drive a gene through a species, we might think of each genome as a national park, an untrammeled space in a nongeographical dimension. A refuge from an increasingly humanized world. I can hate the whine of a mosquito in my tent and still revere the pristine landscape of its genome. Engineering that genome would be like putting a road system through the Gates of the Arctic. There would be some obvious benefits—and something less obvious would be lost forever.

With every new technology, we tend to shoot first and ask questions later. It's a dynamic built into the DNA of our culture, which rewards the intrepid individuals who plant their flag on the virgin coast. Those ice-nine mosquitoes in their negative-pressure vault may end up being hugely important. They may, in fact, be a gift. A living metaphor of interconnectedness and of consequences, they may force us to consider if the time has come to throw out the Age of Exploration model and create a new system of science that rewards wisdom over cleverness.

That's a big ask, and it may seem absurd right now, as we survey the vast genetic frontier stretching away before us. How could we not poke around just a little? But we have a lot of experience with lost frontiers at this point, so perhaps there's still time to ask what we ought to do with this one. What if, after gazing from the decks of their caravels at the towering forests and teeming estuaries of the New World, the early explorers went back to their funders in Europe and said: Sure, we could make the place safe and productive. We could fill it with cities and farms and factories. But here's the thing: It isn't bad the way it is. It's full of mysteries and other ways of being. So . . . this is going to sound crazy, but what if we just left it alone?

BROOKE JARVIS

The Insect Apocalypse Is Here

FROM *The New York Times Magazine*

SUNE BOYE RIIS was on a bike ride with his youngest son, enjoying the sun slanting over the fields and woodlands near their home north of Copenhagen, when it suddenly occurred to him that something about the experience was amiss. Specifically, something was *missing*.

It was summer. He was out in the country, moving fast. But strangely, he wasn't eating any bugs.

For a moment, Riis was transported to his childhood on the Danish island of Lolland, in the Baltic Sea. Back then, summer bike rides meant closing his mouth to cruise through thick clouds of insects, but inevitably he swallowed some anyway. When his parents took him driving, he remembered, the car's windshield was frequently so smeared with insect carcasses that you almost couldn't see through it. But all that seemed distant now. He couldn't recall the last time he needed to wash bugs from his windshield; he even wondered, vaguely, whether car manufacturers had invented some fancy new coating to keep off insects. But this absence, he now realized with some alarm, seemed to be all around him. Where had all those insects gone? And when? And why hadn't he noticed?

Riis watched his son, flying through the beautiful day, not eating bugs, and was struck by the melancholy thought that his son's childhood would lack this particular bug-eating experience of his own. It was, he granted, an odd thing to feel nostalgic about. But he couldn't shake a feeling of loss. "I guess it's pretty human to think that everything was better when you were a kid," he said. "Maybe I didn't like it when I was on my bike and I ate all the bugs, but looking back on it, I think it's something everybody should experience."

I met Riis, a lanky high school science and math teacher, on a hot day in June. He was anxious about not having yet written his address for the school's graduation ceremony that evening, but first, he had a job to do. From his garage, he retrieved a large insect net, drove to a nearby intersection, and stopped to strap the net to the car's roof. Made of white mesh, the net ran the length of his car and was held up by a tent pole at the front, tapering to a small, removable bag in back. Drivers whizzing past twisted their heads to stare. Riis eyed his parking spot nervously as he adjusted the straps of the contraption. "This is not one hundred percent legal," he said, "but I guess, for the sake of science."

Riis had not been able to stop thinking about the missing bugs. The more he learned, the more his nostalgia gave way to worry. Insects are the vital pollinators and recyclers of ecosystems and the base of food webs everywhere. Riis was not alone in noticing their decline. In the United States, scientists recently found the population of monarch butterflies fell by 90 percent in the last 20 years, a loss of 900 million individuals; the rusty-patched bumblebee, which once lived in 28 states, dropped by 87 percent over the same period. With other, less-studied insect species, one butterfly researcher told me, "all we can do is wave our arms and say, 'It's not here anymore!'" Still, the most disquieting thing wasn't the disappearance of certain species of insects; it was the deeper worry, shared by Riis and many others, that a whole insect world might be quietly going missing, a loss of abundance that could alter the planet in unknowable ways. "We notice the losses," says David Wagner, an entomologist at the University of Connecticut. "It's the diminishment that we don't see."

Because insects are legion, inconspicuous, and hard to meaningfully track, the fear that there might be far fewer than before was more felt than documented. People noticed it by canals or in backyards or under streetlights at night—familiar places that had become unfamiliarly empty. The feeling was so common that entomologists developed a shorthand for it, named for the way many people first began to notice that they weren't seeing as many bugs. They called it the windshield phenomenon.

To test what had been primarily a loose suspicion of wrongness, Riis and 200 other Danes were spending the month of June roaming their country's back roads in their outfitted cars. They were part of a study conducted by the Natural History Museum of Denmark, a joint effort of the University of Copenhagen, Aarhus University, and North Carolina State University. The nets would stand in for windshields as Riis and the other volunteers drove through various habitats—urban areas, forests, agricultural tracts, uncultivated open land, and wetlands—hoping to quantify the disorienting sense that, as one of the study's designers put it, "something from the past is missing from the present."

When the investigators began planning the study in 2016, they weren't sure if anyone would sign up. But by the time the nets were ready, a paper by an obscure German entomological society had brought the problem of insect decline into sharp focus. The German study found that, measured simply by weight, the overall abundance of flying insects in German nature reserves had decreased by 75 percent over just 27 years. If you looked at midsummer population peaks, the drop was 82 percent.

Riis learned about the study from a group of his students in one of their class projects. They must have made some kind of mistake in their citation, he thought. But they hadn't. The study would quickly become, according to the website Altmetric, the sixth-most-discussed scientific paper of 2017. Headlines around the world warned of an "insect Armageddon."

Within days of announcing the insect-collection project, the Natural History Museum of Denmark was turning away eager volunteers by the dozens. It seemed there were people like Riis everywhere, people who had noticed a change but didn't know what to make of it. How could something as fundamental as the bugs in the sky just disappear? And what would become of the world without them?

Anyone who has returned to a childhood haunt to find that everything somehow got smaller knows that humans are not great at remembering the past accurately. This is especially true when it comes to changes to the natural world. It is impossible to maintain a fixed perspective, as Heraclitus observed 2,500 years ago: it is not the same river, but we are also not the same people.

A 1995 study, by Peter H. Kahn and Batya Friedman, of the way some children in Houston experienced pollution summed up our blindness this way: "With each generation, the amount of environmental degradation increases, but each generation takes that amount as the norm." In decades of photos of fishermen holding up their catch in the Florida Keys, the marine biologist Loren McClenachan found a perfect illustration of this phenomenon, which is often called "shifting baseline syndrome." The fish got smaller and smaller, to the point where the prize catches were dwarfed by fish that in years past were piled up and ignored. But the smiles on the fishermen's faces stayed the same size. The world never feels fallen, because we grow accustomed to the fall.

By one measure, bugs are the wildlife we know best, the nondomesticated animals whose lives intersect most intimately with our own: spiders in the shower, ants at the picnic, ticks buried in the skin. We sometimes feel that we know them rather too well. In another sense, though, they are one of our planet's greatest mysteries, a reminder of how little we know about what's happening in the world around us.

We've named and described a million species of insects, a stupefying array of thrips and firebrats and antlions and caddis flies and froghoppers and other enormous families of bugs that most of us can't even name. (Technically, the word *bug* applies only to the order Hemiptera, also known as true bugs, species that have tubelike mouths for piercing and sucking—and there are as many as 80,000 named varieties of those.) The ones we think we do know well, we don't: there are 12,000 types of ants, nearly 20,000 varieties of bees, almost 400,000 species of beetles, so many that the geneticist J. B. S. Haldane reportedly quipped that God must have an inordinate fondness for them. A bit of healthy soil a foot square and two inches deep might easily be home to 200 unique species of mites, each, presumably, with a subtly different job to do. And yet entomologists estimate that all this amazing, absurd, and understudied variety represents perhaps only 20 percent of the actual diversity of insects on our planet—that there are millions and millions of species that are entirely unknown to science.

With so much abundance, it very likely never occurred to most entomologists of the past that their multitudinous subjects might dwindle away. As they poured themselves into studies of the life cycles and taxonomies of the species that fascinated them, few thought to measure or record something as boring as their number. Besides, tracking quantity is slow, tedious, and unglamorous work: setting and checking traps, waiting years or decades for your data to be meaningful, grappling with blunt baseline questions instead of more sophisticated ones. And who would pay for it? Most academic funding is short-term, but when what you're interested in is invisible, generational change, says Dave Goulson, an entomologist at the University of Sussex, "a three-year monitoring program is no good to anybody." This is especially true of insect populations, which are naturally variable, with wide, trend-obscuring fluctuations from one year to the next.

When entomologists began noticing and investigating insect declines, they lamented the absence of solid information from the past in which to ground their experiences of the present. "We see a hundred of something, and we think we're fine," Wagner says, "but what if there were a hundred thousand two generations ago?" Rob Dunn, an ecologist at North Carolina State University who helped design the net experiment in Denmark, recently searched for studies showing the effect of pesticide spraying on the quantity of insects living in nearby forests. He was surprised to find that no such studies existed. "We ignored really basic questions," he said. "It feels like we've dropped the ball in some giant collective way."

If entomologists lacked data, what they did have were some very worrying clues. Along with the impression that they were seeing fewer bugs in their own jars and nets while out doing experiments —a windshield phenomenon specific to the sorts of people who have bug jars and nets—there were documented downward slides of well-studied bugs, including various kinds of bees, moths, butterflies, and beetles. In Britain, as many as 30 to 60 percent of species were found to have diminishing ranges. Larger trends were harder to pin down, though a 2014 review in *Science* tried to quantify these declines by synthesizing the findings of existing studies and found that a majority of monitored species were declining, on average by 45 percent.

Entomologists also knew that climate change and the overall degradation of global habitat are bad news for biodiversity in general, and that insects are dealing with the particular challenges posed by herbicides and pesticides, along with the effects of losing meadows, forests, and even weedy patches to the relentless expansion of human spaces. There were studies of other, better-understood species that suggested that the insects associated with them might be declining, too. People who studied fish found that the fish had fewer mayflies to eat. Ornithologists kept finding that birds that rely on insects for food were in trouble: 8 in 10 partridges gone from French farmlands; 50 and 80 percent drops, respectively, for nightingales and turtledoves. Half of all farmland birds in Europe disappeared in just three decades. At first, many scientists assumed the familiar culprit of habitat destruction was at work, but then they began to wonder if the birds might simply be starving. In Denmark, an ornithologist named Anders Tottrup was the one who came up with the idea of turning cars into insect trackers for the windshield-effect study after he noticed that rollers, little owls, Eurasian hobbies, and bee-eaters—all birds that subsist on large insects such as beetles and dragonflies—had abruptly disappeared from the landscape.

The signs were certainly alarming, but they were also just signs, not enough to justify grand pronouncements about the health of insects as a whole or about what might be driving a widespread, cross-species decline. "There are no quantitative data on insects, so this is just a hypothesis," Hans de Kroon, an ecologist at Radboud University in the Netherlands, explained to me—not the sort of language that sends people to the barricades.

Then came the German study. Scientists are still cautious about what the findings might imply about other regions of the world. But the study brought forth exactly the kind of longitudinal data they had been seeking, and it wasn't specific to just one type of insect. The numbers were stark, indicating a vast impoverishment of an entire insect universe, even in protected areas where insects ought to be under less stress. The speed and scale of the drop were shocking even to entomologists who were already anxious about bees or fireflies or the cleanliness of car windshields.

The results were surprising in another way too. The long-term details about insect abundance, the kind that no one really thought existed, hadn't appeared in a particularly prestigious journal and didn't come from university-affiliated scientists, but from a small society of insect enthusiasts based in the modest German city of Krefeld.

Krefeld sits a half-hour drive outside Düsseldorf, near the western bank of the Rhine. It's a city of brick houses and bright flower gardens and a *stadtwald*—a municipal forest and park—where paddle boats float on a lake, umbrellas shade a beer garden, and (I couldn't help noticing) the afternoon light through the trees illuminates small swarms of dancing insects.

Near the center of the old city, a paper sign, not much larger than a business card, identifies the stolid headquarters of the society whose research caused so much commotion. When it was founded, in 1905, the society operated out of another building, one that was destroyed when Britain bombed the city during World War II. (By the time the bombs fell, members had moved their precious records and collections of insects, some of which dated back to the 1860s, to an underground bunker.) Nowadays, the society uses more than 6,000 square feet of an old three-story school as storage space. Ask for a tour of the collections, and you will hear such sentences as "This whole room is Lepidoptera," referring to a former classroom stuffed with what I at first took to be shelves of books but which are in fact innumerable wooden frames containing pinned butterflies and moths; and, in an even larger room, "every bumblebee here was collected before the Second World War, 1880 to 1930"; and, upon opening a drawer full of sweat bees, "It's a new collection, thirty years only."

On the shelves that do hold books, I counted 31 clearly well-loved volumes in the series "Beetles of Middle Europe." A 395-page book that catalogued specimens of spider wasps—where they were collected; where they were stored—of the western Palearctic said "1948–2008" on the cover. I asked my guide, a society member named Martin Sorg, who was one of the lead authors of the paper, whether those dates reflected when the specimens were collected. "No," Sorg replied, "that was the time the author needed for this work."

Sorg, who rolls his own cigarettes and wears John Lennon glasses and whose gray hair grows long past his shoulders, is not a freewheeling type when it comes to his insect work. And his insect work is really all he wants to talk about. "We think details about nature and biodiversity declines are important, not details about life histories of entomologists," Sorg explained after he and Werner Stenmans, a society member whose name appeared alongside Sorg's on the 2017 paper, dismissed my questions about their day jobs. Leery of an article that focused on him as a person, Sorg also didn't want to talk about what drew him to entomology as a child or even what it was about certain types of wasps that had made him want to devote so much of his life to studying them. "We normally give life histories when someone is dead," he said.

There was a reason for the wariness. Society members dislike seeing themselves described, over and over in news stories, as "amateurs." It's a framing that reflects, they believe, a too-narrow understanding of what it means to be an expert or even a scientist —what it means to be a student of the natural world.

Amateurs have long provided much of the patchy knowledge we have about nature. Those bee and butterfly studies? Most depend on mass mobilizations of volunteers willing to walk transects and count insects, every two weeks or every year, year after year. The scary numbers about bird declines were gathered this way, too, though because birds can be hard to spot, volunteers often must learn to identify them by their sounds. Britain, which has a particularly strong tradition of amateur naturalism, has the best-studied bugs in the world. As technologically advanced as we are, the natural world is still a very big and complex place, and the best way to learn what's going on is for a lot of people to spend a lot of time observing it. The Latin root of the word *amateur* is, after all, the word *lover*.

Some of these citizen-scientists are true beginners clutching field guides; others, driven by their own passion and following in a long tradition of "amateur" naturalism, are far from novices. Think of Victorians with their butterfly nets and curiosity cabinets; of Vladimir Nabokov, whose theories about the evolution of Polyommatus blue butterflies were ignored until proved correct by DNA testing more than 30 years after his death; of young Charles Darwin, cutting his classes at Cambridge to collect beetles at Wicken Fen and once putting a live beetle in his mouth because his hands were already full of other bugs.

The Krefeld society is volunteer-run, and many members have other jobs in unrelated fields, but they also have an enormous depth of knowledge about insects, accumulated through years of what other people might consider obsessive attention. Some study the ecology or evolutionary taxonomy of their favorite species or map their populations or breed them to study their life histories. All hone their identification skills across species by amassing their own collections of carefully pinned and labeled insects like those that fill the society's storage rooms. Sorg estimated that of the society's 63 members, a third are university-trained in subjects such as biology or earth science. Another third, he said, are "highly specialized and highly qualified but they never visited the university," while the remaining third are actual amateurs who are still in the process of becoming "real" entomologists: "Some of them may also have a degree from the university, but in our view, they are beginners."

The society members' projects often involved setting up what are called malaise traps, nets that look like tents and drive insects flying by into bottles of ethanol. Because of the scientific standards of the society, members followed certain procedures: They always employed identical traps, sewn from a template they first used in 1982. (Sorg showed me the original rolled-up craft paper with great solemnity.) They always put them in the same places. (Before GPS, that meant a painstaking process of triangulating with surveying equipment. "We are not sure about a few centimeters," Sorg granted.) They saved everything they caught, regardless of what the main purpose of the experiment was. (The society bought so much ethanol that it attracted the attention of a narcotics unit.)

Those bottles of insects were gathered into thousands of boxes, which are now crammed into what were once offices in the upper reaches of the school. When the society members, like entomologists elsewhere, began to notice that they were seeing fewer insects, they had something against which to measure their worries.

"We don't throw away anything, we store everything," Sorg explained. "That gives us today the possibility to go back in time."

In 2013 Krefeld entomologists confirmed that the total number of insects caught in one nature reserve was nearly 80 percent lower than the same spot in 1989. They had sampled other sites, analyzed old data sets, and found similar declines: where 30 years earlier, they often needed a liter bottle for a week of trapping, now a half-liter bottle usually sufficed. But it would have taken even highly trained entomologists years of painstaking work to identify all the insects in the bottles. So the society used a standardized method for weighing insects in alcohol, which told a powerful story simply by showing how much the overall mass of insects dropped over time. "A decline of this mixture," Sorg said, "is a very different thing than the decline of only a few species."

The society collaborated with de Kroon and other scientists at Radboud University in the Netherlands, who did a trend analysis of the data that Krefeld provided, controlling for things like the effects of nearby plants, weather, and forest cover on fluctuations in insect populations. The final study looked at 63 nature preserves, representing almost 17,000 sampling days, and found consistent declines in every kind of habitat they sampled. This suggested, the authors wrote, "that it is not only the vulnerable species but the flying-insect community as a whole that has been decimated over the last few decades."

For some scientists, the study created a moment of reckoning. "Scientists thought this data was too boring," Dunn says. "But these people found it beautiful, and they loved it. They were the ones paying attention to Earth for all the rest of us."

The current worldwide loss of biodiversity is popularly known as the sixth extinction: the sixth time in world history that a large number of species have disappeared in unusually rapid succession, caused this time not by asteroids or ice ages but by humans. When we think about losing biodiversity, we tend to think of the last northern white rhinos protected by armed guards, of polar bears on dwindling ice floes. Extinction is a visceral tragedy, universally understood: There is no coming back from it. The guilt of letting a unique species vanish is eternal.

But extinction is not the only tragedy through which we're living. What about the species that still exist, but as a shadow of what they once were? In *The Once and Future World*, the journalist J. B. MacKinnon cites records from recent centuries that hint at what has only just been lost: "In the North Atlantic, a school of cod stalls a tall ship in midocean; off Sydney, Australia, a ship's captain sails from noon until sunset through pods of sperm whales as far as the eye can see . . . Pacific pioneers complain to the authorities that splashing salmon threaten to swamp their canoes." There were reports of lions in the south of France, walruses at the mouth of the Thames, flocks of birds that took three days to fly overhead, as many as 100 blue whales in the Southern Ocean for every one that's there now. "These are not sights from some ancient age of fire and ice," MacKinnon writes. "We are talking about things seen by human eyes, recalled in human memory."

What we're losing is not just the *diversity* part of biodiversity, but the *bio* part: life in sheer quantity. While I was writing this article, scientists learned that the world's largest king penguin colony shrank by 88 percent in 35 years, that more than 97 percent of the bluefin tuna that once lived in the ocean are gone. The number of

Sophie the Giraffe toys sold in France in a single year is nine times the number of all the giraffes that still live in Africa.

Finding reassurance in the survival of a few symbolic standard-bearers ignores the value of abundance, of a natural world that thrives on richness and complexity and interaction. Tigers still exist, for example, but that doesn't change the fact that 93 percent of the land where they used to live is now tigerless. This matters for more than romantic reasons: large animals, especially top predators like tigers, connect ecosystems to one another and move energy and resources among them simply by walking and eating and defecating and dying. (In the deep ocean, sunken whale carcasses form the basis of entire ecosystems in nutrient-poor places.) One result of their loss is what's known as trophic cascade, the unraveling of an ecosystem's fabric as prey populations boom and crash and the various levels of the food web no longer keep each other in check. These places are emptier, impoverished in a thousand subtle ways.

Scientists have begun to speak of functional extinction (as opposed to the more familiar kind, numerical extinction). Functionally extinct animals and plants are still present but no longer prevalent enough to affect how an ecosystem works. Some phrase this as the extinction not of a species but of all its former interactions with its environment—an extinction of seed dispersal and predation and pollination and all the other ecological functions an animal once had, which can be devastating even if some individuals still persist. The more interactions are lost, the more disordered the ecosystem becomes. A 2013 paper in *Nature*, which modeled both natural and computer-generated food webs, suggested that a loss of even 30 percent of a species' abundance can be so destabilizing that other species start going fully, numerically extinct—in fact, 80 percent of the time it was a secondarily affected creature that was the first to disappear. A famous real-world example of this type of cascade concerns sea otters. When they were nearly wiped out in the northern Pacific, their prey, sea urchins, ballooned in number and decimated kelp forests, turning a rich environment into a barren one and also possibly contributing to numerical extinctions, notably of the Steller's sea cow.

Conservationists tend to focus on rare and endangered species, but it is common ones, because of their abundance, that power the living systems of our planet. Most species are not common, but within many animal groups most individuals—some 80 percent of them—belong to common species. Like the slow approach of twilight, their declines can be hard to see. White-rumped vultures were nearly gone from India before there was widespread awareness of their disappearance. Describing this phenomenon in the journal *BioScience*, Kevin Gaston, a professor of biodiversity and conservation at the University of Exeter, wrote: "Humans seem innately better able to detect the complete loss of an environmental feature than its progressive change."

In addition to extinction (the complete loss of a species) and extirpation (a localized extinction), scientists now speak of *defaunation*: the loss of individuals, the loss of abundance, the loss of a place's absolute animalness. In a 2014 article in *Science*, researchers argued that the word should become as familiar, and influential, as the concept of deforestation. In 2017 another paper reported that major population and range losses extended even to species considered to be at low risk for extinction. They predicted "negative cascading consequences on ecosystem functioning and services vital to sustaining civilization" and the authors offered another term for the widespread loss of the world's wild fauna: *biological annihilation.*

It is estimated that, since 1970, Earth's various populations of wild land animals have lost, on average, 60 percent of their members. Zeroing in on the category we most relate to, mammals, scientists believe that for every six wild creatures that once ate and burrowed and raised young, only one remains. What we have instead is ourselves. A study published this year in the *Proceedings of the National Academy of Sciences* found that if you look at the world's mammals by weight, 96 percent of that biomass is humans and livestock; just 4 percent is wild animals.

We've begun to talk about living in the Anthropocene, a world shaped by humans. But E. O. Wilson, the naturalist and prophet of environmental degradation, has suggested another name: the Eremocine, the age of loneliness.

Wilson began his career as a taxonomic entomologist, studying ants. Insects—about as far as you can get from charismatic megafauna—are not what we're usually imagining when we talk about biodiversity. Yet they are, in Wilson's words, "the little things that run the natural world." He means it literally. Insects are a case study in the invisible importance of the common.

*

Scientists have tried to calculate the benefits that insects provide simply by going about their business in large numbers. Trillions of bugs flitting from flower to flower pollinate some three-quarters of our food crops, a service worth as much as $500 billion every year. (This doesn't count the 80 percent of wild flowering plants, the foundation blocks of life everywhere, that rely on insects for pollination.) If monetary calculations like that sound strange, consider the Maoxian Valley in China, where shortages of insect pollinators have led farmers to hire human workers, at a cost of up to $19 per worker per day, to replace bees. Each person covers 5 to 10 trees a day, pollinating apple blossoms by hand.

By eating and being eaten, insects turn plants into protein and power the growth of all the uncountable species—including freshwater fish and a majority of birds—that rely on them for food, not to mention all the creatures that eat those creatures. We worry about saving the grizzly bear, says the insect ecologist Scott Hoffman Black, but where is the grizzly without the bee that pollinates the berries it eats or the flies that sustain baby salmon? Where, for that matter, are we?

Bugs are vital to the decomposition that keeps nutrients cycling, soil healthy, plants growing, and ecosystems running. This role is mostly invisible, until suddenly it's not. After introducing cattle to Australia at the turn of the 19th century, settlers soon found themselves overwhelmed by the problem of their feces: for some reason, cow pies there were taking months or even years to decompose. Cows refused to eat near the stink, requiring more and more land for grazing, and so many flies bred in the piles that the country became famous for the funny hats that stockmen wore to keep them at bay. It wasn't until 1951 that a visiting entomologist realized what was wrong: the local insects, evolved to eat the more fibrous waste of marsupials, couldn't handle cow excrement. For the next 25 years, the importation, quarantine, and release of dozens of species of dung beetles became a national priority. And that was just one unfilled niche. (In the United States, dung beetles save ranchers an estimated $380 million a year.) We simply don't know everything that insects do. Only about 2 percent of invertebrate species have been studied enough for us to estimate whether they are in danger of extinction, never mind what dangers that extinction might pose.

When asked to imagine what would happen if insects were to

disappear completely, scientists find words like *chaos, collapse, Armageddon.* Wagner, the University of Connecticut entomologist, describes a flowerless world with silent forests, a world of dung and old leaves and rotting carcasses accumulating in cities and roadsides, a world of "collapse or decay and erosion and loss that would spread through ecosystems"—spiraling from predators to plants. E. O. Wilson has written of an insect-free world, a place where most plants and land animals become extinct; where fungi explodes, for a while, thriving on death and rot; and where "the human species survives, able to fall back on wind-pollinated grains and marine fishing" despite mass starvation and resource wars. "Clinging to survival in a devastated world, and trapped in an ecological dark age," he adds, "the survivors would offer prayers for the return of weeds and bugs."

But the crux of the windshield phenomenon, the reason that the creeping suspicion of change is so *creepy,* is that insects wouldn't have to disappear altogether for us to find ourselves missing them for reasons far beyond nostalgia. In October, an entomologist sent me an email with the subject line, "Holy [expletive]!" and an attachment: a study just out from *Proceedings of the National Academy of Sciences* that he labeled, "Krefeld comes to Puerto Rico." The study included data from the 1970s and from the early 2010s, when a tropical ecologist named Brad Lister returned to the rain forest where he had studied lizards—and, crucially, their prey—40 years earlier. Lister set out sticky traps and swept nets across foliage in the same places he had in the 1970s, but this time he and his co-author, Andrés Garcia, caught much, much less: 10 to 60 times less arthropod biomass than before. (It's easy to read that number as *60 percent* less, but it's *sixtyfold* less: where once he caught 473 milligrams of bugs, Lister was now catching just 8 milligrams.) "It was, you know, devastating," Lister told me. But even scarier were the ways the losses were already moving through the ecosystem, with serious declines in the numbers of lizards, birds, and frogs. The paper reported "a bottom-up trophic cascade and consequent collapse of the forest food web." Lister's inbox quickly filled with messages from other scientists, especially people who study soil invertebrates, telling him they were seeing similarly frightening declines. Even after his dire findings, Lister found the losses shocking: "I didn't even know about the earthworm crisis!"

The strange thing, Lister said, is that, as staggering as they are, all the declines he documented would still be basically invisible to the average person walking through the Luquillo rain forest. On his last visit, the forest still felt "timeless" and "phantasmagorical," with "cascading waterfalls and carpets of flowers." You would have to be an expert to notice what was missing. But he expects the losses to push the forest toward a tipping point, after which "there is a sudden and dramatic loss of the rain-forest system," and the changes will become obvious to anyone. The place he loves will become unrecognizable.

The insects in the forest that Lister studied haven't been contending with pesticides or habitat loss, the two problems to which the Krefeld paper pointed. Instead, Lister chalks up their decline to climate change, which has already increased temperatures in Luquillo by 2 degrees Celsius since Lister first sampled there. Previous research suggested that tropical bugs will be unusually sensitive to temperature changes; in November, scientists who subjected laboratory beetles to a heat wave reported that the increased temperatures made them significantly less fertile. Other scientists wonder if it might be climate-induced drought or possibly invasive rats or simply "death by a thousand cuts"—a confluence of many kinds of changes to the places where insects once thrived.

Like other species, insects are responding to what Chris Thomas, an insect ecologist at the University of York, has called "the transformation of the world": not just a changing climate but also the widespread conversion, via urbanization, agricultural intensification, and so on, of natural spaces into human ones, with fewer and fewer resources "left over" for nonhuman creatures to live on. What resources remain are often contaminated. Hans de Kroon characterizes the life of many modern insects as trying to survive from one dwindling oasis to the next but with "a desert in between, and at worst it's a poisonous desert." Of particular concern are neonicotinoids, neurotoxins that were thought to affect only treated crops but turned out to accumulate in the landscape and to be consumed by all kinds of nontargeted bugs. People talk about the "loss" of bees to colony collapse disorder, and that appears to be the right word: affected hives aren't full of dead bees, but simply mysteriously empty. A leading theory is that exposure to neurotoxins leaves bees unable to find their way home. Even hives exposed to low levels of neonicotinoids have been shown to collect less pollen and produce fewer eggs and far fewer queens.

Some recent studies found bees doing better in cities than in the supposed countryside.

The diversity of insects means that some will manage to make do in new environments, some will thrive (abundance cuts both ways: agricultural monocultures, places where only one kind of plant grows, allow some pests to reach population levels they would never achieve in nature), and some, searching for food and shelter in a world nothing like the one they were meant for, will fail. While we need much more data to better understand the reasons or mechanisms behind the ups and downs, Thomas says, "the average across all species is still a decline."

Since the Krefeld study came out, researchers have begun searching for other forgotten repositories of information that might offer windows into the past. Some of the Radboud researchers have analyzed long-term data, belonging to Dutch entomological societies, about beetles and moths in certain reserves; they found significant drops (72 percent, 54 percent) that mirrored the Krefeld ones. Roel van Klink, a researcher at the German Center for Integrative Biodiversity Research, told me that before Krefeld, he, like most entomologists, had never been interested in biomass. Now he is looking for historical data sets—many of which began as studies of agricultural pests, like a decades-long study of grasshoppers in Kansas—that could help create a more thorough picture of what's happening to creatures that are at once abundant and imperiled. So far he has found forgotten data from 140 old data sets for 1,500 locations that could be resampled.

In the United States, one of the few long-term data sets about insect abundance comes from the work of Arthur Shapiro, an entomologist at the University of California, Davis. In 1972 he began walking transects in the Central Valley and the Sierras, counting butterflies. He planned to do a study on how short-term weather variations affected butterfly populations. But the longer he sampled, the more valuable his data became, offering a signal through the noise of seasonal ups and downs. "And so here I am in Year 46," he said, nearly half a century of spending five days a week, from late spring to the end of autumn, observing butterflies. In that time he has watched overall numbers decline and seen some species that used to be everywhere—even species that "everyone regarded as a junk species" only a few decades ago—all but disappear. Shapiro believes that Krefeld-level declines are likely to be happening all over the globe. "But, of course, I don't cover the entire globe," he added. "I cover I-80."

There are also new efforts to set up more of the kind of insect-monitoring schemes researchers wish had existed decades ago, so that our current level of fallenness, at least, is captured. One is a pilot project in Germany similar to the Danish car study. To analyze what is caught, the researchers turned to volunteer naturalists, hobbyists similar to the ones in Krefeld, with the necessary breadth of knowledge to know what they're looking at. "These are not easy species to identify," says Aletta Bonn, of the German Center for Integrative Biodiversity Research, who is overseeing the project. (The skills required for such work "are really extreme," Dunn says. "These people train for decades with other amateurs to be able to identify beetles based on their genitalia.") Bond would like to pay the volunteers for their expertise, she says, but funding hasn't caught up to the crisis. That didn't stop the "amateurs" from being willing to help: "They said, 'We're just curious what's in there, we would like to have samples.'"

Goulson says that Europe's tradition of amateur naturalism may account for why so many of the clues to the falloff in insect biodiversity originate there. (Tottrup's design for the car net in Denmark, for example, was itself adapted from the invention of a dedicated beetle-collecting hobbyist.) As little as we know about the status of European bugs, we know significantly less about other parts of the world. "We wouldn't know anything if it weren't for them," the so-called amateurs, Goulson told me. "We'd be entirely relying on the fact that there's no bugs on the windshield."

Thomas believes that this naturalist tradition is also why Europe is acting much faster than other places—for example, the United States—to address the decline of insects: interest leads to tracking, which leads to awareness, which leads to concern, which leads to action. Since the Krefeld data emerged, there have been hearings about protecting insect biodiversity in the German Bundestag and the European Parliament. European Union member states voted to extend a ban on neonicotinoid pesticides and have begun to put money toward further studies of how abundance is changing, what is causing those changes, and what can be done. When I knocked on the door of de Kroon's office, at Radboud University in the Dutch city Nijmegen, he was looking at some photos from another meeting he had that day: Willem-Alexander, the king of the Netherlands, had taken a tour of the city's efforts to make its riverside a friendlier habitat for bugs.

Stemming insect declines will require much more than this, however. The European Union already had some measures in place to help pollinators—including more strictly regulating pesticides than the United States does and paying farmers to create insect habitats by leaving fields fallow and allowing for wild edges alongside cultivation—but insect populations dropped anyway. New reports call for national governments to collaborate; for more creative approaches such as integrating insect habitats into the design of roads, power lines, railroads, and other infrastructure; and, as always, for more studies. The necessary changes, like the causes, may be profound. "It's just another indication that we're destroying the life-support system of the planet," Lister says of the Puerto Rico study. "Nature's resilient, but we're pushing her to such extremes that eventually it will cause a collapse of the system."

Scientists hope that insects will have a chance to embody that resilience. While tigers tend to give birth to three or four cubs at a time, a ghost moth in Australia was once recorded laying 29,100 eggs, and she still had 15,000 in her ovaries. The fecund abundance that is insects' singular trait should enable them to recover, but only if they are given the space and the opportunity to do so.

"It's a debate we need to have urgently," Goulson says. "If we lose insects, life on Earth will . . ." He trailed off, pausing for what felt like a long time.

In Denmark, Sune Boye Riis's transect with his car net took him past a bit of woods, some suburban lawns, some hedges, a Christmas-tree farm. The closest thing to a meadow that we passed was a large military property, on which the grass had been allowed to grow tall and golden. Riis had received instructions not to drive too fast, so traffic backed up behind us, and some people began to honk. "Well," Riis said, "so much for science." After three miles, he turned around and drove back toward the start. His windshield stayed mockingly clean.

Riis had four friends who were also participating in the study. They had a bet going among them: Who would net the biggest bug? "I'm way behind," Riis said. "A bumblebee is in the lead." His biggest catch? "A fly. Not even a big one."

At the end of the transect, Riis stopped at another parlous road-side spot, unfastened the net, and removed the small bag at its tip. Some volunteers, captivated by what the study revealed about the world around them, asked the organizers for extra specimen bags, so they could do more sampling on their own. Some even asked if they could buy the entire car-net apparatus. Riis, though, was content to peer through the mesh, inside of which he could make out a number of black specks of varying tininess.

There was also a single butterfly, white-winged and delicate. Riis thought of the bet with his friends, for which the meaning of big-ness had not been defined. He wondered how it might be reck-oned. What gave a creature value?

"Is it weight?" he asked, staring down at the butterfly. In the big bag, it looked small and sad and alone. "Or is it grace?"

MATT JONES

No Heart, No Moon

FROM *The Southern Review*

THE SPACE RACE killed the sparrow.

Of course, there were other factors.

There was the decision in '46 by the Brevard Mosquito Control District to slather the Merritt Island salt marshes in DDT dropped aerially from a No. 2 diesel fuel carrier.

Then, because the mosquitoes grew resistant to DDT, there was the application of BHC and dieldrin and malathion.

There was brand name FLIT, a petroleum derivative, the same stuff that was sprayed over the Spanish moss that hung from the rafters of the Rhythm Club in Natchez the night the one-story building went up in flames killing 209 people.

There was Paris green powdered finely over the dikes, the same stuff that killed Parisian rats; the same popular pigment used in the paintings of Cézanne and Van Gogh; the same crystalline powder that, despite its name, gave fireworks their blue hue.

There was the direct and indirect poisoning from various insecticides. The physiological problems. The eggshell thinning. The reproductive failures.

There were the dikes themselves. The impoundments built up along Banana Creek and Banana River. The flooding of the salt marshes to drown out the mud-loving mosquitoes.

There was the railroad causeway that went up just north of Roach Hole in 1963.

There was the loss and degradation of habitat. The disappearance of cordgrass and seashore saltgrass.

There was the invasion of dense sea myrtle and snakes and raccoon and aggressive redwing blackbirds.

There were the controlled burns.

And the flooding. So much flooding.

The Orlando Jetport that opened to the public in '62.

There was the SR 528, otherwise known as the Martin B. Anderson Beachline Expressway: the Bee Line that stretches from the Space Coast all the way to Disney.

And make no mistake: Disney was involved.

But more than anything, the fate of the dusky seaside sparrow was intertwined with the space race that started sometime around '55 with Khruschev, Sputnik, Yuri Gagarin, Kennedy, and that Special Message to Congress on Urgent National Needs in which JFK so boldly declared, "It will not be one man going to the moon—if we make this judgment affirmatively, it will be an entire nation."

That was the real deathblow to the dusky seaside sparrow: man's ambition. The year NASA purchased most of North Merritt Island, where the largest colony of dusky seaside sparrows lived, was the same year John F. Kennedy stood in front of a crowd of 35,000 people at Rice Stadium in Houston, Texas, and said, "We choose to go to the moon in this decade and do the other things, not because they are easy, but because they are hard, because that goal will serve to organize and measure the best of our energies and skills, because that challenge is one that we are willing to accept, one we are unwilling to postpone, and one which we intend to win, and the others, too."

Many ornithologists had observed that these particular birds, no larger than canaries, their feathers black and white, kept a limited range for hunting, nesting, and habitation. They were known for their unique tendency to stay close to home. They were a nonmigratory species. So perhaps—just maybe—the duskies were also responsible for their own demise. Perhaps they were simply either unwilling or unable to adapt. Maybe they simply lacked the boldness that made us humans consider leaving our first and only home.

But as surely as Americans would one day step foot on the moon, so too would the dusky seaside sparrow travel beyond the land it had always known, touching down finally about an hour west across the center of the Sunshine State in what was then being touted as the Vacation Kingdom of the World.

*

After he journeyed to the center of the Earth, but before he traveled 20,000 leagues beneath the sea, the science fiction author Jules Verne set his sights on outer space.

Following the end of the American Civil War, Verne released the fourth book in his Voyages Extraordinaires series: *From the Earth to the Moon.* In it, members of the Baltimore Gun Club attempt to build a cannon so long and powerful it can launch three men to the surface of the moon.

The novel got a lot of things wrong.

For instance, Konstantin Tsiolkovsky, a Soviet rocket scientist and one of the founding fathers of astronautics, critiqued Verne's theory of using a cannon to reach the moon. The explosive force required to achieve the escape velocity necessary to exit Earth's atmosphere would have disintegrated the cannon, the men, and the dreams inside each of their heads. Even so, *From the Earth to the Moon* inspired Tsiolkovsky, and many of his theories were later used to shape the Soviet space program that would eventually send both the first satellite and the first animal into orbit.

And there were plenty of things Verne got right.

One of them was placing the location of his imagined launch site at N 27°7'0", W 82°9'0", an area more commonly known as Florida. In doing so, he planted the Sunshine State in the American subconscious as a sort of gateway between myth and reality —here and there, Earth and moon.

Fools would call it fate that Verne wrote about Florida, but Fate is a town in Texas over 600 miles west of Florida, just a half-hour drive from Dealey Plaza in Dallas, where John F. Kennedy was assassinated. Fate—the town—was also the place where Lee Harvey Oswald's widow got remarried. When you start to pick apart fate —the idea, not the place—you begin to realize it doesn't make sense. Fate is a cannon inside your head capable of shooting you to the moon, but the future is something much different.

Nearly a century after the publication of *From the Earth to the Moon,* a joint NASA–air force team was busy looking for the location of their new launch site. Of the eight sites up for consideration, there was Brownsville, Texas; the White Sands Missile Range in New Mexico; and Cumberland Island, Georgia. The engineers had to consider access to deepwater transport, the populations of surrounding towns, and the type of land available and the cost of it.

The same can be said of Walt Disney, who, before selecting

Orlando as the future site of Walt Disney World, visited investors in Niagara Falls; the St. Louis Riverfront; Marceline, Missouri; and the blighted area of Kansas City known only as Signboard Hill. After settling on Florida, Disney and his team first considered a 6,000-acre parcel of land on East Lake Tohopekaliga and sites in Sebring and New Smyrna Beach. After weighing factors such as weather and access to tourists and price per acre, Disney purchased almost 30,000 acres of Florida swampland between Orlando and Kissimmee by way of various shell corporations.

Nowhere in the official documentation was a quality such as fate admitted into the decision-making process for either NASA or Disney. However, there was the strange fact that, on the same day John F. Kennedy was assassinated—November 22, 1963—Walt Disney reportedly looked out the window of a small twin-propeller plane and decided that the stretch of freshwater lakes and interstate construction below would be the site of his next greatest project. Now, you might be able to call that fate. But the piece of land Disney saw from his window had already had a host of other names.

In fact, it was a tiny island in the center of Bay Lake that Disney was rumored to have fallen in love with when he and his associates flew over it on November 22, 1963, while scouting potential locations. When Disney first purchased this island, it was known as Riles Island, and before that, Idle Bay Isle and Raz Island, and later on, Treasure Island and Discovery Island, and before being abandoned altogether, it would eventually become known as the final resting place of the last dusky seaside sparrow.

The race to the moon was first a race to send a living creature beyond the Kármán Line, that invisible altitude 62 miles above sea level that separates Earth's atmosphere from outer space. Technically, the United States won that first sprint in 1947 when they launched fruit flies on about a one-hour round-trip journey into the thermosphere and back. In the immediate years following the success of that first journey, the US sent up white mustard, scarlet globe radish, Radium Brand spring rye, and wild lily seeds that were all soaked and planted in the ground upon their return. In an alternative history, the road to the moon would be lined by radiation-kissed wildflowers and root vegetables that glow in the dark, but in this version of events, the route to the moon was perilous, the shoulder littered with irradiated perennials and roadkill. If the moon were a goddess hungry for sacrifice, then each failed biological payload was evidence of her insatiable appetite.

The death of Albert I, the first rhesus monkey to be rocketed into outer space, was soon followed by the death of Albert II. There were also dozens of white mice and hamsters, anesthetized cats, dogs, frogs, and fertilized chicken eggs. It was only in 1957 that the Soviet Union successfully carried an animal into orbit. The dog's name was Laika, and at the time of her launch the Soviets had not yet developed the technology to retrieve spacecraft from orbit. She is thought to have died hours after takeoff, but her vessel orbited above Earth for five months. The Germans called her the "She-Hound of Heaven." The Americans preferred "Muttnik."

Only a year after that, the United States launched an intermediate-range ballistic missile manned—or monkeyed—by a navy-trained South American squirrel monkey named Gordo. While Gordo made it into the upper reaches of the thermosphere, past where the International Space Station currently orbits, his capsule's parachute failed to open when he reentered Earth's atmosphere, and he sank into the depths of the Atlantic. Scientists believe Gordo was alive at the time of impact because the in-flight telemetric data being transmitted back to Earth indicated a slight elevation in pulse at splashdown.

While the USSR eventually won the space race in 1961 by sending Yuri Gagarin into orbit, the Americans stole the show again on July 16, 1969, when NASA launched a Saturn V rocket from the Kennedy Space Center in Merritt Island, Florida. Four days and nearly 240,000 miles later, the three-man crew of Michael Collins, Buzz Aldrin, and Neil Armstrong arrived at their destination. Collins piloted the command module *Columbia* as Aldrin and Armstrong descended toward the moon's surface inside the lunar module named after the national bird of the United States: the *Eagle*.

Armstrong's heart rate jumped from 77 beats per minute to 156 beats per minute as Aldrin called out the altitude readings: "750 feet, coming down at 23 degrees . . . 700 feet, 21 down . . . 400 feet, down at 9." When they finally touched down, Armstrong quietly said, "Houston. Tranquility Base here. The *Eagle* has landed."

The dusky seaside sparrow was still stuck back down on Earth. In fact, in that same year, one biologist observed that only thirty

singing male dusky seaside sparrows remained on Merritt Island.
The scientific community had been sounding the alarm about the
disappearance of the dusky for years, but there was little concern
shown beyond the small circle of ornithologists studying Florida's
Atlantic coast. The average sparrow is about as large as a human
heart, though not nearly as important to the survival of actual humans. Perhaps the greatest thing the dusky seaside sparrow had
working against it was that it was not as glorious or impressive as
other species. It was no bald eagle. It was no heart. It was no moon.

What the duskies needed was to develop the same kind of drive
that made Kennedy look toward the stars, that made Disney look
down from the sky at a tiny patch of island in the center of Florida's Bay Lake and say, *There. There's the spot where I'll change the world.*
Either that, or they simply needed support and publicity if they
were going to survive.

Luckily, support did arrive. In 1971, the same year the Walt Disney World Resort opened just 50 miles west of Merritt Island, the
federal government allocated over $2 million to purchase 6,250
acres beside the Indian River to create the Saint Johns National
Wildlife Refuge, home to the only other dusky population that was
known to exist.

Even so, the dusky completely disappeared from Merritt Island
over the course of the 1970s. The habitat west of the refuge, along
the Saint Johns River, was ill managed. The Bee Line Expressway
that stretched between Florida's Space Coast and Orlando was expanded. Residential properties were erected. The marshes were
continually drained, invasive shrubs grew in thick, and ranchers
conducted controlled burns to create more pastureland. Because
the bald eagle had been declared an endangered species in 1967,
and because it had flown all the way to the moon, new environmental protections were put in place and its population steadily
rebounded. However, by 1979 only six duskies remained in Saint
Johns National Wildlife Refuge. Biologists were able to trap only
five for a captive-breeding program that was the last chance to save
the species. There was just one problem: all of the surviving sparrows were male.

While the duskies dwindled on the Florida coast, their new habitat
was being prepared in the center of the state.

Originally intended to debut as a pirates' getaway that would emulate the 1950 Disney film of the same name, Treasure Island was built up with 15,000 cubic yards of soil and 1,000 tons of trees and boulders, transforming it into a tropical oasis where, with the purchase of a Special Adventure ticket, guests could sail across the Seven Seas Lagoon into Bay Lake to spend the day exploring the wreckage of a fictional ship, the *Walrus,* and observing a variety of imported flora and fauna.

Opening to the public on April 8, 1974, Treasure Island became a singular place in the Walt Disney World Resort. There, visitors could spot brilliantly colored macaws and cockatoos from Cap'n Flint's Perch and stumble across vulturine guinea fowl while traipsing through the Indian orchid trees, blue passionflowers, and Chinese gardenias that stretched all the way from Black Dog Bridge to Scavenger Beach.

Despite its being billed as a "tropical island paradise," Treasure Island failed to attract many visitors. So in 1976 Treasure Island closed down for redevelopment. A snack bar was added, along with an aviary, and the pirate references were scrapped. A few months later, Discovery Island opened as not only a tropical destination for Disney tourists, but also a breeding facility for rare birds that would shortly thereafter be accredited by the American Association of Zoological Parks and Aquariums.

The five surviving dusky seaside sparrows that had been captured in Saint Johns National Wildlife Refuge were named after the colors of the identification bands affixed to their tiny legs: Orange Band, White Band, Red Band, Yellow Band, and Blue Band. While some of them would eventually wind up at Disney's Discovery Island, they first traveled to the Santa Fe College Teaching Zoo in Gainesville, Florida. In a cooperative effort with several organizations, including the Wildlife Research Laboratory of the Florida Game and Fresh Water Fish Commission and the Florida Museum of Natural History, the last male duskies were entered into an experimental breeding program, where they were mated with females from a closely related subspecies that was native to the Gulf Coast: the Scott's seaside sparrow. The goal of this program was genetic backcrossing, in which the first generation of female hybrid offspring, being 50 percent dusky, would mate with the remaining pure duskies to produce chicks of 75 percent purity, and so on for several generations, in hope of eventually producing a bird that was at least 90 percent dusky.

While Red Band died of a tumor early in the breeding program, the remaining four duskies made the journey to Discovery Island to continue the crossbreeding experiment. It was just about this time, and halfway across the theme park, that Walt Disney's original dream was being resurrected from the ashes. Eleven years after the Walt Disney World Resort first opened, so too did EPCOT Center.

Originally modeled on Ebenezer Howard's *Garden Cities of Tomorrow,* Disney first imagined EPCOT Center (later renamed Epcot) as a kind of urban utopia, unmarred by crime and poverty and waste, radiating outward in concentric circles of entertainment, shopping, green space, industry, and high-density apartment housing for 20,000 live-in residents. More than anything, Epcot was supposed to function as a model "progress city" that tourists could visit, enjoy, and later try to emulate in their own home towns.

Walt Disney often spoke of Epcot like Kennedy spoke about the moon. He spoke of challenges and new technologies and better tomorrows. He called Epcot a "virgin land" and a place "that will always be in the state of becoming." As something that wasn't quite fate would have it, this turned out to be true, just not in the way Walt Disney had imagined.

Housed in a building known as Spaceship Earth, an 18-story geodesic sphere designed with the help of science fiction writer Ray Bradbury, Epcot wasn't the fully functional live-in city Walt Disney had originally pictured, but it was an attraction that promised to offer guests a glimpse into how something like a future was created.

The attraction itself was composed of a time-machinesque experience that took guests on a 16-minute dark ride through a history populated by animatronic figures. While guests were on the ride, lights and projectors drew their attention to a series of important historical moments—the origin of prehistoric man, the invention of the alphabet, the fall of Rome, the Renaissance, and the Apollo 11 *Eagle* landing on the moon—all of which, much like time itself, quickly faded into darkness and receded into the past as the train of guests crept forward.

In Epcot's early days, the entire journey through time was narrated by Lawrence Dobkin, a voice made famous during the golden age of radio. At the end of the ride, guests would arrive at the top of the track only to be met with a large planetarium full of twinkling stars and galaxies so close they seemed attainable by simple extension of the arm. Then the vehicles turned around and made their descent back into reality, the artificial sky slowly fading from view. This was just about the point in the ride where Lawrence Dobkin's ethereal voice echoed, "Tomorrow's world approaches, so let us listen and learn, let us explore and question and understand. Let us go forth and discover the wisdom to guide great Spaceship Earth through the uncharted seas of the future. Let us dare to fulfill our destiny." For those who didn't have the time or money to purchase the Special Adventure ticket that would have taken them out to Discovery Island, it sure would have been neat to have the silvery tweee-tweee of the dusky piped in through the speakers.

Success stories of imperiled species coming back from the brink of extinction are not unheard of. In fact, the Endangered Species Act, passed by Congress in 1973, played a part in saving the nene, the American peregrine falcon, and what John F. Kennedy once called a bird that "aptly symbolizes the strength and freedom of America": the bald eagle.

The dusky seaside sparrow was among the species protected by the Endangered Species Act. However, the crossbreeding program that took place on Disney's Discovery Island presented an interesting challenge. As reported by the *Chicago Tribune*, the hybrid sparrows, at least in the view of the US Fish and Wildlife Service, were a "nonspecies—deserving neither federal protection nor the right to be released on federal lands."

Between the remaining four duskies—Orange, Yellow, Blue, and White—only five viable hybrid chicks were produced. One was a male of 75 percent purity, while the other four were females that were 25, 50, 75, and 87.5 percent dusky, the last of which was thought by ornithologists to have looked nearly indistinguishable from a purebred bird.

The original duskies and their chicks lived in 8-by-10-foot screened-in cages stocked with clumps of native cordgrass, crickets, and insect larvae. At the top of each cage was a miniature sprinkler system that rained down pockets of mist when the weather was at its hottest. Their faux habitats were hidden from the public eye, tucked away safely between the toucan exhibit and the Discovery Island restrooms. By the time Blue died and then Yellow and then

White, the Discovery Island tour guides spoke in hushed whispers about the only two birds left with any hope of saving the species —Orange and the 87.5-percenter—as a "project for salvation."

It really can't be overstated: 1986 was a year of failed flight.

On January 28, the NASA Space Shuttle orbiter *Challenger* lifted off from Kennedy Space Center in Merritt Island, Florida. On the CNN live broadcast, Tom Mintier had just said, "So the twenty-fifth space shuttle mission is now on the way after more delays than NASA cares to count," when the *Challenger* broke apart over the Atlantic Ocean 73 seconds into its flight. It was the *Challenger*'s 10th launch. There were at least eight delays, from bad weather at the transoceanic abort-landing site in Dakar, Senegal, to a malfunction of the microswitch indicator used to ensure that the hatch was safely locked.

For the people watching, either from their living rooms or the grassy lawns alongside the Florida coast, it appeared as if the orbiter had unexpectedly exploded in a giant fireball. But there was no explosion, at least not in the traditional sense of the word. Despite the fact that history would remember the flight of the *Challenger* as ill-fated, fate had little to do with what happened on that cold January morning in 1986.

The disintegration was actually caused by an O-ring seal that failed shortly after liftoff. During preliminary tests, a number of NASA engineers filed reports and voiced concerns that the O-rings were faulty, even going so far as to redesignate the small rubber parts "Critical 1," knowing that the failure of a part so seemingly minuscule as the O-ring could open an almost indiscernible gap in the infrastructure of the orbiter, through which gases would leak and combust. Both NASA managers and engineers had been aware of this problem since 1977.

The Rogers Commission, charged with investigating what caused the *Challenger* to disintegrate just over a minute into its flight, concluded that the tragedy had little to do with fate and more to do with "go fever," a uniquely human condition rooted in our ability to overlook issues and ignore risks based on a desire to succeed, even in the face of irrefutable danger. The commission eventually surmised that what had happened to the *Challenger* was "an accident rooted in history."

The same could be said of the ill-fated dusky seaside sparrow.

Of the few remaining birds with dusky blood still in them, only Orange Band and the one hybrid female chick—the 87.5-percenter —offered a viable path forward, to redemption, to achieving a dusky that was over 90 percent pure. If there were ever a place where finding salvation seemed not only possible, but also likely, then it would have to be at Disney World, WHERE DREAMS COME TRUE.

Science fiction author Jules Verne has been called a "merchant of dreams" because of books like *Twenty Thousand Leagues Under the Sea* and *Journey to the Center of the Earth*. He trafficked in dream machines, in opulent submarines, elephantine locomotives, and cannons capable of blasting men all the way to the moon. That was the beauty of a Vernian dream: the future was capacious. There was room enough for anyone and everyone. In an 8-by-10-foot bullet-shaped room called a "projectile coach," Verne was able to comfortably fit 50 gallons of brandy, two dogs, six rifles with plenty of ammunition, a dozen small shrubs, a half-dozen chickens, several cases of wine, enough food and water to last an entire year, and three adult men all bound for the moon.

The employees of Disney's Discovery Island could have learned something from Jules Verne, who knew so well how to preserve the future in an 8-by-10 space, dimensions that proved too small for the dusky seaside sparrow. In 1985 Orange had given birth to another 87.5 percent dusky, the sister of the hybrid female with which he ultimately failed to mate. The chick broke its neck against the side of the cage. Charles Cook, the head curator of Discovery Island at the time, said, "It was a terrible misfortune. They fly in a straight line. They take off like a jet fighter. They have to learn their boundaries."

Orange Band died on June 17, 1987, just an hour's drive from Merritt Island, 50 or so miles as the crow flies. He was blind in one eye and estimated to be as old as 14. "At the end," Tom Sander of the Florida *Sun Sentinel* reported, Orange "was treated like royalty, kept in a special eight-by-ten-foot cage, cooled and bathed by a soft artificial rainfall, fed a particularly nutritious diet of crickets, seeds and grubs and protected from infection by a requirement that his human guardians disinfect their boots every time they came to pay a visit."

The sort of rosy ending that Sander imagined has long been a fixture of our storytelling: that everything will be OK. That against all odds, humanity is destined not only for survival, but also for greatness. But if history has shown us anything, it is that there is no such thing as destiny.

There is no such thing as fate.

There is only the future, itself an accident rooted in history, an inadvertent offspring of the past so similar in appearance and design that it is sometimes indistinguishable, even to the trained eye.

KEVIN KRAJICK

The Scientific Detectives Probing the Secrets of Ancient Oracles

FROM *Atlas Obscura*

ONE HOT SPRING day, two scientists began climbing a steep, boulder-strewn ridge near Selcuk, in southwestern Turkey. The few stunted pines sprinkled here and there offered little shade. "Don't touch the rocks with your hands too much," warned their guide. "Scorpions." Several hundred feet up, a dark, narrow opening pierced the slope. They climbed in and descended about 20 feet to the floor of a small, cool cavern. In the indirect sunlight, they could see stalactites lacing the walls, and curving passages, too small to enter, spiraling down. A newly shed snakeskin lay on the floor. At the rear, five natural steps led up to a rock formation resembling a tangle of human bones. They had found their quarry—an oracle of Cybele, earth goddess of Asia Minor, curer of disease, granter of fertility, seer of all things. From prehistoric times until 2,000 years ago, and perhaps longer, people came to this cave to ask Cybele the same questions we ask today. Whom should I marry? How can I make more money? How long will I live? Today, we have therapists and algorithms, risk analysts and actuarial charts. The ancients had oracles.

The scientists were John Hale, an archaeologist from the University of Louisville, and Jelle Zeilinga de Boer, a geologist from Wesleyan University. They had formed themselves into a sort of oracle detective team, seeking out the sites of these ancient prognosticators and attempting to figure out why they are located where they are and what role they played in the ancient world.

Around 50 BC the Roman politician Cicero wrote, "As far as I know, there is no nation whatever, however polished and learned, or however barbarous and uncivilized, which does not believe it is possible that future events maybe be indicated, and understood." Oracles were the most famous and enduring institutions of the ancient world. The best known was Greece's Oracle of Delphi, where, for at least 1,000 years, kings and common pilgrims visited a cave where a priestess leaned over a sacred spring and inhaled from it the breath of the god. Elsewhere, the future was divined via haruspicy, the reading of organs from sacrificed animals; empyromancy, the interpretation of flickering flames; or augury, which involved observing lightning flashes and other phenomena. At Dodona, priests of Zeus were said to hear the future in the rustling leaves of a sacred oak. At Sura, on the Turkish coast, it was in the patterns of fish congregating around a magical whirlpool. On earthquake-prone Mount Garganus in Italy, the method was incubation: A supplicant performed purification rituals, slaughtered a black ram, and slept on its skin in the sanctuary. That night's dream held portents, interpreted by a resident priest. At the Cybele shrine visited by Hale and de Boer, divination apparently involved dice made from the knucklebones of sheep. Thousands have been found at its entrance, along with votive statues, coins, and other offerings.

Almost all these sites had one thing in common. They all were located upon or within extraordinary natural features—deep caverns, strange rock formations, bubbling springs, ancient groves—that apparently had something to do with their powers. Some researchers believe cults connected to some of these sites could go back to well before the rise of civilization, as far as 25,000 years. As religions and belief systems changed, some sacred sites went with them—Cybele shrines were renamed for more familiar Greek gods—at least until Christianity came to dominate the Mediterranean. That was when the Roman emperor Theodosius declared Christianity the state religion, and outlawed oracles, in AD 385. Sites great and small were pillaged, repurposed, or just buried and forgotten. (One exception is the cave on Mount Garganus, now a sanctuary of the Archangel Michael of Monte Sant'Angelo.) Sixteen centuries later, archaeologists started digging.

In the 1990s looters rediscovered the cave of Cybele before archaeologists got there. Years later, Hale and de Boer were led there by the director of the Ephesus Museum in nearby Selcuk,

Cengiz Icten. Outside the cave, Icten poked the dry ground with a walking stick and grinned. Suddenly we saw what he was seeing. Everywhere in the loose soil were pottery sherds, shaped stones, even a corroded bronze coin. "It took many centuries for all this to build up," he said. "This place goes back a long way."

Suddenly a huge, dark shape exploded from the mouth of the cave, almost knocking us all down, and flew into the sky. It was a large owl. As it flapped off, we could hear its babies cooing from some hidden crevice within. "This place does not feel spooky to me," said Hale. "This seems like a growing place. It feels like a revelatory place."

Hale and de Boer got their start in the oracle business at the famed Delphi itself. The archaeologist and geologist had met in 1995, while both were touring ancient ruins elsewhere. They became friends over a bottle of wine and their shared interest in the oracle—a space where their spheres of academic interest overlapped. That night they vowed to solve the mystery.

Hale brought a deep knowledge of ancient history, languages, mythology, and architecture. De Boer supplied expertise in the even more ancient: the origins of rocks, the mechanisms of earthquakes, the workings of volcanoes. He had surveyed the area around Delphi in 1981 for a study of Greek earthquake hazards. "Geology is at the ground level of everything, whether it's biological, archaeological, anthropological, or ecological," he said.

"It's my hope," Hale said, "that by learning about the past, one might find some inspiration about how to live today."

Delphi sits on the slopes of Mount Parnassus, about 75 miles west of Athens. It is thought to have originated as a sanctuary of Gaia, the pre-Greek earth goddess. The Greeks later said it was where Zeus had placed the center of the world. It was also the main abode of Apollo, god of the sun and of prophecy, near where he slew the giant serpent Python. By the fifth century BC, it hosted an elaborate complex of ritual buildings. Over the colonnade of Apollo's temple was carved γνῶθι σεαυτόν. KNOW THYSELF.

According to Greek writer Plutarch (AD 46–120), inside the temple a small, dimly lit underground sanctum enclosed a cleft in the bedrock. The feature exuded a sweet-smelling vapor—the *pneuma,* or "breath of the god." The pneuma, he wrote, was produced by "natural underground forces," and was emitted "as if from a spring." Once a month, a priestess, or Pythia, went through elaborate purification rituals, sat in a special chair, and hung her head over the chasm to inhale the pneuma. Then she began speaking in a strange, disembodied voice. Questioners were admitted.

The Pythia's answers could be cryptic or unwelcome, but they were always taken seriously. Matters of business, marriage, treaties, and wars were undertaken on her counsel. In legend, it was the Pythia who told Oedipus he would kill his father and marry his mother. Some time before 399 BC, one Chaerephon asked the Pythia if anyone was wiser than his friend Socrates. "No," said the oracle—either confirming the philosopher's greatness or denying the very existence of human wisdom. According to fourth-century BC historian Herodotus, in 546 BC, the Lydian king Croesus sacrificed 3,000 animals, burned piles of valuables, and sent a huge treasure to honor the oracle. Then he sent a messenger to ask whether he should attack his rival, Cyrus of Persia. The priestess replied that if he did, he would "destroy a mighty empire." Croesus attacked and was defeated. By some accounts, he was given a last-minute reprieve from being burned alive, and sent a messenger to the oracle to ask why it had betrayed him. The Pythia replied: "Croesus ought, if he had been wise, to have sent again and inquired which empire was meant, that of Cyrus or his own; but if he neither understood what was said, nor took the trouble to seek for enlightenment, he has only himself to blame for the result." Her message was simple. Know thyself.

What kept people coming back to oracles—outside, of course, of the universal desire for certainty about the future? Perhaps the seers did, indeed, have a good track record in predicting the future. One reason for this could be that oracles were the greatest intelligence-gathering and -disseminating agencies of their day. According to a 1956 history of oracles by historian H. W. Parke, temple officials often subjected powerful people to days or even weeks of questioning before allowing them to consult the oracle. This meant that those officials had deep access to political developments, military strategies, and economic trends—sometimes from opposite sides of a conflict. This may have helped them make informed judgments that they could pass on to the priest or priestess. Another factor is that oracles often favored the most generous tippers—who were, thanks to their wealth, probably more likely to prevail in an economic or military conflict anyway. Croesus notwithstanding.

Then, there is the third theory: the uncanny. "There is a hell of a lot more around us than we know about," said de Boer, a die-hard scientific empiricist with little patience for speculation. "When people ask me, 'Do you know how it all worked?' I have to say no. There are some things we will never know."

It should be noted, though, that something can seem extranatural without being outright supernatural. The investigation of many such phenomena is often known as geomythology, or the study of how natural processes—from earthquakes and volcanic eruptions to floods and eclipses—get encoded in religious stories, mythology, and folklore. In the case of Delphi, it had been speculated that the pneuma was some gas or vapor, emitted from a natural chasm or spring, with psychoactive effects.

In the late 19th and early 20th centuries, Delphi was rediscovered by archaeologists. At the time, scholars denounced the whole idea of the pneuma, supernatural or geological, as a myth or even a hoax. Excavation of the ruins revealed no obvious cleft or cave where the oracular sanctum might have been. There was also no obvious sign of volcanism that would account for the release of gases. According to Parke, some researchers believed that the priestess was inspired by sitting over a hole filled with burning marijuana. Another, more recent, theory has it that the Pythia was high from chewing the toxic leaves of the oleander tree, or inhaling their smoke.

De Boer, however, had examined the area closely in his initial survey, with a geologist's eye. To the east of Delphi, he spotted an earthquake fault exposed by a modern road cut, and followed it on foot to near the temple complex. "It was beautifully expressed on the surface," he said. To the west lay a known fault, striking in the same direction. And if you connected the ends, the thread clearly ran under the temple, though that part was obscured by rocky debris and the buildings themselves. De Boer had read Plutarch, and connected that cleft with what he saw in the ground. It was not a geological smoking gun, exactly, but it was a geomythological lead. "Present-day humans are pretty arrogant when they think the ancients could not have observed things clearly," he said.

Later, in the late 1990s, de Boer and Hale visited Delphi together and, among other things, dug up Greek government geological maps showing that the limestone in the area was laced with tarry petrochemicals. There was no evidence of geothermal features, but a slow slippage of the fault could create enough heat to vaporize those deposits. They also found traces of a second fault, almost perpendicular to the first, also below the temple floor. This intersection would have created an ideal vent for underground gases. There was no spring, as Plutarch suggested, but Hale and de Boer found evidence of a drain, and, uphill, some still-running spring water. They sampled this water and chiseled out pieces of travertine, a chalky rock that forms when chemical-laden spring waters react with air. In both they found traces of hydrocarbon gases.

One of the gases in the flowing water is ethylene, a substance used in the early 20th century as an anesthetic, and still widely used in the chemical industry. In small doses, it is said to induce an out-of-body euphoria and a release of inhibition. In the interest of science, of course, Hale and a couple of friends in Louisville got hold of a tank of ethylene, opened the valve in a backyard garden shed about the size of the alleged inner sanctum, and took turns, well, huffing it. Hale is pretty sure this was legal. They lost the feeling in their hands and feet, and began seeing the world as if from outside. "Very strange, but not scary," said Hale. The next logical step? Predicting the outcome of the next Kentucky Derby.

Starting in 2001, the team published a series of scientific papers laying out the case that the Delphic Oracle operated exactly as described, and that much of it could be explained scientifically. Though not everyone bought into all their conclusions, many scholars were converted. Modest fame followed, and a book. There was just one problem. Every garden-shed prediction about the Kentucky Derby was dead wrong.

Hale likes pointing out that the Greek words for "prophet" and "madness"—*mantos* and *mania*—come from a common root. "When Plato considered the Delphic Oracle, he said that the priestess was never of any use when she was in her right mind. But when she was mad, she benefited all mankind," said Hale. "That is a beautiful thought. It tells us that there are special places on Earth that shape human belief."

Emboldened by their work at Delphi, Hale and de Boer looked farther afield. Southwest Turkey was a logical starting point. In the centuries before Christ, the Greeks had colonized the region. At other times there were Hittites, Lydians, Persians, and Romans. There are multicultural ruins all over the place, including the oracular Greek temples of Klaros and Didyma, nearly as important as Delphi in their time, if not quite as well known. According to inscriptions dating back as far as 600 BC, rulers from as far away as present-day Russia and Mauritania consulted these oracles about plagues, labor disputes, and religious crises. Locals asked about planting crops, money matters, or, in one case, whether to embark on piracy. (Didyma approved.) So one spring they set out to explore as many of them as they could, gather samples, and devise, if possible, a unifying theory of ancient oracles.

To reach Klaros, we drove through wooded hills and farmland near the Aegean coast to a small valley, where we followed a dirt road through the lemon groves. The road entered a swampy area and dead-ended against a rocky wall. From out of high reeds rose a set of broad steps leading to a great stone platform. Only a few columns and walls still stood, but in the remains of the sanctuary were fragments of a sculpture of Apollo said to have once been over 20 feet high. Countless names and inscriptions had been carved into the ruins—possibly the greatest surviving collection of ancient Greek inscriptions in a single place. Artifacts found around the temple foundations date at least as far back as 1200 BC. "People obviously sensed early on that there was something special about this place," said Hale.

The stone slabs that once formed the main temple floor had been hauled away. This had exposed a basement labyrinth once hidden within the platform—and, Hale and de Boer suspected, Klaros's oracular secrets. The labyrinth, excavated by French archaeologists in the 1980s, led from the front steps to two chambers in the rear, all now filled waist-deep with stagnant water.

According to an AD 18 description by Roman aristocrat Tacitus, prophecies here were offered only on certain nights: "A priest, after hearing merely the number and names of the clients, [went] down into a cave; there he [drank] from a secret fountain." The priest went into a trance, and then cried out his prophecies from an unseen corner. About 50 years after Tacitus, Pliny the Elder noted that these priests served only one-year terms—possibly, he noted, because the fountain "inspires wonderful oracles, but shortens the life of the drinker."

For decades prior to the site's excavation, researchers had looked for such a cave in nearby hillsides. The discovery of the labyrinth and these chambers suggested that the oracular cave was, in fact, embodied by the temple itself. It even looked like the structure had been repeatedly expanded and elaborated around it, much like at Delphi.

"Let's try and get a feel for the oracular experience," said Hale. In bathing suits and water shoes—we knew ahead what to expect —we descended four steps into the watery labyrinth entrance. The water was warm, opaque with algal scum, and alive with frogs and turtles. I tried not to think too hard about what else might be down there, rubbing against our legs. We waded through six turns to reach the chambers. At least we had the open sky above us. The experience would have been a lot spookier for ancient pilgrims following the priest. It would probably have been pitch-black, the tunnel barely shoulder-wide with the ceiling at head-height. At the end of the labyrinth, we came to a room covered with surviving stone arches, and with stone benches along the walls. Here questioners must have waited to hear prophecies.

Beyond this room was a rectangular inner chamber where archaeologists had found a circular hole in the floor, containing water—the secret fountain, apparently. Perhaps the water source had shifted and begun to overflow, or maybe rain had filled the old basement. In any case, we could not see the hole and had to feel for it with our feet—carefully. A caretaker had warned us it went down at least 20 feet. Off to one side, we finally felt it. It was covered by what felt like a modern metal grate.

Hale and de Boer suspected this spring was effervescing hydrocarbon gases like those at Delphi—maybe even a lot more potent, if the priests were dying prematurely. "In a closed space, it would be like sniffing gasoline, only worse," suggested de Boer. "Of course, not very healthy."

By feel, we fed a long, plastic hose down through the grate. A big plastic syringe was attached to the other end, and we sucked up water samples from the depths. These samples would go back to de Boer's lab. Our next target would be Didyma.

Pilgrims originally reached Didyma by walking 10 miles from the coastal city of Miletus along the Sacred Way, a stone path flanked by sphinxes, fountains, and tombs. Remnants of it can still be seen from a parallel, far less impressive asphalt road. The temple lies on the outskirts of the small village of Didim, where a century of excavations has revealed a stupendous building with a multistory central court, much bigger than Klaros—much bigger than the

Parthenon, actually, which would fit comfortably inside it. Like Delphi, Didyma is said to have featured a spring above which a priestess sat. The spring is thought to have dried up after Persian invaders burned and looted the place in 493 BC. It miraculously returned, supposedly, after Alexander the Great passed through some 150 years later. Today its exact location has been lost.

When we arrived the place was crawling with tourists. This didn't bother Hale or de Boer. They were looking for something very specific: the site of a now-vanished little house in the center, where a priestess "receive[d] the god by imbibing the vapor of the water," according to the fourth-century AD writer Iamblichus. In the courtyard, we spotted three round, well-like structures, all currently dry. Any of them could have been the spring. De Boer speculated that they all could have been the spring; maybe it had dried up and popped up elsewhere periodically, he said, due to natural shifts in underground waterways. "They must have moved the well from time to time to keep up," he said.

In the absence of any water within the temple itself, de Boer went to a well in front, where pilgrims are thought to have purified themselves before entering. It held plenty of water, as well as coins that people had tossed much more recently. We poked the hose down and sucked up some water. "Second or third best, but better than nothing," said de Boer.

Some months later, de Boer called me with the results: the water at both Klaros and Didyma contains ethylene, along with other hydrocarbon gases including methane and ethane. Ever the cautious scientist, he said he needed to go back for more investigation.* But, he said, "This gives us a good indication that a similar process was going on at all these places."

We tried to investigate other oracular haunts over the next few days, but time had shifted the landscape and muddled evidence at many. Patara, once a seaside city said to host an oracle of Apollo, was sunken into silt and underbrush, leaving little to see. At Sura, ancient people had once bought kebabs that they tossed into a strange whirlpool—possibly a freshwater spring blending with the sea below the tide line—and priests told the future by observing

---

* This, sadly, would not happen. Jelle Zeilinga de Boer passed away before he got a chance to go back.

the fish that gathered for the feast. We found the ruins of the temple there and a nearby spring, but the shoreline had long since receded, and the site itself was mired in a swamp. We hunted for Acharaca, a long-lost cave dedicated to the god of the underworld Pluto and his queen Persephone. There, sacrificial bulls led in were said to simply drop dead. Near the rumored site, we sniffed hydrogen sulfide—a sign that perhaps the bulls were victims of volcanic gases. Locals told stories of collapsed underground vaults, and warned of venomous snakes. Hale and de Boer, patient but weary, resolved to return some other time.

We also ventured to the ancient town of Hierapolis, whose ruins sprawl out over a mountain slope. Pure white terraces, formed of travertine-type minerals precipitating from dozens of chemical-laden springs, mantle the mountainside. The waters, used as a spa in ancient times and still available for wading or swimming, are said to treat high blood pressure, skin disease, rheumatism, and eye problems.

In the ruins above the spa was the place de Boer and Hale had come to see: a small oracle called the Plutonium. "It was supposedly an entrance to hell," said Hale. Ancient authors write of a dense, deadly vapor issuing from it. Sacrificial animals such as sparrows were thrown in and quickly died. "Mysteriously taken by the god," said Hale. The apparent portal is a tiny arched doorway cut into a cliff next to the ruins of a modest Apollonian temple. Except for a head-size gap, the door was closed off with a recent-looking block-and-mortar job. On a nearby wall was a partly obliterated Greek inscription. Hale struggled to make it out: "dreams . . . earth . . . oracle . . ."

As we approached, we were struck with a terrible stench. It was hard to say if this came from the cave or two small dead porcupines on the marble pavement nearby. An elderly lady was sweeping up a pile of dead birds with a broom, like it was her regular job.

De Boer said that in the 1980s Turkish scientists had shown the vapor to be largely carbon dioxide, which can come from volcanic sources. Heavier than air, it can kill by displacing oxygen wherever it pools. They also identified whiffs of sulfuric acid and a few other asphyxiants and poisons. Ancient accounts said that while the sacrificial animals died in the Plutonium, the priests could go in and out unscathed. "I believe they had bladders of air under their robes," said de Boer. "It must have made people really afraid."

Since our trip, German scientists have shown that carbon dioxide around the Plutonium tends to pool up in the cool of the night, creating a lethal layer a couple of feet thick. The concentration of the gas falls rapidly with height—so it could quickly suffocate animals close to the ground, while humans could wade through, safe and unperturbed.

In poking around before the trip, I had come across a vaguely sourced article in *Omni* magazine that asserted that two vacationing Australians had entered the Plutonium recently and disappeared. An equally vague website claimed, "Many people throughout history who went past the mouth of the cave never returned." We were undeterred as we eyed that head-sized gap.

I volunteered. "Promise you'll wave or something, if you're passing out," said Hale, as I prepared to look in. "Oh, sure. If I look like I'm slumping, just grab my legs and pull," I said.

My head just fit through. A hot, humid billow burned my eyes. I held my breath and blinked as my vision adjusted to the darkness. In a cell-like room, a square shaft descended about six feet down, where a narrow black cleft curved to the right and out of sight. A dark shape lay on the floor, unidentifiable. Gasping for air, I pulled my head out.

Later, Fettah Anli, the friendly owner of the nearby Hal-Tur Hotel, told us he had grown up playing among the ruins. He was quite sure no one had ever disappeared in the Plutonium, but he did say that locals had once hung a sign over the door saying DEVIL'S HOLE, and lowered small dogs and other unfortunate animals to their deaths for the entertainment of paying visitors.

Fifteen years ago, he went on, a tourist from New Zealand—he remembered the man's name was Thomas—went swimming in a mineral-water pool nearby, and decided to explore a narrow underground feeder channel. "After he swam in, his wife kept waiting for him," said Anli. "But he did not come out. Then she started screaming."

It took three days for a backhoe to reach Thomas's body. He had become wedged in a tight spot 40 feet into the cave, and apparently drowned. It seemed that repeated retellings of the story had morphed his journey to the underworld into yet another myth.

J. B. MACKINNON

You Really Don't Want to Know What It's Like to Be a Right Whale These Days

FROM *The Atlantic*

WE MIGHT BEGIN with a way of killing a whale that next to no one today would find acceptable. In the autumn of 1385, far enough back in time that we tend to think of humans then and now almost as different species, a whale beached near the southern tip of Greenland. Among the Norse settlers who gathered around the animal was a recent arrival named Björn Einarsson, an Icelandic chieftain who, on a return trip from Norway, had found his home country bound in its namesake ice and was forced to carry on. With winter on the horizon, the newcomers had been struggling to procure enough food.

Einarsson has left the impression across six centuries of a man who expects to have good luck. Sure enough, as the settlers butchered the whale, they found a spearhead embedded in its flesh. By Norse law, fully half the whale belonged to whomever had struck the ultimately fatal blow, and Einarsson recognized the spearhead's mark. It belonged to Ólafur Ísfirðingur, who lived in the same district of Iceland as Einarsson himself. Ísfirðingur obviously could not be reached in a timely manner (a typical law of the era spells out that the finder of a dead whale must alert the hunter if he can "travel there twice on the day concerned, if he leaves early in the morning"), which meant that Einarsson could claim the "shooter's share" of the whale meat on a promise to pay his neighbor later. We can imagine much rejoicing among the misplaced sailors.

Between the lines, we can also read snatches of the whale's story, which is considerably more of a downer. Early Norse whaling mainly involved the spear-drift technique: spear a whale, then hope that it drifts ashore nearby so that you can get your share. Unfortunately, the process could be a slow one. The whale that put food on Einarsson's table in late-autumn Greenland had most likely been speared during the Icelandic summer. When the end came, it had swum about a thousand miles across several months, in pain, terribly injured, slowly dying.

Spear-drift whaling was eventually replaced by more efficient means of killing whales. Along the way, we grew distant enough from stranded-in-Greenland-variety food desperation to agree that wounding an animal and leaving it to die across hours, let alone months, is not OK. That troublesome vegan at the dinner party would agree, but so, too, would members of the NRA, whose hunter's code of ethics calls on them to "insure clean, sportsmanlike kills."

In acknowledging the brutality of certain hunting practices, we drew a baseline of human responsibility for the welfare of wild animals: We should not cause them undue suffering. In that, there has been plenty of room for controversy (a meat-is-murder activist, a leg-hold trapper, and a catch-and-release fly fisherman walk into a bar . . .). Still, the baseline matters. It matters more and more, in fact, because everyday life for a growing roster of wild creatures has become so unpleasant, on our watch and by our hands, that their suffering calls for consideration by reasonable people. One of those animals happens to be a kind of whale, the North Atlantic right whale, which lives along the eastern seaboard of the United States and Canada. As one researcher put it: "We're not actually going out and sticking them with a piece of steel anymore. We're just ruining their lives."

Americans have trouble seeing right whales, and not because the leviathans live in remote places. I recently spent a day dipping in and out of every stretch of public seashore I could find along the Massachusetts coast between Boston and Cape Cod. An endangered species, North Atlantic right whales number about 450 in total, but one-quarter of the entire population had made the seasonal pilgrimage to Cape Cod Bay, and right whales spouted at every stop. (There are three species of right whale, separated by vast tracts of ocean; here, *right whale* refers to the North Atlantic variety unless otherwise noted.)

Would-be whale watchers knew where to go: there's an app. Yet the whales were easy to miss, low black hummocks among low black waves. Right whales don't plunge forward, arcing and diving; they plug along the surface, back and forth in boustrophedon style, then vanish for long periods. Watching people watch for whales without the patience for their subtle presence, and with genuine disappointment that they weren't more showy and exciting, made the shore feel like a dividing line between utterly different worlds.

That division is an illusion, the marine biologist Scott Kraus told me when we met in his Boston waterfront office, just 30 miles of abominable traffic away. Kraus, the vice president for research at the New England Aquarium, pulled up an image on his computer screen of the US East Coast, covered with a matrix of lines that crowded into the Atlantic. "That's the shipping," Kraus said. "Then we add fishing." He clicked on another slide, which laid a new matrix on top of the previous one. And on he went, adding undersea pipelines and cables, coastal wind-energy fields, and military operations until the image looked like a quintuple-exposure photograph of a map of Manhattan. If you layered in the general public's use of the ocean, he said, "I can't even imagine. It would blow it out."

Nearly a decade ago, Kraus nicknamed the right whale the "urban whale." The Atlantic coast of North America is visible by night from space as a cluster of light whose intensity is rivaled only by parts of western Europe. That its influence extends into the ocean should surprise no one, and yet it does. "We keep looking to it as a way to expand civilization," said Kraus. "You know, 'The whole ocean is empty there. We can do anything!'"

Another convenient delusion is our tendency to see endangered species as evolution's snowflakes: unique expressions of life's diversity, but also high maintenance and generally unable to roll with the punches. It's hard to make this case with right whales. They evolved at least 4 million years ago, twice as deep in the past as our own genus. An adult may weigh 80 tons and reach 45 feet in length, fully a quarter of it accounted for by a great anvil of a head and a sweeping, downturned mouth. Individuals can live 100 years and possibly many more—few have the opportunity to find out.

They take their name from having been the "right" whale to

hunt, because of the value of their blubber and baleen, and as such, they'd already been driven to rarity by the time of the American Revolution. Yet they do not die easy. The intentional killing of right whales was banned in 1935, but in March of that year, it took a group of fishermen—apparently not up to speed on international law—six hours, seven hand-thrown harpoons, and 150 rifle rounds to kill a 32-foot calf off Fort Lauderdale, Florida.

If right whales are threatened with extinction, it's not from a lack of grit. It's because their home—which spans 2,000 miles of coastline from southern Canada to northern Florida and cannot be described as small or niche—is one of the most human-modified and -influenced regions on Earth. With due respect to Kraus, the North Atlantic right whale is not so much the urban whale as the Anthropocene whale.

So-called "whale huggers" face an unusual problem when it comes to the species' welfare: stories of right-whale suffering are often so awful that no one wants to hear them. ("Inevitably you put your head inside the head of the whale and say, 'What's this feel like?' And it's not a good place to be," said one.) Fortunately, a suite of less-awful, more-bearable ills is also available for discussion. We can save the worst for later.

One of the first people to start thinking about how we make whales miserable, as opposed to how we kill them, was the marine-acoustics scientist Chris Clark, now retired as a graduate professor of Cornell University. In the 1990s, with Cold War tensions subsiding, Clark was selected as the US Navy's marine-mammal scientist.

Using the navy's underwater listening posts, he was able to tune in to singing fin whales—second only to the blue whale in size—across a patch of sea larger than Oregon. In a data visualization he later created, the singing whales wink on and off: hotspots that arise, spread their sonic glow, and fade. Then enormous flares ripple across the entire space. That's the acoustic imprint of a seismic air gun, used to probe for oil and gas deposits under the seafloor. "This was an epiphany," Clark said. He had witnessed the way that human-made sounds could overwhelm, at enormous scales, whales' ability to hear and be heard in the ocean.

I asked for his opinion about what day-to-day life is like for right whales now, two decades later. "Acoustic hell," Clark replied. "Humans can't comprehend the magnitude of the insult that we pour into the ocean." While no one can say how an animal experiences its world, there are clues that Clark is correct. When the 9/11 attacks took place in 2001, researchers from the New England Aquarium happened to be in the Bay of Fundy, just across the US border into Canada, testing right-whale feces for stress hormones. Over the following days, boat traffic abruptly dropped off. The scientists were struck by how clearly they could hear whale calls through their equipment, as though they'd been standing beside a freeway that fell silent and could suddenly hear birdsong. The whale stress levels measured in those quiet waters were the lowest by far that were recorded across four summers of sampling.

Noise is what biologists refer to as a "sublethal" impact, meaning it doesn't directly cause death. The list of sublethal impacts has grown long, however. Right whales have the highest prevalence of infection with giardia and cryptosporidium, mainly from sewage and agricultural manure runoff, ever recorded in any mammal. In humans, these cause the diseases known as beaver fever and crypto, respectively, which involve debilitating digestive complaints. No one knows what problems, if any, they cause in right whales.

The whales are similarly exposed to an alphabet soup of chemicals (DDT, PCBs, PAHs, etc.), oil and gas, flame retardants, pharmaceuticals, pesticides—all the effluvia of civilization. Then there are blooms of red tide and other toxic algae, which can cause paralysis and death in humans, and are increasingly common. One study found paralytic shellfish poisoning in the feces of all 16 right whales it sampled. Again, no one can say what effect these pollutants might be having on right whales.

We do know that it is getting harder for them to find food. Right whales' diet is composed mainly of flea-like plankton called copepods, a predator-to-prey size ratio comparable to you or I meeting our daily needs by eating specks of dust. We have approximately zero idea how giant whales find tiny plankton in an enormous ocean, but we do know that the distribution of copepods has become even more unpredictable as the climate changes.

It's worth pointing out here that right whales seem to be at least as capable of feeling as other animals that we care about (dogs, cats, horses, etc.). They respond—sometimes with irritation— even to being gently touched with a pole at the base of their tail by researchers. In one such case, a whale promptly sank beneath the waves, then rose, belly up, to lift the scientists' boat out of the water, rock it back and forth (one crew member tumbled into the ocean), and set the vessel back down before going about its business. You don't need a PhD in biology to get the message the whale was sending, or to admire its self-restraint.

You also don't need to be an expert to interpret the fact that whales off Iceland appear to be more common where whaling is permitted (but rare) than within the near-shore protected zone where whale-watching boats are everywhere. Whales sleep differently than we do (they put one hemisphere of their brain to sleep at a time), and yet we can bet, as researchers do, that a noisy, busy, polluted ocean is a more challenging place to sleep than the ocean as it used to be, the ocean that some whales are old enough to still remember.

"These whales are suffering from the effects of multiple cumulative impacts," said Rosalind Rolland, a senior scientist at New England Aquarium. She points to the stark contrast between North Atlantic right whales and southern right whales, a closely related species that lives in a far less humanized environment. Rolland once traveled to the Auckland Islands, 300 miles south of New Zealand's mainland, to do a visual health assessment on southern right whales. "They were fat, they were happy, they had no skin lesions, they were curious. It was dealing with a completely different animal," she told me.

As you've probably guessed, North Atlantic right whales don't look quite so good. The health of every class of right whale—male and female, young and old—has visibly worsened over the past three decades. They are thinner, more heavily infested with whale lice, more marked by lesions and scars. The research that showed this, led by Rolland, also found that the more sickly a female right whale's appearance, the less likely that she would successfully give birth to a calf.

For an endangered species, a lack of births is a kind of death, and this year, for the first time since reliable record-keeping began nearly 30 years ago, no calves at all were born in the right-whale population. The animals' welfare may now be so poor, their suffering so serious, that sublethal impacts have turned lethal.

The traditional view of animal welfare has been that we're responsible for animals under our care—pets, livestock, zoo animals, laboratory animals. Concerns about wild-animal welfare have typically

ended at the question of what we put them through as we kill
them.

That began to change in the 1990s. In Britain, some members
of London's Institute of Zoology had come to recognize the Anthropocene concept, though the term would not enter popular
usage until the 21st century. "Much of the Earth's surface is now
under human control, partial control, or influence, and this inevitably often affects the fate of wild animals," three of them wrote in
a pioneering 1994 paper. "We have a degree of responsibility for
their welfare."

A quarter century later, Jennifer Jacquet, an assistant professor
in environmental studies and animal studies at New York University, is more emphatic. "Is there wild-animal suffering anymore? I
mean, are there wild animals? Is there suffering in the wild, or are
there just human-altered animals suffering from human-caused
harms?"

Like any emerging school of thought, wild-animal welfare
makes visible what was previously overlooked, or ignored. For example, you might reasonably imagine that clearing wild-animal
habitats for houses, or farms, or auto malls, involves a respectable
trade-off: human needs versus animal inconvenience. A 2017 study
of land clearing in Australia addresses what is actually involved.
Turns out the animals do not simply pack their bags and make a
fresh start somewhere new. "The clear scientific consensus is that
most, and in some cases all, of the individuals present at a site will
die as a consequence of that vegetation being removed, either immediately or in a period of days to months afterward," the authors
write.

They lay out the suffering in exhaustive detail: Animals are
crushed, impaled, or lacerated. Some are buried alive. They endure internal bleeding, broken bones, spinal damage, eye injuries,
head injuries. Limbs are lost. "Degloving"—partial skinning, alive
—occurs. Those that flee their homes (many are surprisingly re-
luctant to do so) are often run over on nearby roads, entangled in
fences, die of exposure, or are made easy prey for predators. You
don't really want to hear this, but tree-dwelling species may cower
in their holes up to the moment they pass through the sawmill or
the wood-chipping machine. You don't really want to hear that
koala bears may starve when land is cleared—"an issue that has
surprisingly not generated much discussion." By the authors' esti-

mate, 50 million mammals, birds, and reptiles ultimately die each year due to land clearing in two Australian states alone.

Wild-animal welfare has remained out of mind for so long partly because the harms involved are often indirect or unintentional—no one chose to disrupt the right whales' food supply, or give them chronic stress disorders. The sheer scale of potential responsibility also encourages willful blindness. David Fraser, the research chair in animal welfare at the University of British Columbia, calls the hurt we heap on wildlife through our day-to-day machinations "the huge, neglected issue of the century." So far, wild-animal welfare has largely remained an academic concern, with some thinkers inevitably proposing that we become what one called "compassionate gods," knitting sweaters for elk during deadly cold snaps and preparing meat substitutes for predators so that they, too, can go vegan. But the argument that humans should prevent all of life's cruelties to other species is distinct from the one that suggests we have at least some responsibility for suffering that we cause.

The questions involved are large, and numerous. Consider the pied flycatcher, a European songbird that resembles a miniaturized penguin. In the past, the flycatcher's migration and nesting season coincided with the spring outbreak of caterpillars that the birds feed to their young. Due to climate change, peak caterpillar has been taking place earlier and earlier across the past 20 years, while the flycatchers show no sign of moving up their migration. Research in the Netherlands found that where the caterpillar cycle has shifted, pied flycatcher populations have declined by 90 percent. Given that fear, hunger, anxiety, and so on normally occur in nature, at what point does our own contribution of misery trigger a duty of care? How capable of suffering is a pied flycatcher, anyway? Does the fact that the species is not yet in danger of extinction ease the situation's moral urgency? Are we called upon to find some boots-on-the-ground solution—there isn't an obvious one—to flycatcher famine, or just chalk it up as another reminder of the ethical burden of climate change as a whole?

"The point is not to overwhelm people and make them feel completely despondent or guilty. The point is to do what we can," said Jacquet. "I don't think it's going to be solved overnight, but I hope that the moral compass moves a bit."

We might start with those cases that most clearly link wild-animal suffering to species endangerment, and to human responsibility.

Behold the right whale: at least moderately intelligent, and plainly able to experience physical and psychological pain. Nearing extinction, and dependent on a habitat so radically altered by human actions that it has been described by fisheries scientists as "arguably domesticated." And, lest we forget, probably suffering every damned day.

We don't only harm right whales indirectly, of course. We also kill them outright. Half of known right-whale deaths since the 1980s have direct human causes, and that number does not include whales whose bodies are never found or are found in such a state of decomposition that no cause of death can be determined. The true percentage is certainly higher, and rising.

There are two main ways that we kill right whales: by running them over with boats (known as "ship strike"), and through entanglement in fishing gear. From the traditional conservation perspective, which is mainly concerned with the survival of species, ship strike and entanglement are comparable crises: both leave rare whales equally dead. Introduce the idea of wild-animal welfare and there is, unfortunately, more to the story.

It isn't clear why whales fail to avoid being hit by boats, but ship strike is probably no more inexplicable than the fact that people get run over by cars: suddenly a large object looms out of the background of everyday life to smite you. The good news is that regulators have had some success moving shipping lanes and applying vessel-speed restrictions in parts of right-whale habitat. Doing so, however, does little to address whales' welfare—the "acoustic hell," the sublethally lethal ocean. In this case, conservation that fails to consider welfare is paradoxical. We can end up saving whales only to leave them swimming in a sea of pain.

At least victims of ship strike usually don't suffer a lot: death comes swiftly. Now consider the ways that frontline wildlife veterinarians described to me the results of entanglement:

"The injuries are some of the worst we've seen to any animal."

"One of the grossest abuses of wild-animal sensibility in the modern world."

"Barbaric in the extreme."

"If this was an entangled cow that you drove past on your way to work, wrapped up in line and with line cutting into it—no one would be able to drive by."

I have been holding back, as I said I would, from the darkest reaches of this story. But let me now make at least one case of deepest right-whale suffering unavoidable. The whale in question never received one of the affable names applied to well-known individuals, such as Kleenex, or Churchill, or Porter. Instead, it was only ever known by its number in a long-standing photo-identification catalogue: #2030.

Read a media report of a whale bound up in ropes or nets and you are likely to think that this is one unlucky whale, akin to the sea turtle with a plastic straw up its nose. In fact, entanglement is not rare or even unusual for right whales; it is a constant: 85 percent of them have scars from past entanglements, as many as a quarter may acquire new scars in any given year, and both numbers seem to be increasing.

Since 2010, nearly 40 percent of known right-whale deaths—15 whales in total—have been caused by entanglement in fishing gear. It's the leading cause of mortality. (Remember that many losses go unrecorded; the real figure may be twice as high.) Once again, it's not like the whales are somehow hopelessly maladaptive—the fraction that dies from entanglements is minuscule compared to the number of entanglements overall. It's just that ropes and fishing gear are nearly unavoidable.

Across large spans of right-whale habitat, lobster and crab fishing is so intensive that even some fishers compare it to farming. There can be so many buoys marking underwater lobster-pot lines that a boat at the surface can't travel in a straight line—as one Cape Cod fisherman put it to me, "It's ridiculous. There's pots everywhere. You can't move." "Ropeless" crab and lobster pots already exist that could replace vertical lines in the water with buoys that release from the seafloor only when the fishers come to retrieve their catch. But such solutions involve costs for the industry, which in turn are passed along to the consumer. We prefer our seafood to come at a reasonable price.

In 1999, right whale #2030 was seen at Cultivator Shoal, about a hundred miles east of Cape Cod. Just 12 years old but already 45 feet in length, she was entangled in a gill net and its rigging, which she had picked up sometime earlier. Perhaps she had run into the gear at night, or while diving in dark water, or simply because she was focused on other things. In any case, she responded as many right whales do, spinning in an effort to free herself. She ended up bound in loops of rope and tightly wound netting. She looked something like a draft horse in heavy harness.

The next time #2030 was spotted, she was in the Bay of Fundy. A 10-day effort to cut away the fishing gear still left her with tight coils around her flippers connected by one taut line across her back and two under her belly. On the 11th day she left the bay and vanished into open ocean. She was later found floating, dead, off Cape May, New Jersey.

Michael Moore was involved in the postmortem. A senior marine biologist with Woods Hole Oceanographic Institution and veterinarian for the International Fund for Animal Welfare's marine-mammal rescue and research team, Moore had started his career on Icelandic whale boats in 1983, where he researched the efficacy of explosive harpoons. He found they were very efficient. The typical time to death was usually no more than a few minutes.

That finding stuck with him when he began, in 1998, to examine the bodies of entangled right whales. "I've always had this really gut sense of horror in terms of what these animals go through," he told me. Each case is broadly similar, but uniquely awful in the details. The rope across #2030's back had cut down through six inches of blubber to the flesh below, then started to work its way along her body. In the whaling days, the removal of a dead whale's blubber came to be called flensing. Whale #2030 had been flensed alive. She had been "degloved." By the time she died, she had an open-wound saddle across her back that measured more than four feet wide.

Her cause of death was listed as "massive traumatic injury." Starvation played a part, too. Later tests estimated that the gear hanging off #2030 had increased her body's drag through the water by 161 percent, costing her more energy than she was able to take in as food. As her condition deteriorated, so many orange and white whale lice spread over her body that she appeared to have turned a different color.

Whale #2030's skeleton hangs in the Museum of the Earth in Ithaca, New York. If you know what to look for, you can see where funky, cauliflowery bone growths arose in the body's attempt to protect itself from its bonds. You can see smooth grooves worn into the joints and bones of its flippers, where the fishing ropes sawed back and forth.

Whale #2030 was first seen entangled on May 10, 1999, and was found dead on October 20 that year. When the end came, she had swum about a thousand miles across several months, in pain, terribly injured, slowly dying. In other words, her death was every bit as terrible as that of the spear-drift whale found by Norse settlers in Greenland in 1385, far enough back in time that we tend to think of humans then and now almost as different species.

BILL MCKIBBEN

How Extreme Weather Is
Shrinking the Planet

FROM *The New Yorker*

THIRTY YEARS AGO, this magazine published "The End of Nature," a long article about what we then called the greenhouse effect. I was in my twenties when I wrote it, and out on an intellectual limb: climate science was still young. But the data were persuasive, and freighted with sadness. We were spewing so much carbon into the atmosphere that nature was no longer a force beyond our influence—and humanity, with its capacity for industry and heedlessness, had come to affect every cubic meter of the planet's air, every inch of its surface, every drop of its water. Scientists underlined this notion a decade later when they began referring to our era as the Anthropocene, the world made by man.

I was frightened by my reporting, but at the time it seemed likely that we'd try as a society to prevent the worst from happening. In 1988 George H. W. Bush, running for president, promised that he would fight "the greenhouse effect with the White House effect." He did not, nor did his successors, nor did their peers in seats of power around the world, and so in the intervening decades what was a theoretical threat has become a fierce daily reality. As this essay goes to press, California is ablaze. A big fire near Los Angeles forced the evacuation of Malibu, and an even larger fire, in the Sierra Nevada foothills, has become the most destructive in California's history. After a summer of unprecedented high temperatures and a fall "rainy season" with less than half the usual precipitation, the northern firestorm turned a city called Paradise into an inferno within an hour, razing more than 10,000 buildings and killing at least 63 people; more than 600 others are missing. The authorities brought in cadaver dogs, a lab to match evacuees' DNA with swabs taken from the dead, and anthropologists from California State University at Chico to advise on how to identify bodies from charred bone fragments.

For the past few years, a tide of optimistic thinking has held that conditions for human beings around the globe have been improving. Wars are scarcer, poverty and hunger are less severe, and there are better prospects for wide-scale literacy and education. But there are newer signs that human progress has begun to flag. In the face of our environmental deterioration, it's now reasonable to ask whether the human game has begun to falter—perhaps even to play itself out. Late in 2017, a United Nations agency announced that the number of chronically malnourished people in the world, after a decade of decline, had started to grow again—by 38 million, to a total of 815 million, "largely due to the proliferation of violent conflicts and climate-related shocks." In June 2018, the Food and Agriculture Organization of the UN found that child labor, after years of falling, was growing, "driven in part by an increase in conflicts and climate-induced disasters."

In 2015, at the UN Climate Change Conference in Paris, the world's governments, noting that the Earth has so far warmed a little more than 1 degree Celsius above preindustrial levels, set a goal of holding the increase this century to 1.5 degrees Celsius (2.7 degrees Fahrenheit), with a fallback target of two degrees (3.6 degrees Fahrenheit). This past October, the UN's Intergovernmental Panel on Climate Change published a special report stating that global warming "is likely to reach 1.5 C between 2030 and 2052 if it continues to increase at the current rate." We will have drawn a line in the sand and then watched a rising tide erase it. The report did not mention that, in Paris, countries' initial pledges would cut emissions only enough to limit warming to 3.5 degrees Celsius (about 6.3 degrees Fahrenheit) by the end of the century, a scale and pace of change so profound as to call into question whether our current societies could survive it.

Scientists have warned for decades that climate change would lead to extreme weather. Shortly before the IPCC report was published, Hurricane Michael, the strongest hurricane ever to hit the Florida Panhandle, inflicted $30 billion worth of material damage and killed 45 people. President Trump, who has argued that global warming is "a total, and very expensive, hoax," visited Florida to survey the wreckage, but told reporters that the storm had not caused him to rethink his decision to withdraw the United States from the Paris climate accords. He expressed no interest in the IPCC report beyond asking "who drew it." (The answer is 91 researchers from 40 countries.) He later claimed that his "natural instinct" for science made him confident that the climate would soon "change back." A month later, Trump blamed the fires in California on "gross mismanagement of forests."

Human beings have always experienced wars and truces, crashes and recoveries, famines and terrorism. We've endured tyrants and outlasted perverse ideologies. Climate change is different. As a team of scientists recently pointed out in the journal *Nature Climate Change,* the physical shifts we're inflicting on the planet will "extend longer than the entire history of human civilization thus far."

The poorest and most vulnerable will pay the highest price. But already, even in the most affluent areas, many of us hesitate to walk across a grassy meadow because of the proliferation of ticks bearing Lyme disease, which have come with the hot weather; we have found ourselves unable to swim off beaches, because jellyfish, which thrive as warming seas kill off other marine life, have taken over the water. The planet's diameter will remain 8,000 miles, and its surface will still cover 200 million square miles. But the Earth, for humans, has begun to shrink, under our feet and in our minds.

Climate change, like *urban sprawl* or *gun violence,* has become such a familiar term that we tend to read past it. But exactly what we've been up to should fill us with awe. During the past 200 years, we have burned immense quantities of coal and gas and oil—in car motors, basement furnaces, power plants, steel mills—and as we have done so, carbon atoms have combined with oxygen atoms in the air to produce carbon dioxide. This, along with other gases like methane, has trapped heat that would otherwise have radiated back out to space.

There are at least four other episodes in the Earth's half-billion-year history of animal life when CO_2 has poured into the atmosphere in greater volumes, but perhaps never at greater speeds. Even at the end of the Permian Age, when huge injections of CO_2 from volcanoes burning through coal deposits culminated in "the

Great Dying," the CO_2 content of the atmosphere grew at perhaps a tenth of the current pace. Two centuries ago, the concentration of CO_2 in the atmosphere was 275 parts per million; it has now topped 400 per million and is rising more than 2 parts per million each year. The extra heat that we trap near the planet every day is equivalent to the heat from 400,000 bombs the size of the one that was dropped on Hiroshima.

As a result, in the past 30 years we've seen all 20 of the hottest years ever recorded. The melting of ice caps and glaciers and the rising levels of our oceans and seas, initially predicted for the end of the century, have occurred decades early. "I've never been at . . . a climate conference where people say 'that happened slower than I thought it would,'" Christina Hulbe, a New Zealand climatologist, told a reporter for *Grist* last year. This past May, a team of scientists from the University of Illinois reported that there was a 35 percent chance that, because of unexpectedly high economic growth rates, the UN's "worst-case scenario" for global warming was too optimistic. "We are now truly in uncharted territory," David Carlson, the former director of the World Meteorological Organization's climate-research division, said in the spring of 2017, after data showed that the previous year had broken global heat records.

We are off the literal charts as well. In August I visited Greenland, where, one day, with a small group of scientists and activists, I took a boat from the village of Narsaq to a glacier on a nearby fjord. As we made our way across a broad bay, I glanced up at the electronic chart above the captain's wheel, where a blinking icon showed that we were a mile inland. The captain explained that the chart was from five years ago, when the water around us was still ice. The American glaciologist Jason Box, who organized the trip, chose our landing site. "We called this place the Eagle Glacier because of its shape," he said. The name, too, was five years old. "The head and the wings of the bird have melted away. I don't know what we should call it now, but the eagle is dead."

There were two poets among the crew, Aka Niviana, who is Greenlandic, and Kathy Jetnil-Kijiner, from the low-lying Marshall Islands, in the Pacific, where "king tides" recently washed through living rooms and unearthed graveyards. A small lens of fresh water has supported life on the Marshall Islands' atolls for millennia, but as salt water intrudes, breadfruit trees and banana palms wilt and die. As the Greenlandic ice we were gazing at continues to melt, the water will drown Jetnil-Kijiner's homeland. About a third of the carbon responsible for these changes has come from the United States.

A few days after the boat trip, the two poets and I accompanied the scientists to another fjord, where they needed to change the memory card on a camera that tracks the retreat of the ice sheet. As we took off for the flight home over the snout of a giant glacier, an eight-story chunk calved off the face and crashed into the ocean. I'd never seen anything quite like it for sheer power—the waves rose 20 feet as it plunged into the dark water. You could imagine the same waves washing through the Marshalls. You could almost sense the ice elevating the ocean by a sliver—along the seafront in Mumbai, which already floods on a stormy day, and at the Battery in Manhattan, where the seawall rises just a few feet above the water.

When I say the world has begun to shrink, this is what I mean. Until now, human beings have been spreading, from our beginnings in Africa, out across the globe—slowly at first, and then much faster. But a period of contraction is setting in as we lose parts of the habitable Earth. Sometimes our retreat will be hasty and violent; the effort to evacuate the blazing California towns along narrow roads was so chaotic that many people died in their cars. But most of the pullback will be slower, starting along the world's coastlines. Each year, another 24,000 people abandon Vietnam's sublimely fertile Mekong Delta as crop fields are polluted with salt. As sea ice melts along the Alaskan coast, there is nothing to protect towns, cities, and native villages from the waves. In Mexico Beach, Florida, which was all but eradicated by Hurricane Michael, a resident told the *Washington Post,* "The older people can't rebuild; it's too late in their lives. Who is going to be left? Who is going to care?"

In one week at the end of last year, I read accounts from Louisiana, where government officials were finalizing a plan to relocate thousands of people threatened by the rising Gulf ("Not everybody is going to live where they are now and continue their way of life, and that is a terrible, and emotional, reality to face," one state official said); from Hawaii, where, according to a new study, 38 miles of coastal roads will become impassable in the next few decades; and from Jakarta, a city with a population of 10 million, where a rising Java Sea had flooded the streets. In the first days of 2018, a nor'easter flooded downtown Boston; dumpsters and cars floated through the financial district. "If anyone wants to question global warming, just see where the flood zones are," Marty Walsh, the mayor of Boston, told reporters. "Some of those zones did not flood thirty years ago."

According to a study from the United Kingdom's National Oceanography Centre last summer, the damage caused by rising sea levels will cost the world as much as $14 trillion a year by 2100, if the UN targets aren't met. "Like it or not, we will retreat from most of the world's non-urban shorelines in the not very distant future," Orrin Pilkey, an expert on sea levels at Duke University, wrote in his book *Retreat from a Rising Sea*. "We can plan now and retreat in a strategic and calculated fashion, or we can worry about it later and retreat in tactical disarray in response to devastating storms. In other words, we can walk away methodically, or we can flee in panic."

But it's not clear where to go. As with the rising seas, rising temperatures have begun to narrow the margins of our inhabitation, this time in the hot continental interiors. Nine of the ten deadliest heat waves in human history have occurred since 2000. In India, the rise in temperature since 1960 (about 1 degree Fahrenheit) has increased the chance of mass heat-related deaths by 150 percent. The summer of 2018 was the hottest ever measured in certain areas. For a couple of days in June, temperatures in cities in Pakistan and Iran peaked at slightly above 129 degrees Fahrenheit, the highest reliably recorded temperatures ever measured. The same heat wave, nearer the shore of the Persian Gulf and the Gulf of Oman, combined triple-digit temperatures with soaring humidity levels to produce a heat index of more than 140 degrees Fahrenheit. June 26 was the warmest night in history, with the mercury in one Omani city remaining above 109 degrees Fahrenheit until morning. In July a heat wave in Montreal killed more than 70 people, and Death Valley, which often sets American records, registered the hottest month ever seen on our planet. Africa recorded its highest temperature in June, the Korean Peninsula in July, and Europe in August. The *Times* reported that, in Algeria, employees at a petroleum plant walked off the job as the temperature neared 124 degrees. "We couldn't keep up," one worker told the reporter. "It was impossible to do the work."

This was no illusion; some of the world is becoming too hot for humans. According to the National Oceanic and Atmospheric Administration, increased heat and humidity have reduced the amount of work people can do outdoors by 10 percent, a figure that is predicted to double by 2050. About a decade ago, Australian and American researchers, setting out to determine the highest survivable so-called wet-bulb temperature, concluded that when temperatures passed 35 degrees Celsius (95 degrees Fahrenheit) and the humidity was higher than 90 percent, even in "well-ventilated shaded conditions," sweating slows down, and humans can survive only "for a few hours, the exact length of time being determined by individual physiology."

As the planet warms, a crescent-shaped area encompassing parts of India, Pakistan, Bangladesh, and the North China Plain, where about 1.5 billion people (a fifth of humanity) live, is at high risk of such temperatures in the next half century. Across this belt, extreme heat waves that currently happen once every generation could, by the end of the century, become "annual events with temperatures close to the threshold for several weeks each year, which could lead to famine and mass migration." By 2070, tropical regions that now get one day of truly oppressive humid heat a year can expect between 100 and 250 days, if the current levels of greenhouse-gas emissions continue. According to Radley Horton, a climate scientist at the Lamont-Doherty Earth Observatory, most people would "run into terrible problems" before then. The effects, he added, will be "transformative for all areas of human endeavor—economy, agriculture, military, recreation."

Humans share the planet with many other creatures, of course. We have already managed to kill off 60 percent of the world's wildlife since 1970 by destroying their habitats, and now higher temperatures are starting to take their toll. A new study found that peak-dwelling birds were going extinct; as temperatures climb, the birds can no longer find relief on higher terrain. Coral reefs, rich in biodiversity, may soon be a 10th of their current size.

As some people flee humidity and rising sea levels, others will be forced to relocate in order to find enough water to survive. In late 2017 a study led by Manoj Joshi, of the University of East Anglia, found that, by 2050, if temperatures rise by two degrees a quarter of the Earth will experience serious drought and desertification.

The early signs are clear: São Paulo came within days of running out of water last year, as did Cape Town this spring. In the fall, a record drought in Germany lowered the level of the Elbe to below 20 inches and reduced the corn harvest by 40 percent. The Potsdam Institute for Climate Impact Research concluded in a recent study that, as the number of days that reach 86 degrees Fahrenheit or higher increases, corn and soybean yields across the US grain belt could fall by between 22 and 49 percent. We've already over-pumped the aquifers that lie beneath most of the world's bread-baskets; without the means to irrigate, we may encounter a repeat of the 1930s, when droughts and deep plowing led to the Dust Bowl—this time with no way of fixing the problem. Back then, the Okies fled to California, but California is no longer a green oasis. A hundred million trees died in the record drought that gripped the Golden State for much of this decade. The dead limbs helped spread the waves of fire, as scientists earlier this year warned that they could.

Thirty years ago, some believed that warmer temperatures would expand the field of play, turning the Arctic into the new Midwest. As Rex Tillerson, then the CEO of Exxon, cheerfully put it in 2012, "Changes to weather patterns that move crop production areas around—we'll adapt to that." But there is no rich topsoil in the far North; instead, the ground is underlaid with permafrost, which can be found beneath a fifth of the northern hemisphere. As the permafrost melts, it releases more carbon into the atmosphere. The thawing layer cracks roads, tilts houses, and uproots trees to create what scientists call "drunken forests." Ninety scientists who released a joint report in 2017 concluded that economic losses from a warming Arctic could approach $90 trillion in the course of the century, considerably outweighing whatever savings may have resulted from shorter shipping routes as the Northwest Passage unfreezes.

Churchill, Manitoba, on the edge of the Hudson Bay, in Canada, is connected to the rest of the country by a single rail line. In the spring of 2017 record floods washed away much of the track. OmniTrax, which owns the line, tried to cancel its contract with the government, declaring what lawyers call a "force majeure," an unforeseen event beyond its responsibility. "To fix things in this era of climate change—well, it's fixed, but you don't count on it being the fix forever," an engineer for the company explained at a media briefing in July. This summer, the Canadian government re-opened the rail at a cost of $117 million dollars—about $190,000 per Churchill resident. There is no reason to think the fix will last, and every reason to believe that our world will keep contracting.

All this has played out more or less as scientists warned, albeit faster. What has defied expectations is the slowness of the re-sponse. The climatologist James Hansen testified before Congress about the dangers of human-caused climate change 30 years ago. Since then, carbon emissions have increased with each year except 2009 (the height of the global recession) and the newest data show that 2018 will set another record. Simple inertia and the human tendency to prioritize short-term gains have played a role, but the fossil fuel industry's contribution has been by far the most damag-ing. Alex Steffen, an environmental writer, coined the term *preda-tory delay* to describe "the blocking or slowing of needed change, in order to make money off unsustainable, unjust systems in the meantime." The behavior of the oil companies, which have pulled off perhaps the most consequential deception in mankind's his-tory, is a prime example.

As journalists at InsideClimate News and the *Los Angeles Times* have revealed since 2015, Exxon, the world's largest oil company, understood that its product was contributing to climate change a decade before Hansen testified. In July 1977, James F. Black, one of Exxon's senior scientists, addressed many of the company's top leaders in New York, explaining the earliest research on the green-house effect. "There is general scientific agreement that the most likely manner in which mankind is influencing the global climate is through carbon-dioxide release from the burning of fossil fuels," he said, according to a written version of the speech that was later recorded, and obtained by InsideClimate News. In 1978, speak-ing to the company's executives, Black estimated that a doubling of the carbon dioxide concentration in the atmosphere would in-crease average global temperatures by between 2 and 3 degrees Celsius (5.4 degrees Fahrenheit), and as much as 10 degrees Cel-sius (18 degrees Fahrenheit) at the poles.

Exxon spent millions of dollars researching the problem. It out-fitted an oil tanker, the *Esso Atlantic*, with CO_2 detectors to measure how fast the oceans could absorb excess carbon, and hired mathe-maticians to build sophisticated climate models. By 1982, they had concluded that even the company's earlier estimates were probably too low. In a private corporate primer, they wrote that heading off global warming and "potentially catastrophic events" would "require major reductions in fossil fuel combustion."

An investigation by the *LA Times* revealed that Exxon executives took these warnings seriously. Ken Croasdale, a senior researcher for the company's Canadian subsidiary, led a team that investigated the positive and negative effects of warming on Exxon's Arctic operations. In 1991 he found that greenhouse gases were rising due to the burning of fossil fuels. "Nobody disputes this fact," he said. The following year, he wrote that "global warming can only help lower exploration and development costs" in the Beaufort Sea. Drilling season in the Arctic, he correctly predicted, would increase from two months to as many as five months. At the same time, he said, the rise in the sea level could threaten onshore infrastructure and create bigger waves that would damage offshore drilling structures. Thawing permafrost could make the earth buckle and slide under buildings and pipelines. As a result of these findings, Exxon and other major oil companies began laying plans to move into the Arctic, and started to build their new drilling platforms with higher decks, to compensate for the anticipated rises in sea level.

The implications of the exposés were startling. Not only did Exxon and other companies know that scientists like Hansen were right; they used his NASA climate models to figure out how low their drilling costs in the Arctic would eventually fall. Had Exxon and its peers passed on what they knew to the public, geological history would look very different today. The problem of climate change would not be solved, but the crisis would, most likely, now be receding. In 1989 an international ban on chlorine-containing man-made chemicals that had been eroding the Earth's ozone layer went into effect. Last month, researchers reported that the ozone layer was on track to fully heal by 2060. But that was a relatively easy fight, because the chemicals in question were not central to the world's economy, and the manufacturers had readily available substitutes to sell. In the case of global warming, the culprit is fossil fuel, the most lucrative commodity on Earth, and so the companies responsible took a different tack.

A document uncovered by the *LA Times* showed that, a month after Hansen's testimony, in 1988, an unnamed Exxon "public af-

fairs manager" issued an internal memo recommending that the
company "emphasize the uncertainty" in the scientific data about
climate change. Within a few years, Exxon, Chevron, Shell, Amoco,
and others had joined the Global Climate Coalition, "to coordinate business participation in the international policy debate" on
global warming. The GCC coordinated with the National Coal Association and the American Petroleum Institute on a campaign,
via letters and telephone calls, to prevent a tax on fossil fuels, and
produced a video in which the agency insisted that more carbon
dioxide would "end world hunger" by promoting plant growth.
With such efforts, it ginned up opposition to the Kyoto Protocol,
the first global initiative to address climate change.

In October 1997, two months before the Kyoto meeting, Lee
Raymond, Exxon's president and CEO, who had overseen the science department that in the 1980s produced the findings about
climate change, gave a speech in Beijing to the World Petroleum
Congress, in which he maintained that the Earth was actually cooling. The idea that cutting fossil fuel emissions could have an effect
on the climate, he said, defied common sense. "It is highly unlikely
that the temperature in the middle of the next century will be
affected whether policies are enacted now, or twenty years from
now," he went on. Exxon's own scientists had already shown each
of these premises to be wrong.

On a December morning in 1997 at the Kyoto Convention
Center, after a long night of negotiation, the developed nations
reached a tentative accord on climate change. Exhausted del-
egates lay slumped on couches in the corridor, or on the floor in
their suits, but most of them were grinning. Imperfect and limited
though the agreement was, it seemed that momentum had gathered behind fighting climate change. But as I watched the dele-
gates cheering and clapping, an American lobbyist, who had been
coordinating much of the opposition to the accord, turned to me
and said, "I can't wait to get back to Washington, where we've got
this under control."

He was right. On January 29, 2001, nine days after George W.
Bush was inaugurated, Lee Raymond visited his old friend Vice
President Dick Cheney, who had just stepped down as the CEO of
the oil-drilling giant Halliburton. Cheney helped persuade Bush
to abandon his campaign promise to treat carbon dioxide as a
pollutant. Within the year, Frank Luntz, a Republican consultant

for Bush, had produced an internal memo that made a doctrine of the strategy that the GCC had hit on a decade earlier. "Voters believe that there is no consensus about global warming within the scientific community," Luntz wrote in the memo, which was obtained by the Environmental Working Group, a Washington-based organization. "Should the public come to believe that the scientific issues are settled, their views about global warming will change accordingly. Therefore, you need to continue to make the lack of scientific certainty a primary issue in the debate."

The strategy of muddling the public's impression of climate science has proved to be highly effective. In 2017 polls found that almost 90 percent of Americans did not know that there was a scientific consensus on global warming. Raymond retired in 2006, after the company posted the biggest corporate profits in history, and his final annual salary was $400 million. His successor, Rex Tillerson, signed a $500 billion deal to explore for oil in the rapidly thawing Russian Arctic, and in 2012 was awarded the Russian Order of Friendship. In 2016 Tillerson, at his last shareholder meeting before he briefly joined the Trump administration as secretary of state, said, "The world is going to have to continue using fossil fuels, whether they like it or not."

It's by no means clear whether Exxon's deception and obfuscation are illegal. The company has long maintained that it "has tracked the scientific consensus on climate change, and its research on the issue has been published in publicly available peer-reviewed journals." The First Amendment preserves one's right to lie, although, in October, New York State Attorney General Barbara D. Underwood filed suit against Exxon for lying to investors, which *is* a crime. What is certain is that the industry's campaign cost us the efforts of the human generation that might have made the crucial difference in the climate fight.

Exxon's behavior is shocking, but not entirely surprising. Philip Morris lied about the effects of cigarette smoking before the government stood up to Big Tobacco. The mystery that historians will have to unravel is what went so wrong in our governance and our culture that we have done, essentially, nothing to stand up to the fossil fuel industry.

There are undoubtedly myriad intellectual, psychological, and political sources for our inaction, but I cannot help thinking that the influence of Ayn Rand, the Russian émigré novelist, may have played a role. Rand's disquisitions on the "virtue of selfishness" and unbridled capitalism are admired by many American politicians and economists—Paul Ryan, Tillerson, Mike Pompeo, Andrew Puzder, and Donald Trump, among them. Trump, who has called *The Fountainhead* his favorite book, said that the novel "relates to business and beauty and life and inner emotions. That book relates to . . . everything." Long after Rand's death, in 1982, the libertarian gospel of the novel continues to sway our politics: Government is bad. Solidarity is a trap. Taxes are theft. The Koch brothers, whose enormous fortune derives in large part from the mining and refining of oil and gas, have peddled a similar message, broadening the efforts that Exxon-funded groups like the Global Climate Coalition spearheaded in the late 1980s.

Fossil fuel companies and electric utilities, often led by Koch-linked groups, have put up fierce resistance to change. In Kansas, Koch allies helped turn mandated targets for renewable energy into voluntary commitments. In Wisconsin, Scott Walker's administration prohibited state land officials from talking about climate change. In North Carolina, the state legislature, in conjunction with real-estate interests, effectively banned policymakers from using scientific estimates of sea-level rise in the coastal-planning process. Earlier this year, Americans for Prosperity, the most important Koch front group, waged a campaign against new bus routes and light-rail service in Tennessee, invoking human liberty. "If someone has the freedom to go where they want, do what they want, they're not going to choose public transit," a spokeswoman for the group explained. In Florida, an anti-renewable-subsidy ballot measure invoked the "Rights of Electricity Consumers Regarding Solar Energy Choice."

Such efforts help explain why, in 2017, the growth of American residential solar installations came to a halt even before March 2018, when President Trump imposed a 30 percent tariff on solar panels, and why the number of solar jobs fell in the United States for the first time since the industry's great expansion began, a decade earlier. In February, at the Department of Energy, Rick Perry —who once skipped his own arraignment on two felony charges, which were eventually dismissed, in order to attend a Koch brothers event—issued a new projection in which he announced that the United States would go on emitting carbon at current levels through 2050; this means that our nation would use up all the planet's remaining carbon budget if we plan on meeting the 1.5-degree target. Skepticism about the scientific consensus, Perry told the media in 2017, is a sign of a "wise, intellectually engaged person."

Of all the environmental reversals made by the Trump administration, the most devastating was its decision, last year, to withdraw from the Paris accords, making the United States, the largest single historical source of carbon, the only nation not engaged in international efforts to control it. As the *Washington Post* reported, the withdrawal was the result of a collaborative venture. Among the antigovernment ideologues and fossil fuel lobbyists responsible was Myron Ebell, who was at Trump's side in the Rose Garden during the withdrawal announcement, and who, at Frontiers of Freedom, had helped run a "complex influence campaign" in support of the tobacco industry. Ebell is a director of the Competitive Enterprise Institute, which was founded in 1984 to advance "the principles of limited government, free enterprise, and individual liberty," and which funds the Cooler Heads Coalition, "an informal and ad-hoc group focused on dispelling the myths of global warming," of which Ebell is the chairman. Also instrumental were the Heartland Institute and the Koch brothers' Americans for Prosperity. After Trump's election, these groups sent a letter reminding him of his campaign pledge to pull America out. The CEI ran a TV spot: "Mr. President, don't listen to the swamp. Keep your promise." And despite the objections of most of his advisers, he did. The coalition had used its power to slow us down precisely at the moment when we needed to speed up. As a result, the particular politics of one country for one half-century will have changed the geological history of the Earth.

We are on a path to self-destruction, and yet there is nothing inevitable about our fate. Solar panels and wind turbines are now among the least expensive ways to produce energy. Storage batteries are cheaper and more efficient than ever. We could move quickly if we chose to, but we'd need to opt for solidarity and coordination on a global scale. The chances of that look slim. In Russia, the second-largest petrostate after the United States, Vladimir Putin believes that "climate change could be tied to some global cycles on Earth or even of planetary significance." Saudi

Arabia, the third-largest petrostate, tried to water down the recent IPCC report. Jair Bolsonaro, the newly elected president of Brazil, has vowed to institute policies that would dramatically accelerate the deforestation of the Amazon, the world's largest rain forest. Meanwhile, Exxon recently announced a plan to spend $1 million —about a hundredth of what the company spends each month in search of new oil and gas—to back the fight for a carbon tax of $40 a ton. At a press conference, some of the IPCC's authors laughed out loud at the idea that such a tax would, this late in the game, have sufficient impact.

The possibility of swift change lies in people coming together in movements large enough to shift the zeitgeist. In recent years, despairing at the slow progress, I've been one of many to protest pipelines and to call attention to Big Oil's deceptions. The movement is growing. Since 2015, when 400,000 people marched in the streets of New York before the Paris climate talks, activists—often led by indigenous groups and communities living on the front lines of climate change—have blocked pipelines, forced the cancellation of new coal mines, helped keep the major oil companies out of the American Arctic, and persuaded dozens of cities to commit to 100 percent renewable energy.

Each of these efforts has played out in the shadow of the industry's unflagging campaign to maximize profits and prevent change. Voters in Washington State were initially supportive of a measure on last month's ballot which would have imposed the nation's first carbon tax—a modest fee that won support from such figures as Bill Gates. But the major oil companies spent record sums to defeat it. In Colorado, a similarly modest referendum that would have forced frackers to move their rigs away from houses and schools went down after the oil industry outspent citizen groups 40 to 1. This fall, California's legislators committed to using only renewable energy by 2045, which was a great victory in the world's fifth-largest economy. But the governor refused to stop signing new permits for oil wells, even in the middle of the state's largest cities, where asthma rates are high.

New kinds of activism keep springing up. In Sweden this fall, a one-person school boycott by a 15-year-old girl named Greta Thunberg helped galvanize attention across Scandinavia. At the end of October, a new British group, Extinction Rebellion—its name both a reflection of the dire science and a potentially feisty response—announced plans for a campaign of civil disobedience. Last week, 51 young people were arrested in Nancy Pelosi's office for staging a sit-in, demanding that the Democrats embrace a "Green New Deal" that would address the global climate crisis with policies to create jobs in renewable energy. They may have picked a winning issue: several polls have shown that even Republicans favor more government support for solar panels. This battle is epic and undecided. If we miss the two-degree target, we will fight to prevent a rise of three degrees, and then four. It's a long escalator down to hell.

Last June, I went to Cape Canaveral to watch Elon Musk's Falcon 9 rocket lift off. When the moment came, it was as I'd always imagined: the clouds of steam venting in the minutes before launch, the immensely bright column of flame erupting. With remarkable slowness, the rocket began to rise, the grip of gravity yielding to the force of its engines. It is the most awesome technological spectacle human beings have produced.

Musk, Jeff Bezos, and Richard Branson are among the billionaires who have spent some of their fortunes on space travel—a last-ditch effort to expand the human zone of habitability. In November 2016, Stephen Hawking gave humanity a deadline of a thousand years to leave Earth. Six months later, he revised the timetable to a century. In June 2017 he told an audience that "spreading out may be the only thing that saves us from ourselves." He continued, "Earth is under threat from so many areas that it is difficult for me to be positive."

But escaping the wreckage is, almost certainly, a fantasy. Even if astronauts did cross the 34 million miles to Mars, they'd need to go underground to survive there. To what end? The multimillion-dollar attempts at building a "biosphere" in the Southwestern desert in 1991 ended in abject failure. Kim Stanley Robinson, the author of a trilogy of novels about the colonization of Mars, recently called such projects a "moral hazard." "People think if we fuck up here on Earth we can always go to Mars or the stars," he said. "It's pernicious."

The dream of interplanetary colonization also distracts us from acknowledging the unbearable beauty of the planet we already inhabit. The day before the launch, I went on a tour of the vast grounds of the Kennedy Space Center with NASA's publicaffairs officer, Greg Harland, and the biologist Don Dankert. I'd been warned beforehand by other NASA officials not to broach the topic of global warming; in any event, NASA's predicament became obvious as soon as we climbed up on a dune overlooking Launch Complex 39, from which the Apollo missions left for the moon, and where any future Mars mission would likely begin. The launchpad is a quarter of a mile from the ocean—a perfect location, in the sense that, if something goes wrong, the rockets will fall into the sea, but not so perfect, since that sea is now rising. NASA started worrying about this sometime after the turn of the century, and formed a Dune Vulnerability Team.

In 2012 Hurricane Sandy, even at a distance of a couple of hundred miles, churned up waves strong enough to break through the barrier of dunes along the Atlantic shoreline of the Space Center and very nearly swamped the launch complexes. Dankert had millions of cubic yards of sand excavated from a nearby air force base, and saw to it that 187,000 native shrubs were planted to hold the sand in place. So far, the new dunes have yielded little ground to storms and hurricanes. But what impressed me more than the dunes was the men's deep appreciation of their landscape. "Kennedy Space Center shares real estate with the Merritt Island Wildlife Refuge," Harland said. "We use less than ten percent for our industrial purposes."

"When you look at the beach, it's like eighteen-seventies Florida —the longest undisturbed stretch on the Atlantic Coast," Dankert said. "We launch people into space from the middle of a wildlife refuge. That's amazing."

The two men talked for a long time about their favorite local species—the brown pelicans that were skimming the ocean, the Florida scrub jays. While rebuilding the dunes, they carefully bucket-trapped and relocated dozens of gopher tortoises. Before I left, they drove me half an hour across the swamp to a pond near the Space Center's headquarters building, just to show me some alligators. Menacing snouts were visible beneath the water, but I was more interested in the sign that had been posted at each corner of the pond explaining that the alligators were native species, not pets. PUTTING ANY FOOD IN THE WATER FOR ANY REASON WILL CAUSE THEM TO BECOME ACCUSTOMED TO PEOPLE AND POSSIBLY DANGEROUS, it went on, adding that, if that should happen, THEY MUST BE REMOVED AND DESTROYED.

Something about the sign moved me tremendously. It would have been easy enough to poison the pond, just as it would have been easy enough to bulldoze the dunes without a thought for the tortoises. But NASA hadn't done so, because of a long series of laws that draw on an emerging understanding of who we are. In 1867 John Muir, one of the first Western environmentalists, walked from Louisville, Kentucky, to Florida, a trip that inspired his first heretical thoughts about the meaning of being human. "The world, we are told, was made especially for man—a presumption not supported by all the facts," Muir wrote in his diary. "A numerous class of men are painfully astonished whenever they find anything, living or dead, in all God's universe, which they cannot eat or render in some way what they call useful to themselves." Muir's proof that this self-centeredness was misguided was the alligator, which he could hear roaring in the Florida swamp as he camped nearby, and which clearly caused man mostly trouble. But these animals were wonderful nonetheless, Muir decided—remarkable creatures perfectly adapted to their landscape. "I have better thoughts of those alligators now that I've seen them at home," he wrote. In his diary, he addressed the creatures directly: "Honorable representatives of the great saurian of an older creation, may you long enjoy your lilies and rushes, and be blessed now and then with a mouthful of terror-stricken man by way of dainty."

That evening, Harland and Dankert drew a crude map to help me find the beach, north of Patrick Air Force Base and south of the spot where, in 1965, Barbara Eden emerged from her bottle to greet her astronaut at the start of the TV series *I Dream of Jeannie.* There, they said, I could wait out the hours until the predawn rocket launch and perhaps spot a loggerhead sea turtle coming ashore to lay her eggs. And so I sat on the sand. The beach was deserted, and under a near-full moon I watched as a turtle trundled from the sea and lumbered deliberately to a spot near the dune, where she used her powerful legs to excavate a pit. She spent an hour laying eggs, and even from 30 yards away you could hear her heavy breathing in between the whispers of the waves. And then, having covered her clutch, she tracked back to the ocean, in the fashion of others like her for the past 120 million years.

REBECCA MEAD

The Story of a Face

FROM *The New Yorker*

ABBY STEWART SOMETIMES thought that she was born to
be a teacher. At the small college in Colorado where she was an
instructor in the biology department, she enjoyed preparing lec-
tures for business or history majors who were simply fulfilling a
requirement by taking her course in anatomy and physiology. She
worked hard to prove to these students that they should still care
about biology, and one way she captured their attention was by
describing natural phenomena that, at first glance, might seem
peculiar. She revealed that clown fish—like Nemo, in the Pixar
movie—are hermaphrodites, starting out as male but sometimes
becoming female as they mature. And she talked about how fungi
send out subterranean tentacles, which fuse with those of other
fungi. When you eat a mushroom, Abby told her students, you are
eating combined sex organs.

Abby wasn't much older than her students—she had recently
turned thirty—and she had a playfulness that made her seem even
younger. It was fun to freak the kids out a bit. But the point was to
convey to them that nature was not black and white. It contained
infinite shades of gray.

One afternoon last December, Abby took a walk through Muir
Woods, north of San Francisco, with her mother, Bette. (Some
names have been changed.) Abby wandered through the redwood
groves in jeans and hiking boots, stopping to remark on a burl
here or a root system there. As she and her mother chatted beneath the scented pines, the resemblance between them was striking: they had the same fair hair; the same soft, pinkish complex-

ion; the same vivid blue eyes. But Abby, who is about six feet tall, towered above her more delicately framed mother, and bent to her in a gentle, attentive way.

The last time they had been to Muir Woods together, Bette recalled, Abby was six months old; they'd been on a family trip to the Bay Area from their hometown, in the South. Abby laughed, saying that she obviously couldn't remember that day—this visit might as well be her first. But for most of the hike she was pensive. The next day, in San Francisco, she would be undergoing a seven-hour operation at the hands of Dr. Jordan Deschamps-Braly, a craniofacial surgeon who specializes in a process called facial feminization. His practice serves transgender patients like Abby, who was designated male at birth, and publicly identified as a man before she began transitioning, almost two years ago. Abby would submit to an array of surgical procedures—on her brow, chin, jaw, nose, and throat—that would leave her looking subtly altered: like her own cousin, or her sister.

The surgery would, in another sense, unwind time. It would give Abby the face that she might now have if the baby who was once carried around Muir Woods had been spared the unwelcome ravages of puberty—if testosterone had not thickened her brow, sunk her eyes deeper, and weighted her jaw. As Abby saw it, testosterone had blighted her with an Adam's apple that, no matter how long she grew her curly hair, or how soft her skin became from hormone-replacement therapy, irremediably read to a stranger as male.

Although Abby had never undergone a surgical procedure more serious than a tonsillectomy, she had not been nervous in the three months since her first consultation with Deschamps-Braly. The technicalities of the procedures fascinated her. She wished that the whole surgery could be recorded on video, and had requested that someone on the medical team take photographs. But after she and Bette boarded a bus back to San Francisco, where they were staying in a hotel, she looked out the window at the twilight stealing over the folds of the valley, and her mood became somber: now it really was the night before.

The bus driver, speaking over a microphone, offered the passengers a platitude: "I hope you've had a good day, and made some memories that will last a lifetime." Abby gave an ironic sigh. Her phone buzzed: it was a text from Sofia, her girlfriend of the past three years, who had just arrived at the San Francisco airport. "Tomorrow's going to go by in the blink of an eye, literally," it read. "You'll get to fall asleep and wake up the way you were supposed to be born." A more down-to-earth addendum followed: "Plus some swelling."

Abby put her phone away, and looked over at Bette. "Thank you for bringing me, Mom," she said.

"You're welcome, sweetie," Bette replied.

When Jordan Deschamps-Braly was six or seven years old, he was building a model airplane and gashed his arm open with a knife. He was rushed, screaming, to the emergency room, and his father made the unconventional suggestion that it might calm the boy to watch the surgeon stitch him up. Jordan fell quiet, captivated by the procedure. Just as some kids fixate on being firefighters or police officers, he knew from then on that, when he grew up, he wanted to wear blue scrubs and a surgical mask.

Deschamps-Braly was born in 1979, in Oklahoma, to an affluent family that had lived in Ada, a small town southeast of Oklahoma City, for generations. His grandfather had owned a cattle ranch. His parents, George and Dania, were lawyers who ran a small joint practice. Jordan, an only child, went to the local high school, and then attended the University of Oklahoma in Oklahoma City, as both an undergraduate and a medical student. Having waited for years to try surgery, Jordan, who had always liked working with his hands, found plastic surgery particularly gratifying. As a general surgeon, you might open someone's abdomen and repair his bowels or excise her colon cancer, but after you closed the patient up there was no visible trace of your work beyond a little scar—and, hopefully, a recovered patient. With plastic surgery, your artistry was the point. He applied for one of the specialty's highly competitive residencies. His backup plan was to return to school and train as an architect.

He won a fellowship in craniofacial surgery at Milwaukee Children's Hospital, in Wisconsin. He learned to treat kids with such conditions as hydrocephaly, which causes the skull to expand to an extraordinary size, and Treacher Collins syndrome, which results in drooping eyes and an unusually small chin. He worked with Arlen Denny, an eminent surgeon who had trained in France under Paul Tessier, an even more eminent surgeon. Tessier, who died in

2008, was known as the father of craniofacial surgery. In the 1960s, he mastered such innovations as transorbital surgery—the separation of the face from the skull. He practiced the technique on cadavers: cutting the skin along the hairline from ear to ear, and then peeling the forehead away from the skull and the eye sockets.

Deschamps-Braly avidly studied the history of craniofacial surgery, learning about other pioneers, including René Le Fort, who, in the late 19th century, categorized the types of skull fractures that might be caused by blunt trauma. Le Fort's methods would today be considered out of bounds: he tossed cadavers from the roof of a medical school in Paris to see where their faces broke on impact, or smashed their faces in with clubs and analyzed the results. Deschamps-Braly was thrilled by such commitment, which resulted in a catalogue of what are known as Le Fort fractures—the fault lines along which surgeons now deliberately break and reshape the bones of their patients.

Deschamps-Braly wrote to a surgeon in Paris who had studied under Tessier, and asked if he might train there. He was accepted, and moved to France in 2011. He came to see that, however skilled plastic surgeons were at manipulating soft tissue, their effects would be limited if they did not also address the foundation of the face: the bones beneath the skin. During his rotations in Oklahoma, he had been fascinated by the work of oral surgeons, who performed elective surgeries not just to correct a patient's bite but to alter his or her appearance, and to correct what the patient saw as deficits—bringing forward a recessive jaw, or reducing an overly prominent one. After several months in Paris, Deschamps-Braly moved to Zurich, where he trained with an expert in corrective jaw surgery. On Thursday evenings, he returned to Paris on the bullet train, so that he could perform surgery on Friday and spend the weekend with his girlfriend, Maya—a model, and a student of political science and contemporary literature at the Sorbonne.

In 2012 Deschamps-Braly moved back to the United States, with Maya, and the following year they married. He was working at Children's Hospital Oakland, in California, while also establishing a private practice in craniofacial surgery. One day in 2013, Deschamps-Braly learned that another surgeon trained by Tessier, Douglas Ousterhout, who was in his late 70s, was looking for someone to take over his practice, in San Francisco. Ousterhout was known for a niche specialty, facial feminization: reshaping the

faces of people who felt that they looked excessively masculine.
Ousterhout had transformed the faces of more than 1,400 trans
patients, and had published extensively, in scientific journals and
in a handbook for lay readers. Deschamps-Braly had never worked
on a trans patient, but he was excited by the prospect. The two
surgeons met for dinner and hit it off. Ousterhout agreed to delay
his retirement long enough to teach Deschamps-Braly his singular
techniques.

Ousterhout told Deschamps-Braly about how he came to his specialty. In 1982, a colleague approached him with a challenge. The
colleague was a practitioner of what was then known as genital
sex-reassignment surgery, and is now called genital gender-confir-
mation surgery. At the time, the procedure was popularly regarded
as the defining one for transgender patients. The colleague had
performed genital surgery on a patient some years earlier, and the
patient, who subsequently got breast implants, now had the body
she desired. But she remained troubled by the masculine traits in
her face, particularly her brow ridge. She hated it, and wore bangs
to cover it up, but despite her best efforts she thought that she
still had a man's face. The colleague asked Ousterhout, a highly
regarded cranio-maxillofacial surgeon, if he could help.

Ousterhout had never thought much about the broad structural
differences between masculine and feminine faces. When he was
performing reconstructive surgery on a child whose bone plates
had fused incorrectly, his aim was to give the patient's brain room
to develop; when fixing a cleft palate, his goal was to ensure that
the child could eat and breathe and speak. The finer distinctions
of gender were of little concern. But after being presented with his
colleague's surgical problem, Ousterhout went across town to the
Arthur A. Dugoni School of Dentistry, which had a renowned collection of human skulls that had been gathered, mostly from au-
topsies, by an orthodontist. Ousterhout spent hours there, taking
measurements of the head from infancy to adulthood—observing,
for example, how a masculine jaw developed nubbins at the corners, squaring the face, and noting the more pointed quality of a
feminine chin.

Armed with this research, and with information from physical-
anthropology textbooks, he operated on the trans patient. He
did not reduce her brow ridge, which he considered to be within

the bounds of feminine physiognomy, but instead added medical-grade plastic into the concavity above it, giving her a smoother, more rounded profile. The surgery, which took four hours, was aggressive. When it was done, the nurses joked darkly with the patient that the doctor must have punched her really hard. But after the full recovery period of a month, the transformation was impressive. Minute changes in the brow—a matter of a millimeter or two—had brought about dramatic results. The patient felt that she looked pretty and feminine: like herself, or the self she wanted to be.

Word spread in the trans community of Ousterhout's work, and a trickle of patients turned into a steady stream. He started attending conferences and gatherings for trans people, giving presentations and offering consultations. His patients invariably had means, because the procedure, unlike genital surgery, was considered cosmetic by insurers and therefore not covered by them. Most insurance companies still classify facial feminization, which can cost as much as $60,000, as an elective surgery. A few insurance companies, such as Blue Cross Blue Shield of Massachusetts, now cover some facial-feminization procedures.

Over the years, Ousterhout added several elements to his repertoire. First he began offering nose jobs, having realized that, when patients' brows were altered, it threw their noses into new, and sometimes unflattering, relief. Many of his innovations were developed in response to patients' requests. After an airline pilot whose forehead he had feminized complained about the size of her jaw, he devised a technique for reducing the lower jaw without damaging the sensitive bundle of nerves that extend to the chin. For those who wanted a smaller chin, Ousterhout developed a method of excising sometimes more than a centimeter of bone from the lower face.

The anthropologist Eric Plemons spent a year observing Ousterhout's practice, and recently published a book, *The Look of a Woman: Facial Feminization Surgery and the Aims of Trans-Medicine.* He argues that Ousterhout not only honed a set of techniques; he also developed a theory of gender difference. Ousterhout came to believe that, for trans patients, the most meaningful surgical intervention they could undergo was not genital but facial surgery. Few people you meet see your genitals, but everyone sees your face, and instantly makes assumptions about your gender, based on a subconscious assessment of your features. (Trans men typically have an easier time signaling their gender: testosterone therapy induces the growth of a beard, or the development of male-pattern baldness, and though trans men are sometimes of smaller stature, a short man is hardly viewed as remarkable, in the way that a very tall woman can be.)

Ousterhout initially sought to bring his patients within the middle of the femininity range that he had established through his research into facial shapes. But as he became known as the leading authority in facial feminization—a field that was rapidly being populated by other surgeons—his surgical interventions became more extensive. He gradually came to believe that he should try to make his patients look not just like average women but like beautiful women. In part, this was to counterbalance common masculine traits that a trans patient cannot alter, such as the size of her hands. But Ousterhout's decision also had the effect of upholding certain cultural assumptions about what is beautiful or feminine. As Plemons, who is trans, writes, "*Feminine* is a term in which biological femaleness and aesthetic desirability collapse." At the very least, Ousterhout wished to enable his patients to open the door to the UPS guy in their sweatpants, without the armor of makeup or careful hair styling, and be perceived as female. But he also believed that he had the ability to give his patients a face that emulated a feminine ideal.

Not everyone in the trans community sees facial feminization as offering unalloyed benefits. "Passing" can be a fraught notion for trans people, much as it has been for people of color. For some, facial feminization is seen as bolstering restrictive stereotypes while stigmatizing gender nonconformity. And given that the surgery is too expensive for most trans women, it has been criticized as perpetuating what Plemons calls "an embodied form of woman that was idealized by many but available only to a few." The sociologist Heather Laine Talley, in her 2014 book, *Saving Face: Disfigurement and the Politics of Appearance*, has argued that "facial feminization relies on and reproduces essentialized notions about what distinguishes a male face from a female face." From this perspective, facial feminization may offer the individual who undergoes it a reprieve from prejudice, but it may also reinforce the broader oppressive structures that leave trans people disproportionately vulnerable to discrimination and violence.

The debate about the extent to which facial feminization reinforces regressive stereotypes is not limited to academics; it has also permeated popular culture. Caitlyn Jenner's acknowledgment, in 2015, of having undergone facial feminization brought unprecedented attention to the process. Laverne Cox, one of the stars of *Orange Is the New Black*, who is also a trans activist, objects to the imposition of conventional beauty standards on trans women. On her blog, Cox has argued, "There are many trans folks because of genetics and/or lack of material access who will never be able to embody these standards. More importantly many trans folks don't want to embody them, and we shouldn't have to to be seen as ourselves and respected as ourselves." A few years ago, Cox launched the hashtag #transisbeautiful, explaining on her blog that she wanted to "celebrate all those things that make trans folk uniquely trans." She has spoken of being grateful that, by the time she could afford facial-feminization surgery, she no longer wished to undergo it.

Trans people who do not feel inclined, or able, to wage a political battle—who just want to avoid feeling self-conscious when out in public or dismayed when looking in a mirror—may especially desire facial-feminization surgery. *Gender dysphoria* is the technical term for that intense feeling of not-rightness. The most recent update to the *Diagnostic and Statistical Manual of Mental Disorders,* which was published in 2017, contains an important revision, stipulating that gender nonconformity should not be categorized as a mental disorder—a stigma that the trans community has long sought to counter. But the manual also indicates that an individual's distress over feelings of gender nonconformity can be classified as a disorder, which can be managed with a range of potential treatments, including therapy, hormones, and surgery. Gender dysphoria can be understood as a severe dislocation between one's inward sense of self and one's outward appearance. Many women are unhappy with their bodies—in Western consumer society, to be unhappy with some aspect of your body might almost be thought of as a normative condition of womanhood. But hating your thighs is not the same as feeling that your thighs do not belong to you.

Trans people can experience dysphoria about more than just their faces; they can detest their shoulders, or their arms, or their genitals. Some of Ousterhout's patients had already undergone genital surgery; others planned to have it afterward. But many of his patients felt that facial surgery was enough. And, for all his patients, having more feminine faces meant that they were less likely to find themselves the focus of invasive, prurient interest, from strangers or even from friends, about the state of their genitalia. This deeply personal matter would more easily remain private. It would be nobody's business but theirs, and their intimate partners'.

Some of Ousterhout's patients came back for multiple interventions. His very first trans patient, from 1982, returned at one point to have a cleft in her chin removed: she considered it to be another deformity that read as masculine, like Tom Brady's. The final procedure that Ousterhout performed before retiring was a second reduction of her Adam's apple. Ousterhout liked to listen to a classical radio station as he operated, and as he was finishing the procedure his favorite piece of music, Wagner's "The Flight of the Valkyries," came on—an auspicious conclusion to his surgical career. The patient, who is now approaching seventy, intends, upon her death, to donate her skull to Jordan Deschamps-Braly.

Abby Stewart first called Deschamps-Braly's office last fall, and when she was told that a surgical appointment had opened up for later in the year, she flew from Denver to San Francisco for a consultation. They met at his office, in an art deco tower close to Union Square. On the window ledge behind his desk, Deschamps-Braly keeps a model of a skull that delineates, with bronze rods, the proportions of the golden ratio. Unlike many surgeons, Deschamps-Braly does not conduct consultations on Skype, insisting that only by seeing and touching a patient's face can he offer a proper assessment. Nor does he show before-and-after images on his website, feeling that they betray patients' privacy. Deschamps-Braly was not the first surgeon Abby had visited, but she felt confident in his manner, and also in his surgical recommendations. Deschamps-Braly liked Abby, too—he found her personable, admired her intelligence, and knew that her background in science would give her a realistic understanding of what was possible. He felt that it was as important for him to choose his patients as for them to choose him.

Abby had been on hormones for more than a year by the time she visited Deschamps-Braly, and their effects had been profound. Not only had her skin softened and her cheeks filled out; she had developed breasts, and her hips had widened. For many years, Abby had been somewhat overweight and out of shape, but after she moved to Colorado, in 2014, she became a frequent hiker and rock climber. Getting to know other women who enjoyed bouldering had helped her come to feel that there was a model of femininity to which she could comfortably conform: strong, athletic, rangy.

Despite the physical changes wrought by the hormones, Abby continued to suffer from a profound self-consciousness about her face. She felt that when she was seen from the front she looked persuasively feminine, and even striking, with abundant hair that framed her face, and wide-set eyes. But when she turned her head she looked far more masculine: the bossing of her brow showed in profile, as did the length of her jaw. She was so conscious of her Adam's apple that she tucked her chin down, to conceal it, and refrained from turning her head, looking to the side with only her eyes. With her dipped head and her inhibited range of motion, her mannerisms became those of a demure Victorian.

Abby's self-consciousness in the company of others was nothing compared with the unhappiness she felt when faced with her own reflection. Whenever she passed a mirror, she saw the ghost of her former self, and it appalled her. Though Ousterhout had developed his procedures on the premise that his trans patients wished to move through the world without attracting unwelcome notice, Abby's desire to undergo the process was more interior. The person whose reaction to her face she most wanted to change was herself.

Abby was not among those trans people who knew from early childhood that their gender identity did not line up with the gender they'd been assigned at birth. As a child, she was aware only that something was off. She was a shy kid, and for a long time she had a stammer, but she was an excellent student. Although she lived in a conservative part of the South, her parents were relatively liberal, and her childhood was quite sheltered. When she was taunted in middle school with the label of *gay*—then the all-purpose slur for any form of gender nonconformity—she knew only that the word meant "happy."

When Abby eventually discovered what *gay* meant, she knew that it did not apply to her. Yes, she was shy and emotional, attributes that are not stereotypically masculine, but she was definitely attracted to women. Her interest in them had an additional dimension, though. When she played video games, she often chose a female avatar. She wasn't alone in doing that—boys she knew also did it, because it was more fun to watch a virtual figure in a bodysuit—but later she recognized that her motivation was different. She really wanted to *be* the girl. Bette, Abby's mother, later racked her brain, trying to find clues she had missed about Abby's gender identity, but she came up blank. She knew only that she had an unusually sensitive and intuitive child. Whereas Abby's younger brother came home from school and talked about his day for just a few minutes, Abby sat down with Bette for an hour or two every night, and shared all sorts of things. What a beautiful young man, Bette thought: one day, he is going to make some woman very happy, because he thinks like a woman.

Abby went to college in the Midwest, where she fell in with a group of progressive students, and joined campaigns for LGBTQ rights. In her sophomore year, she had a relationship with a female student. She felt that many of her peers secretly questioned the romance, and were waiting for her to come out as gay. She resented the expectation, while feeling uncomfortable about why the suggestion made her so uncomfortable. Abby forestalled a reckoning, telling herself that, for many young adults, feeling uncomfortable with one's body was just part of life.

After graduating from college, she began a PhD program in biology, but found that laboring alone in a laboratory was not for her. She completed a master's degree, and eventually moved to Colorado, both to seek out teaching work and to have the space to figure out her identity. Many trans women go through a masculinization phase before coming out as trans—sometimes by joining the military, sometimes by bulking up through exercise—in the vain hope that, by embracing an extreme of masculinity, they will find relief from the pressing sense of not-rightness. Abby didn't do this, exactly, but she did become fitter and healthier. She met Sofia, a college student studying math, through rock climbing, and in many ways she felt better about herself than she had in years. But the things that were going well in her life made starker the things that were not.

Abby had begun to find a framework for understanding her identity, in part because of the emergence of trans celebrities. And on the internet trans people chronicled their transitions or posted videos about their personal histories. For Abby, it was empowering to discover that she was not alone with these thoughts. Finally, she mustered the wherewithal to acknowledge to herself that she was trans. One Saturday morning in February 2016, she wrote a twelve-page entry in her journal, and that evening she gave it to Sofia, who read it in bed, with Abby anxiously curled at her side. When Sofia finished it, she told Abby that she loved her, and would support her.

Abby was the kind of person who, once she put her mind to something, was all in. In April, she began hormone therapy. The first effects were mental: she felt a remarkable clearing of the mind. (Other trans women describe a sense of relief that comes from eliminating their masculine sex drive, which had previously interrupted their thoughts like a car alarm that couldn't be shut off.) Abby's skin became more vulnerable to bruising. She was thrilled to see the muscles of her back melt away, revealing the sculptural plates of a woman's scapula.

In the summer of 2016, she went home to see her parents. When she announced that she had something to tell them, Bette thought that Abby was going to tell them she was gay. (Abby later noted, dryly, "I am.") When Bette heard that Abby was trans, she went through a period of mourning, and she worried about Abby's safety and happiness, especially after she read of high rates of suicide among trans people. But the more reading that Bette did, the more she became convinced that the statistics were largely the result of trans people being marginalized by their communities, or rejected by their families. Bette made every effort to understand her child, and, if out of habit she sometimes referred to Abby by the masculine pronoun, she was contrite, and was forgiven.

In the fall of 2016, Abby legally changed her name. Early in her transition, she experimented with wearing makeup and dresses —and she immortalized her first blowout on Instagram—but she preferred not to put on lipstick and foundation, and was most at ease in jeans or workout clothes. She transitioned at work when the new semester started, in January 2017, and was grateful that the institution was supportive. She also felt that she should be a role model for students who might be questioning their own gender identity. She accepted this as a responsibility to her community, but it was a lot to take on while she was relearning how to move through the world: how to dress, how to carry herself, how to modulate her voice.

By the summer of 2017, Abby had crossed an invisible line, and most new people she met correctly gauged her gender, calling her "Ma'am" or "Miss." It was exhilarating to go into a restaurant with Sofia and hear the maître d' address them as "ladies." Anxiety over which public bathroom to use—the subject of legal victories during the presidency of Barack Obama, and of disheartening reversals during the presidency of Donald Trump—abated with time. But Abby remained intensely pained by the vestiges of masculinity in her face. She scrutinized her appearance for changes wrought by estrogen with the obsessive focus of a teenager preparing for a date. In many ways, she *felt* like a teenager—as if she were again in the throes of puberty, though this time on her own terms.

For Abby, transitioning medically was accompanied by some sorrowful choices. She loved children, and had always wanted to have her own. The hormones she was taking would cause irreversible sterility, and she had opted not to freeze and store sperm, as some trans women do, because the long-term cost seemed prohibitive. She believed that she might eventually adopt a child, but when she thought about the fact that she would never experience pregnancy she felt a great sense of loss. But, she decided, if she could not give birth to a child, she could engineer her own rebirth.

About a week before Abby underwent facial-feminization surgery, she experienced one of those dreams that stay with you long after waking, like a prophecy. In it, Sofia sketched a portrait of Abby, and held up the finished work. The woman in the image was extremely feminine, like a model, and Abby felt a jolt of recognition. She thought, That's me.

Deschamps-Braly performs surgery three days a week, and spends the remaining two days on consultations and post-op care. In his very limited spare time, he likes to visit the racing track, to drive his Porsche. He enjoys the speed, but what he loves most is the sense of control—of mastering the interplay of acceleration and turning. Although racing is stressful, he finds it oddly relaxing. As in surgery, you have to contend with many moving pieces, and execute each motion smoothly and without interruption.

On the morning of Abby's surgery, Deschamps-Braly drove from his home, which is close to his office, to the California Pa-

cific Medical Center's Davies Campus, between Lower Haight and
the Castro. He arrived just before seven. Abby was already there,
on a gurney, in a violet-colored medical gown, her hair brushed
back from her face. Deschamps-Braly was dressed more like an
artist than like a doctor, wearing a chunky marled-wool cardigan,
narrow black jeans, and pointy black shoes with silver buckles, his
curly dark hair askew. He chooses not to wear a suit, believing that
it could be intimidating to patients. He carries a Prada briefcase.

Deschamps-Braly stood next to Abby on the gurney, and stroked
her arm. He assured her that, once they entered the operating
room, she would fall asleep quickly, and rouse seven or eight hours
later without any sense of time having passed. "Easy-peasy, like
falling off a log," he said, adding that the first stage of recovery
"shouldn't be painful." Because of swelling, he noted, "it will be
horribly uncomfortable, but not much pain."

He asked if she had any musical requests before she went under. Tupac, Abby said. Any particular Tupac song? he asked. That
was a joke, she said. In the event, the Beatles were singing "Here
Comes the Sun" when Abby was wheeled into the operating room.
On the wall, a lightboard displayed her X-rays, and several other
images. There were two professionally lit photographs of Abby,
one from the front and one in profile, which looked like unusually
glamorous mug shots. (They had been taken in Deschamps-Braly's
office.) Next to them were the same images, slightly altered with
Photoshop, to show what Abby should look like post-op. There
were also schematic diagrams indicating what surgical alterations
Deschamps-Braly needed to make: a centimeter off her chin, two
millimeters off her nose.

Deschamps-Braly, who by now was wearing blue scrubs and maroon Crocs, bent toward Abby and whispered in her ear as the
anesthesiologist, Caroline Dejean, put her to sleep, and the others on the medical team readied surgical implements and counted
screws and wires.

A physician's assistant, Zhanna Byalaya, lifted Abby's head and
placed it gently on a silicone ring, like a pillow. Abby's statuesque
shoulders were bared, and she looked like a funerary memorial of
herself. To prepare her face for surgery, Deschamps-Braly razored
away a band of her hair. An adrenaline solution was injected along
the hairline, to minimize bleeding, and her face was swabbed with
the disinfectant Betadine, which gave her features an orange glow.

With a few sutures, Deschamps-Braly sewed Abby's eyelids shut, to ensure that her eyes remained moist and protected.

Deschamps-Braly rolled up to the gurney on his preferred stool, which was labeled DR. D'S THRONE and marked with a skull-and-crossbones sticker. (Apparently, it's hard to find images of skulls alone.) The first procedure to be performed on Abby was a reduction of her Adam's apple. Deschamps-Braly made a small incision just under Abby's chin, to minimize the visibility of the inevitable scar, then pulled the skin apart to make her voice box visible. He scraped away at the larynx with his instruments, but after a while things got tricky: the cartilage had hardened into bone. "I don't think we can go any further without putting her voice at risk," he said, closing up the incision. (Deschamps-Braly has found that trans men are much less concerned with acquiring an Adam's apple than trans women are with getting rid of one. Nevertheless, he and Ousterhout jointly developed a procedure for building a new Adam's apple for trans men. In a detail that even a Hollywood scriptwriter might deem too much, it is made from cartilage extracted from the patient's rib.)

Around half past eight, Deschamps-Braly moved on to Abby's forehead. In a single deft movement with a scalpel, he sliced from the middle of her hairline to above her right ear. He repeated this unzipping motion on the left side. Then, working along the incision, he separated the skin and the subcutaneous layer of her forehead from her skull and peeled it toward her nose, as if he were removing the rubbery skin of a mango from the yellow flesh inside. Blood pooled at Abby's ears, and it was suctioned away. As Deschamps-Braly folded the loosened skin forward, over her closed eyes, he could see the bony contours of Abby's brow, and the bossing above her eyes.

The mood in the operating room was upbeat—intense and focused but also informal. Deschamps-Braly and Dejean talked about upcoming vacation plans. (He was headed to Paris, she was going to New York City.) When he operates on children, as he does about three times a month, there is a greater sense of urgency: their small bodies cannot easily handle a considerable loss of blood, and it's important to finish surgery quickly. With an operation like Abby's, there is a different pressure—that of achieving excellent craftsmanship.

Once Abby's brow had been exposed, at about 9:00 a.m., Deschamps-Braly began reducing her eye socket, the top of which was exposed to the air, with the top of her eyeballs visible. The procedure, for which he used a small drill, would give Abby more open, less hooded eyes. Deschamps-Braly got up to check the diagrams on the lightboard, then returned and used the same drill to burr away the upper part of her forehead, above and around the bony protrusion that covered her sinus cavities and gave her brow a distinctive masculine ridge.

Some surgeons prefer to reduce the forehead only by burring it, but Deschamps-Braly, like Ousterhout before him, is committed to procedures that yield more dramatic results—drama, in the context of facial feminization, being measured in millimeters, or in fractions thereof. Deschamps-Braly marked the problem area, six centimeters by four centimeters, and using a reciprocating saw he sliced the piece of brow bone off, placing it on a side table. In a few minutes, he would reshape the bone and then reattach it. For the moment, though, he resumed burring her forehead bone, grinding certain areas until they were only a millimeter thick. To examine his handiwork, he gently restored the skin of Abby's forehead to its proper place, smoothing it down with his hand to see whether the more rounded shape was emerging. "It's better," he said.

Deschamps-Braly then turned to the brow bone on his side table. It was the color of raw squid. In some cases, the piece could be restored intact, but angled in a way that produced a flatter profile. Abby's brow was quite prominent, however, so Deschamps-Braly cut the bone into four pieces, with the plan of reconnecting them into a more refined shape. It was important that the edges be carefully aligned. When operating on the skulls of children, he could be less precise: young bodies easily generate more bone, filling in minute gaps left by a surgeon. But with adult patients Deschamps-Braly aimed for something closer to marquetry. When he had shaped the four pieces to his satisfaction, he joined them with stainless steel wires, then placed the reconfigured object back on Abby's brow. He reconnected it swiftly to her skull with the twisting of more wires.

It was now ten thirty. Deschamps-Braly began work on a scalp advancement—bringing Abby's hairline forward by five millimeters. He loosened her scalp from her skull, exposing the front part of her head. Abby's skull, a vulnerable, bloody orb, looked like the head of a newborn. With three tiny sutures, Deschamps-Braly reattached the scalp, moved slightly forward, to her skull. Having folded Abby's forehead back up, so that her face was visible again, he sliced a ribbon of flesh off the top of her brow, then stitched along her hairline, rejoining the scalp and the forehead.

He left the operating room for a break in the cafeteria, where he had a Coke and some peanut-butter crackers. Byalaya washed blood and bone dust from Abby's hair, braided it, and wrapped it in a towel, as if they were at a luxury spa.

When Deschamps-Braly returned to the operating room, he began work on Abby's jaw and chin. He cut a trench inside her mouth where her cheek joined her gum, avoiding the bundle of nerves, like electrical cables in a basement, that supplied sensation to her chin. Inserting the reciprocating saw through this passageway, he burred away Abby's jawbone. Then, inserting a drill through an incision under her jawline, Deschamps-Braly cut off the angular corners at the rear of her jaw.

Deschamps-Braly always sought to get a patient's face within range of what he had determined, based on Ousterhout's research, to be the feminine ideal. The hairline and the brow ridge, Deschamps-Braly felt, should be five to six centimeters apart; the eyebrows should be about 65 millimeters from the tip of the nose; the nasal septum should be about 15 millimeters from the upper lip. (In the typical resting female face, more of the upper teeth are exposed than in a resting male face.) Deschamps-Braly wanted the distance from the tips of Abby's incisors to the base of her chin to be about 40 millimeters, which meant that he needed to reduce the length of her lower face by about a centimeter. He would also bring her chin forward by 6 millimeters, to create the desired harmony. Byalaya pulled down the skin of Abby's lower face, and Deschamps-Braly used a sterilized pencil to draw a T shape onto the bone of her chin. Then he began cutting along his pencil marks.

There was a burning smell as Deschamps-Braly extracted a thick crescent of bone and set it aside. Abby had asked to keep it. He set about restoring her chin with a custom-measured titanium plate, cinching the bone together around the excision to close up the gap and thus recontour the chin's shape. The chin finished, Deschamps-Braly took another short break, then returned to do the most pedestrian procedure, a rhinoplasty, which he started at about 1:30 p.m.

The nose job was also, arguably, the most purely aesthetic intervention that Deschamps-Braly was making. A well-shaped nose might be a desirable thing in a woman, or a man, but it is not necessarily a marker of gender. Deschamps-Braly, however, had no qualms about admitting that, though his official brief was to make his patients look more feminine, his goal was also to make them look prettier. In addition to the techniques he had learned from Ousterhout, he had added one of his own: fat injections to the cheeks, which give a fuller, more youthful aspect to the face. In recent decades, fat injections have been widely adopted by conventional cosmetic surgeons, as a way of making patients look younger without giving them the stretched appearance that can result from a face-lift. In trans patients of any age, fat injections can help feminize the face—a feminine face is typically less chiseled than a masculine one. They can also beautify a patient, giving her a plumped, dewy look.

Abby had decided against fat injections. With Sofia, she had spent long hours talking about stereotypes of femininity, and she was ambivalent about her desire to conform to them to the extent that she did. She thought that conventional ideas of beauty were oppressive to women and men alike. In the end, she decided to alter only the features over which she experienced dysphoria, and to forgo interventions that were simply for the sake of aesthetics. Her cheeks did not bother her, so there was no need for added fat. But Deschamps-Braly had convinced her that a rhinoplasty would harmonize the adjustments to the jaw, the chin, and the forehead. She wasn't happy that getting a nose job might be seen as an expression of vanity. For Abby, a major reason for wishing to look decisively feminine was a desire for personal safety: she was aware that violence and hostility against trans people remained prevalent. She thought that there was a strange cultural dichotomy at work: individuals were celebrated for self-actualization but judged superficial if they crossed some imperceptible threshold.

After Deschamps-Braly made an incision between Abby's nostrils and peeled back her skin to reveal the structure of her nose, Byalaya handed him a small hammer, and he chipped away. "Can I get a ruler?" he asked. Having trimmed off a piece of cartilage, he slid it back into her nose, to the side of her septum. In case she didn't like the results of the rhinoplasty, another surgeon could use the cartilage for a restoration. It was like keeping a spare tire in the trunk.

Deschamps-Braly finished up just before three in the afternoon, snapping off his surgical bib as the Beatles sang "With a Little Help from My Friends." He removed the sutures from Abby's eyes, and crouched beside the gurney to survey her profile. Abby's face would soon be largely covered up, and would remain that way for the next week: plaster would be put over her nose, and her forehead and jaw would be wrapped in compression bandages. But for a moment, before the worst of the swelling set in, Deschamps-Braly could see the face that Abby was going to have: a smooth forehead, a delicate chin, an aquiline nose.

He left the surgical theater and descended to the ground floor, where Bette and Sofia awaited. "Nothing unusual happened," he reported.

In Deschamps-Braly's experience, patients who underwent facial feminization often felt an almost immediate relief from dysphoria. When the plaster and bandages came off, a week after surgery, they recognized their new faces as themselves. What varied was the patients' self-image. Deschamps-Braly had some patients who had no desire to hide their trans-ness, including Adrian Roberts, a DJ and party host who identifies as nonbinary trans-feminine, and who goes by the pronoun *they*. Roberts wanted to be seen as more femme, but not as female. Deschamps-Braly gave Roberts a different forehead, nose, and chin, and fat injections. Roberts said, "More than your body, your face is what makes you *you*." Another patient, Autumn Trafficante, a programmer and a trans activist, viewed Deschamps-Braly as an artist, and told him that he should do whatever he thought best with her face. Trafficante had previously presented as a model-handsome man, with a scruff of beard and a ripped torso. When Deschamps-Braly handed her a mirror, a week after surgery, she broke down in tears of relief: for the first time, she registered her face as female. On her Instagram account, Trafficante offered a visual history of her transition, culminating in sloe-eyed glamour shots of herself in lingerie.

Abby was fascinated by images like these, but she was more modest in temperament. Her Instagram account, which she kept private, showed her and Sofia hiking or climbing in the mountains, ruddy and grinning in the sunlight. When she scrolled back to look at herself in 2015, before she had begun to transition, she could see that the changes wrought by the hormones alone were remarkable. But she had sought facial-feminization surgery in order to stop thinking about her gender, not to draw attention to it.

Abby was well versed in gender theory, and could talk at length about the ways in which gender identity is culturally constructed. She knew that trans individuals occupied contested ground, and that there were people who would argue that, for three decades, she had enjoyed the benefit of male privilege. She questioned, though, what privilege lay in having been obliged to conform to a gender expression she loathed and rebuked. Abby was thoughtful and sincere in her commitment to feminism and intersectionality —the idea that gender issues cannot be divorced from matters of class, race, and disability—and she wanted to be helpful to other trans people who might not have had the same educational and cultural advantages she'd had. She had even written a guide about the practical aspects of transitioning, such as changing your name and updating legal documents, for others who might be following a similar path. She was determined to be an activist and an ally in the trans community. She also wanted to just be herself.

When Abby arrived at the hospital for her surgery, she had an experience that underscored the urgency of her need. She was dressed in jeans and a plaid shirt, knowing that she would soon be changing into a surgical gown, and she wasn't carrying a handbag. As instructed, she wore no makeup or jewelry, and had her hair tied back. As she entered the building and rode the elevator to the surgical ward, two hospital employees separately referred to her as "Sir"—something that had not happened to her for months. At the time, she was too focused on the coming operation to pay much attention, but later she reflected on the occurrence. Having stripped away all the clues that she usually provided to indicate how she wanted to be perceived, she was left with bare physical characteristics that read as male. When you take everything else away, you have just your face. That, for her, confirmed that she had made the right choice.

After Abby's surgery, she was moved to a private room. She felt profoundly weak, and when nurses encouraged her to stand she could hardly rise to her feet. She could not speak above a whisper, and she was grateful that Sofia so intuitively responded to her needs, knowing when she needed to rest, and coaxing her to drink milk from a straw. It was spooky how in sync they were.

Little of Abby's face was visible, and what was visible was very swollen, but whenever Sofia or Bette held up a mirror, or took a photo, Abby could see that her eyes were already different: wider, more open. But Abby was not concentrating on her face; she was concentrating on her recovery. She was in a great deal of discomfort—she felt pressure on her forehead, and numbness in her scalp and jaw. She couldn't shower, or breathe through her nose. When she was released from the hospital, the next afternoon, and went back to her hotel, she was too self-conscious to go far from her room.

Eight days after the surgery, she returned to Deschamps-Braly's office to have the plaster and bandages removed: the "great unveiling," as he called it. She lay on a reclining medical chair in his consulting room, and he carefully removed the dressing. Although Abby had heard the stories of patients who were emotionally overwhelmed at the sight of their new faces, that was not her experience. For her, it was an anticlimax. She was in survival mode: she just wanted to heal, and to get a good night's sleep again. She had a moment of seeing herself and thinking, That's the new me, but mostly she wanted to know when she could take her next pain pill.

Abby was relieved when she was told that she could go home to Colorado. Deschamps-Braly gave her a document to take to the airport, which explained why she no longer looked like the photograph on her driver's license. He also gave her a transparent plastic container covered with a biohazard sticker and a word scrawled with a Sharpie: BONES.

In mid-January, after six weeks of recuperating at home, Abby returned to teaching. At first, she was embarrassed about the lingering puffiness around her jawline and on her forehead, although the latter, especially, was barely noticeable to the casual observer. Before the surgery, she had told a few close colleagues exactly what she would be doing over the break, but mostly she used a phrase that Sofia had come up with: she had undergone a corrective surgery.

Before the operation, Sofia had been worried about Abby's coming transformation, fearing that she might no longer recognize the face she had fallen in love with. And Abby's face *had* changed, and would change more over the next six months, which is how long Deschamps-Braly had told her it would take for everything to settle. Abby's jaw was less angular, her chin more tapered. Her eyebrows arched above expressive eyes, and her brow, despite the residual puffiness, was smooth. In some ways, she looked like an entirely different person, but she was also instantly recognizable: she had the same ready smile, and radiated the same quiet intelligence. Sofia said that she had fallen in love with Abby's old face, and she was in love with Abby's new face. But what she meant was that she was in love with Abby.

Abby herself was still adjusting to life with her reborn face. She was not spending much time outdoors, having been told to avoid the sun as she healed. She felt sluggish, and was eager to get back to exercising. But her facial dysphoria had vanished. The reduction of her Adam's apple was such a relief—she had immediately stopped tucking her chin down and had gone back to turning her head. She would run her hand over her throat and think, Holy crap, it's gone. In the months before the operation, she had become anxious and withdrawn, and had avoided making eye contact when she was out and about. Now she was back to striking up conversations with strangers in the grocery store. And whenever she glimpsed herself in the mirror she no longer saw the ghost of her former self.

Abby kept the transparent container in a drawer in her bedroom. She had yet to figure out exactly what to do with the pieces of bone that Deschamps-Braly had removed, and thought that she might have them turned into jewelry. Perhaps she'd put the fragments in a vial, and wear it on a chain around her neck. The largest piece was the section that he had carved so deftly from her chin. It was the shape of a waxing moon—nature's perfect, elegant curve. When Abby held it in her hand, it already looked like a piece of jewelry: precious, beautiful, superfluous.

MOLLY OSBERG

How to Not Die in America

FROM *Splinter*

ON THE SECOND Tuesday in June, I start to feel fluish. If this is 2016 and I'm still a freelance writer, I'm losing money immediately on the assignments I can't complete because my vision is blurry and my thoughts are erratic. If this is 2013, I am soon taken off the roster at the café where I work.

I am out of my mind with anxiety as I hobble to the clinic, sweating, and pay $60 for cough syrup, $300 for the 10-minute visit (if I even have that in the bank; it's about a week's worth of my earnings slinging coffee). Once I realize I can't keep down the cough syrup and start spitting up bile, maybe I'm so feverish and broke I stay in bed without realizing the bacteria I've inhaled is more lethal than the flu. So perhaps I just up and die right there.

But let's say I somehow make it to the hospital. A friend drives me, because a 15-minute ambulance ride can cost nearly $2,000, which I don't have. I'm struggling financially and I've fallen behind on my ACA payments. My friend realizes in the car I'm not making any sense, and that's because my organs have already begun to shut down. My temperature is well over 100. When the doctors can't figure out what's wrong, they submit me to a credit check before advanced treatment.

My credit is awful. I have a massive, unpaid bill from a few years back when someone made international calls on my stolen phone. Maybe, because of this, I'm transferred to a public hospital, where there aren't 20-odd specialists to arrange an "unusual" surgery. Doctors are required to stabilize a patient, but they aren't required to, say, stabilize a patient just long enough to keep them breathing and take them to another hospital with a full infectious disease wing to do something risky. So maybe that's when I die, before they even figure out what's wrong, because I'm not the type of patient whose financial health can support an elaborate, life-saving procedure.

But even if the hospital could be convinced to ignore my distinct lack of liquidity, in one of these alternate timelines I don't have a parent with the time and language skills and resources to come down to New York and negotiate with doctors who need a legal surrogate to parse a series of difficult options. It's not like I can do it myself, in a medically induced coma. And already, two days after being admitted, I am racking up bills for anesthesia, the input of six specialists, radiology, and antibiotics that come to nearly $30,000. And without the treatment, which costs an additional $12,705 for just for a few hours of the surgeon's time, I am dead.

Let's imagine, though, that I get lucky and my mother makes it to New York in time. She demands they do anything within their power to save me and puts up for the surgery, using her own credit. She convinces the paper-pushers she's good for the bills. I am, after all, her only child. I'm in the ICU for 10 days; the baseline cost can be up to $10,000 a night, which doesn't include the ventilators, the sensors, the multiple IV drips jacked directly into my neck.

By the time I'm out of the hospital, we have been billed $642,650.76. If this is a few years earlier, the well-regarded medical center where I have just spent nearly a month is flat-out refusing requests for financial aid, sending bills for emergency surgery to a collection agency that puts liens on the homes of patients' families and forces them to foreclose. I'm probably not aware this is happening until I'm back in my apartment, on a three-times-daily schedule of antibiotic IV treatments, which have to be administered by a home nurse. She's expensive.

In this version of the story, I have survived, but been without a paycheck for the better part of the summer. Around the time I run out of oxycodone and start waking up in tears, completely paralyzed by pain, the medical bills have begun to pile up on my stoop. I am woozy, spending long, featureless days in bed, trying to remember what kind of person I had been before I went under, and I need help to raise my scooped-out torso from bed. I can't lift anything or cook for myself or walk more than a block. Maybe

I fester for awhile in a rehab center, in the absence of there being anyone readily available to make sure I don't waste away.

Never mind recovering physically or financially in any of these scenarios: I can't imagine surviving emotionally, fielding calls from collections agents, facing eviction, waiting for the pain meds to hit so I can keep at a futile job search with an IV still dangling from my side. I am 29 years old, with no preexisting conditions before this moment, and I am unemployed and exhausted and in pain all the time.

Of course, this is not what happened to me. I am not one of the 28 million Americans who are completely uninsured, or one of the 45,000 people who die every year for lack of coverage. I am not one of the three-quarters of US citizens who don't have access to paid sick leave, and I don't live in one of the 45 states without short-term disability plans. I'm not one of the 30 percent of insured consumers who are slapped with hefty surprise bills after a hospital visit. Which is why I did not become one of the millions of people who default on their medical debt every year, regularly making healthcare bills the leading cause of bankruptcy in this country.

Instead, on that second Tuesday in June 2017, I found myself in what I worry could be a fleeting moment in my life, one in which the institutions around me find it advantageous to protect rather than screw me. I find it baffling that, since my illness, well-meaning people have repeatedly referred to me as a "survivor," as if the fact that I got to go on with my life had to do with some inherent moral strength, rather than the material forces put in motion long before I got sick.

That whole week last June, I worked from home, assuring my editor I'd be back in the office shortly. "I haven't been this sick since I was, like, eight years old dude!" I told her over Slack, the same day I would be put into quarantine. We have generous time-off policies in our newsroom, thanks in large part to our union. I was told to rest and get better. I didn't.

A major feature of a person's 20s is that while you're ostensibly old enough to take care of yourself, you haven't really lived through enough to be cautious. As someone with a pack-a-day habit, I got a little sick every year, and my response was to sleep (or work, or drink) through it until the issue somehow resolved. Before 2017 I don't think I'd been to a doctor in about five years

—though as was later reiterated to me by one chagrined specialist after another, my abysmal life choices up to that point didn't end up making much of a difference.

Either from feverishness or denial, the conviction that I was simply a little sick remained well past any logical point. I have health insurance now, but I've spent portions of my life without it, and with my organs sputtering and a spiking fever I wasn't exactly thinking clearly. I thought of the friend stuck with thousands of dollars in bills, picked up by an ambulance after a bike accident. Call it a reflexive reaction to living in a country where one-third of the population delays medical care out of concern for the cost.

After a weekend of still feeling off-kilter, I went to a clinic for cough syrup, which I quickly found I couldn't keep down. Two days later, sweaty and half out of my mind, I stumbled back to the same clinic. I'd tried to brush my teeth and ended up spewing bile. The doctor didn't even examine me: "The only way you're leaving here is in an ambulance," she told me. She wanted to call the hospital: I protested, still of the opinion that a hospital visit would be too pricey. I called a friend, who called a car, and we drove to the closest ER.

The sheer number of doctors involved in emergency medical care—especially when the cause of illness is unknown—can be staggering. Often, once a person is sick enough not to know what's going on, an army of costly specialists are called in. Doctors with the highest out-of-network markups are the ones patients are unlikely to choose themselves: pathologists, anesthesiologists, emergency medical doctors.

I encountered them all when I was admitted to a Brooklyn hospital, though I only remember a fraction of the people who tried to puzzle out what was sending my body into septic shock. According to my medical records I saw six specialists in 40 hours. There was the anesthesiologist who assuaged my terror when I was put under for a bronchoscopy (I was afraid I'd wake up with a camera down my larynx), the infectious-disease doctors who asked me about my sex life, how often I got high, my last period, whether I'd been anywhere near livestock.

For a while the theory was that I had toxic shock; they poked around for a tampon, didn't find one. There was a mass near my lung, but it wasn't cancer, despite what one doctor seemed to think. I was put into quarantine and texted a friend, desperately cheerful: "They're treating me like a patient in world war Z right now." The joke was short-lived when I started playing scenes from those films in my head. I added: "I'm jusyrinf nor to panic everyone like I'm dying lol."

I was having an awful lot of trouble breathing. My liver and kidneys had begun to shut down, and I was informed something was off with a valve in my heart. My hands and feet turned bright red. The unspecified infection winding through my bloodstream was similar to one common with intravenous drug users, so I testified to a stone-faced doctor at length about the kinds of drugs I do, and how often, and how recently. I got the distinct sense he thought I was lying. He asked about my mental health, which I found odd.

But when I was finally released into a small private hospital room, I wondered if I could actually believe myself. The room's green carpet and green walls looked all wrong, oddly monochromatic: I thought maybe I was going crazy, maybe I had in fact taken drugs. I was suddenly very afraid to be alone.

What I know about the next week is mostly pieced together from what I've been told and what's listed on some of the 100-plus insurance claims filed on my behalf. At some point, my retired mother arrived, and I told her that while I was glad she was there, she didn't have to come: it wasn't like I was dying (an objectively false statement I have no recollection of making). My boyfriend called my boss and told her not to expect me back at work anytime soon.

A week ago, I had been a healthy person in her 20s drinking in a friend's new apartment in Fort Greene. Now doctors tried to give me updates on the tests they were doing and I couldn't understand what they were talking about, even as I could objectively sense how corroded my processing power had become. I spent months of my life at the mercy of people with the capacity—and perhaps more importantly, the incentive—to care for me.

The hospital consortium I was admitted to has declined, of course, to comment on what options patients have when they can't immediately cough up cash. My surgeon, a lovely man who appeared completely elated by my recovery—and took a photo, before he operated, of the Ping-Pong-sized hole in my lung—told me recently that while the surgery he performed on me was perhaps "unusual," my health issue was "certainly high risk." He doesn't get involved in insurance-related issues, obviously; he has paper-pushers for that. "What I do know," he wrote in an email, "is that insurance can limit and make our lives difficult, as they commonly deny patients access and healthcare they deserve."

Sherry Glied, the dean of New York University's school of public service, put it a little more bluntly: under even the lowest Affordable Care Act tier, she says, I'd be paying the out-of-pocket maximum—a little over $6,000—and perhaps I'd be limited to in-network doctors. (The hospital I visited has maintained a relationship with an Affordable Care Act–affiliated insurance provider since 2014.) But depending on an individual hospital's policy, and my credit score, and where I happened to land, whisking me to a second hospital with a dedicated lung surgeon to be treated by some of the country's best infectious-disease doctors might have appeared to be more trouble than it was worth.

The surgeon at the second hospital told my mother I had maybe a 50–50 chance of surviving the surgery, but that if they didn't do it, I would certainly die. She signed all the paperwork and sent my boyfriend home and, though she's not a drinker, went directly to the bar, where she waited for the surgeons to remove a not-insignificant portion of my lung.

As they found when they tested the necrotic tissue removed from my chest cavity, I had contracted a relatively rare form of strep—the septic shock shutting down my organs had been kicked off by a ravenous bacteria that incubated in my lung before bursting a hole clear through it. In topical manifestations, this form of strep is referred to by some journalists as the "flesh-eating disease," feeding as it does on the fatty tissues right beneath the skin. In my case, it wormed its way into an organ and began to kill everything around it. The mortality rate for infections like these can be anywhere between 40 percent and 70 percent in adults. Many people lose limbs or parts of their face. Others require breathing machines or dialysis for the rest of their lives.

In the weeks that followed, medical residents—all *General Hospital*-level attractive, to my great surprise—clustered around with clipboards, asking hundreds of questions to place my illness and recovery within the continuum of their previous reports. "I'm just trying to understand how you got this," said one, after a 15-minute interrogation about my sex life. Months later, my infectious-disease doctor, who wore a different pair of brogued leather shoes every time I saw him, would tell me with a half smile it was actually probably just bad luck.

To compound the sheer sense of unreality and deeply unearned good fortune, I read a CDC study recently noting that for some reason, patients who contract similar kinds of severe strep infections are much likelier to survive during summer months.

When I was woken up from an induced coma a few days later — about a week and a half after I'd first gone home sick — I couldn't understand what everyone was fucking crying about. I slowly discovered the humming tubes coming out of my nose and torso and bladder, mostly by way of hospital staff adjusting or emptying their contents. There was a whole bouquet of IVs in my neck. But my friends knew all about the machines I was hooked up to. They had complicated inside jokes about them I didn't understand, nicknames for (and crushes on) my ICU nurse.

Recovering in a hospital like that, you only really do two things: try to follow directions that seem offensively simple until you try them, and hungrily anticipate the schedule of pain medications — the latter of which were being injected directly into my bloodstream in icy spurts. People kept telling me to *breathe* into whatever contraption was attached to my face.

A few days in, I tried to watch cheerful TV shows provided by a friend, and found them too upsetting to stomach — a side effect either of the fever or the drugs. How could Rick take such obvious pleasure in damaging Morty? What existential despair kept Michael Bluth with a family that so abused him? I watched one of the later *Star Wars* and I'm so ashamed, but I cried. People visited me in shorts and tank tops, dripping wet, and complained about a New York high-summer humidity I could only vaguely place in my mind.

There was a lot of manpower seemingly required to staff my recovery, nurses gliding into the room and turning me over, checking my vitals every two hours. A chaplain in a floor-length floral skirt stalked into my room and asked me how I found meaning. (I told her, tragically, "work." She beamed.) When I couldn't finish the Ensure I was supposed to be chugging three times every day, one of the orderlies suggested I stash it and make smoothies when I got home — after all, that stuff isn't cheap.

One of the most enduring clichés about illness — that it's nearly impossible to imagine yourself well — is true. My sense of inertia had atrophied just as much as my muscles, and it took some time to realize, in the lethargic and infantilizing routine of the hospital, how much had been happening, elsewhere, on my behalf. The silent, softly lit medical center some 30 stories above street level was designed to feel insulated from the world—the doctors positioned as if my care were their only concern—but of course that wasn't true at all. By the time I left the ICU for a double room I had started to seriously worry about my job, another concern left over from my years of precarious work.

The paperwork had already been filed on my behalf: my paychecks were being delivered to my apartment, the doctor's statements had been forwarded to the insurance company and the state. My coworkers sent a very nice card. I was incredulous to find it had been signed by our CEO. Unsurprisingly, emergency time off that comes with some sort of paycheck is vastly more common in white-collar industries. New York State's short-term disability program paid about half of my salary; my employer paid the other.

It deserves to be said that while most everyone in the hospital I encountered was gentle and caring, and while my insurance paid for nearly everything, both institutions—like all in the medical industry—have in many other cases ruined people's lives. The medical center where I spent nearly a month was investigated a few years ago for taking charity money but refusing to give financial aid to low-income patients. My insurance provider has argued innumerable cases against claimants: they denied a $400,000 treatment for breast cancer this past November, just to name one.

For reasons I will never fully understand, the occult transmissions between the doctors and the administrators and the insurance company left me responsible for $2,654.42 out of the $648,221.53 billed. And the paychecks I was getting came in handy for the copays and out-of-pocket costs, sure. But they also helped me weather what was essentially an extended convalescence.

Shortly after I was moved from the ICU, my boyfriend explained that my retired mother had canceled her foreseeable plans and would be staying with us indefinitely. I thought the two of them were out of their minds. But patiently my boyfriend, who was already struggling from taking (unpaid) time off, and who had sunk his own resources into preparing the apartment for my eventual return, told me what the doctors had been telling the two of them: for a month or two after I was discharged, I would have trouble walking and need to be monitored almost constantly for signs of a returning infection. Taking flights of stairs, cooking, lifting anything, even standing long enough to shower was for the moment out of the question. I would probably be in a lot of pain, with the incision in my back where they'd scooped everything out. And three times a day, I'd need to take a home-administered antibiotic through an IV drip that was impossible for me to rig up myself.

Once the IVs had blown through all the veins in my forearms and I could shuffle down the hall with some help, I went home. I was 25 pounds lighter. Outside for the first time all summer, I watched a woman amble down the street with a large suitcase and was in total awe, unable to imagine moving in such a casual way. I spent most of the next few weeks in bed; if no one was around to prop me up, I just lay there on my back. I had vivid, unsubtle dreams about all my friends growing bored with my lack of ambition. I couldn't even read for more than an hour or two without exhausting myself.

The PICC line, a semipermanent IV that runs from the upper arm into the heart, pumping antibiotics into the blood, would have cost between $1,300 and $3,000 without insurance to surgically insert. A nurse came by my apartment twice a week to check it for signs of infection and change the dressing. She told me she was in a reggae band, in addition to being a "healer." She spoke approvingly of the crystals my more superstitious friends had left around. The company she worked for billed my insurance provider just under $7,000 for services and supplies.

Around mid-August they took the PICC line out. My white blood cell count went back to normal; I was cleared to drink by far the best beer of my life; the scabs left over from the drainage-tube punctures in my torso fell off. I was pleased the surgical incision had missed my favorite tattoo, and less pleased when streaks of my hair turned gray and started coming out in clumps, or when my nails fell off—a months-delayed reminder of that time my body was preparing to die. I went back to work toward the end of the summer. I gave consent to include my miraculous recovery in a medical journal.

I assume people think surviving illness changes you because there's something inherently character-building about pain. But what happened wasn't a struggle, in the sense that through perseverance I overcame something difficult. For a fairly brief but unexpected period of my life, I lost my capacity to work, to advocate for myself, to navigate life and all its frictions. For much of my illness —intubated, drugged, feverish—I simply wasn't there. In my absence, there were more than 25 individual doctors and specialists, an army of nurses, the friends who loved me enough to take care of my affairs while I was under, some unknowable number of insurance agents sitting behind desks in another state, silently placing checkmarks next to my claims.

I am lucky not for surviving the infection, but for being a member of a shrinking class of Americans whose lives can absorb a trauma of this magnitude, and for whom being thrown, insensible, into the system is actually a good thing. When people refer to me as a "survivor," which they do often, they're correct, but it's not what they think it means: It has already been decided, especially now that it's again fashionable to claim that healthcare is not a right, who is a designated survivor in this country. It has also been decided who is not.

JOSHUA ROTHMAN

Why Paper Jams Persist

FROM *The New Yorker*

BUILDING 111 ON the Xerox engineering campus, near Rochester, New York, is vast and labyrinthine. On the social media site Foursquare, one visitor writes that it's "like Hotel California." Conference Room C, near the southwest corner, is small and dingy; it contains a few banged-up whiteboards and a table. On a frigid winter afternoon, a group of engineers gathered there, drawing the shades against the late-day sun. They wanted to see more clearly the screen at the front of the room, on which a computer model of a paper jam was projected.

The jam had occurred in Asia, where the owners of a Xerox-manufactured printing press were trying to print a book. The paper they had fed into the press was unusually thin and light, of the sort found in a phone book or a Bible. This had not gone well. Midway through the printing process, the paper was supposed to cross a gap; flung from the top of a rotating belt, it needed to soar through space until it could be sucked upward by a vacuum pump onto another belt, which was positioned upside down. Unfortunately, the press was in a hot and humid place, and the paper, normally lissome, had become listless. At the apex of its trajectory, at the moment when it was supposed to connect with the conveyor belt, its back corners drooped. They dragged on the platform below, and, like a trapeze flier missing a catch, the paper sank downward. As more sheets rushed into the same space, they created a pile of loops and curlicues—what the jam engineers called a "flower arrangement."

"It's the worst-case scenario," Erwin Ruiz, the leader of the paper-jam team, said. In the study of paper jams, Ruiz has found his Fountain of Youth: he is 50 but looks almost two decades younger. Born in Brooklyn, he grew up in Puerto Rico before going to graduate school in Rochester, where he is now a fixture of the city's wintertime indoor beach-volleyball scene. Wearing designer sneakers, hip-hugging maroon trousers, a trim plaid shirt rolled to the elbows, and elegant stubble, he began to pace in front of a whiteboard.

Bruce Thompson, the computer modeler who sat at the head of the table, had spent days creating a simulation of the jam. "We're dealing with a highly nonlinear entity moving at a very high speed," he said. On the screen, his wireframes showed a sheet of paper in midflight. He called up a shadowy slow-motion video made inside the press. "There's a good inch before the vacuum takes effect," he observed.

The team began to consider their options. The most obvious fix would have been to buffet the paper upward from below using a device called an air knife. This was off limits, however, because the bottom side was coated with loose toner. "An air knife will just blow the toner right off," Ruiz said. Another possibility was to place "fingers"—small, projecting pieces of plastic—where they could support the corners as they began to droop. "That might create a higher jam rate on different paper shapes," an engineer said—it could be a "stub point." A mystified silence descended.

A mechanical engineer named Dave Breed pointed toward the upside-down conveyor belt. "The vacuum pump actually works by pulling air through holes in the belts," he said. "So what is the pattern of those holes relative to the corners? Maybe there's no suction there."

On the whiteboard, Ruiz sketched a diagram of the conveyor belt—the VPT, or vacuum-paper transport—showing the holes through which the suction operated. "Optimize belt pattern," he wrote.

"If my understanding of air systems is right," Breed went on, "then the force that gets a sheet moving isn't really pressure—it's *flow*."

Thompson nodded, miming the pushing of air away from himself with his hands. "It's flow," he concurred.

"Could we somehow create more acquisition flow?" Breed asked.

By this point, Ruiz appeared to be vibrating. "Here's a stupid idea," he said. "Bernoulli!" Bernoulli's principle, discovered in 1738, entails that fast-moving air exerts less air pressure than slow-moving air. Because the top side of an airplane wing is curved, while the underside is flat, the air above moves faster than the air below, and the wing rises. "If you have jets of air shooting above the corners, the airflow will lower the pressure, and they'll lift," Ruiz said. Using the flat of his hand, he mimed the paper levitating like a wing.

"We could take the output from the vacuum pump and port it around to make it the air source for your Bernoulli," Breed said.

"Stupid idea number seven!" Ruiz said, grinning triumphantly. The whiteboard now contained an elaborate diagram of rollers, conveyors, vacuum pumps, air knives, air jets, stub points, and fingers. "Jets on corners to lift with Bernoulli," Ruiz wrote. Outside, the wind howled. Lake-effect snow had begun to dust the parking lot. The engineers were aglow: conspirators who'd just planned the perfect crime.

Late in *Oslo*, J. T. Rogers's recent play about the negotiation of the Oslo Accords, diplomats are finalizing the document when one of them reports a snag: "It's stuck in the copy machine and I can't get it out!" The employees in Mike Judge's 1999 film *Office Space* grow so frustrated with their jam-prone printer that they destroy it with a baseball bat in a slow-motion montage set to the Geto Boys' "Still." (Office workers around the country routinely reenact this scene, posting the results on YouTube.) According to the *Wall Street Journal*, printers are among the most in-demand objects in "rage rooms," where people pay to smash things with sledgehammers; Battle Sports, a rage-room facility in Toronto, goes through 15 a week. Meanwhile, in the song "Paper Jam" John Flansburgh, of the band They Might Be Giants, sees the jam as a stark moral test. "Paper jam, paper jam," he sings. "It would be so easy to walk away."

Unsurprisingly, the engineers who specialize in paper jams see them differently. Engineers tend to work in narrow subspecialties, but solving a jam requires knowledge of physics, chemistry, mechanical engineering, computer programming, and interface design. "It's the ultimate challenge," Ruiz said.

"I wouldn't characterize it as annoying," Vicki Warner, who

leads a team of printer engineers at Xerox, said of discovering a new kind of paper jam. "I would characterize it as almost exciting." When she graduated from the Rochester Institute of Technology, in 2006, her friends took jobs in trendy fields, such as automotive design. During her interview at Xerox, however, another engineer showed her the inside of a printing press. All Xerox printers look basically the same: a million-dollar printing press is like an office copier, but 24 feet long and 8 feet high. Warner watched as the heavy, pale-gray double doors swung open to reveal a steampunk wonderland of gears, wheels, conveyor belts, and circuit boards. As in an office copier, green plastic handles offer access to the "paper path"—the winding route, from "feeder" to "stacker," along which sheets of paper are shocked and soaked, curled and decurled, vacuumed and superheated. "Printers are essentially paper torture chambers," Warner said, smiling behind her glasses. "I thought, This is the coolest thing I've ever seen."

There are many loose ends in high-tech life. Like unbreachable blister packs or awkward sticky tape, paper jams suggest that imperfection will persist, despite our best efforts. They're also a quintessential modern problem—a trivial consequence of an otherwise efficient technology that's been made monumentally annoying by the scale on which that technology has been adopted. Every year, printers get faster, smarter, and cheaper. All the same, jams endure.

Gutenberg invented his printing press around 1440; the modern paper jam was invented around 1960. During most of the years in between, jamming was impossible, because printing was done one sheet at a time. Traditional presses lowered inked type onto individual sheets of paper; their successor, the rotary drum, was hand-fed. In 1863 an inventor and newspaper editor named William Bullock created the Bullock press, which was fed by a single roll of paper several miles long. Bullock's press revolutionized the printing industry by vastly increasing printing speeds. Sadly, in 1867 Bullock's leg was caught in the press; it became gangrenous, and he died. There are jams worse than paper jams.

The Bullock press was one of the first presses with a paper path, but by today's standards its path was simple. The most complex step, the composition of type, happened off-line; if a printer wanted to change the type, he had to stop the press to reset it.

The holy grail of printing—a paper path that incorporated composition, and so could produce different pages, one after another
—remained inconceivable. The creation of a miniature press, for
use in offices, was an even wilder dream.

The solution was xerography, invented by Chester Carlson,
the physicist cofounder of Xerox, in 1938. In xerography, static
electricity quickly and precisely manipulates electrostatically sensitive powdered ink—aka toner. As the term *photocopier* suggests, a
xerographic machine is less like a traditional printer and more
like a darkroom. Using an early Xerox machine required placing
an original under a glass pane, reflecting light off it onto a stati-
cally charged photosensitive plate, using the charged plate to draw
toner from a tray, transferring the toned image to plain paper,
and then melting the toner into the paper in a miniature electric
oven. (Between the charging of the plate and the ding of the oven,
or "fuser," each copy took around three minutes to make.) The
Xerox 914, introduced in 1959, automated this process. Caressed
by sultry secretaries in advertisements, it resembled an instrument
console from the Starship *Enterprise* and shipped with a fire extin-
guisher, in case its heating elements set the paper alight; seven
plain-paper copies per minute trundled through its paper path.
Between 1960 and 1979, the 914 earned Xerox around $40 billion
—funding, among other things, the construction of the corporate
campus in Rochester, and jump-starting the development of the
personal computer, at Xerox PARC, in California. (Xerox failed
to capitalize on the PC revolution; recently, Fujifilm announced
plans to acquire a majority stake in the company.) Today, not all
Xerox printers are xerographic; many are ink-jets, which work by
converting an image into a waveform, then using the waveform to
control an ink nozzle. Almost all, however, follow the template of
the 914 paper path: feeder, printer, fuser, stacker.

Jams emerge from an elemental struggle between the natural
and the mechanical. "Paper isn't manufactured—it's processed,"
Warner said, as we ambled among the copiers in a vast Xerox show-
room with Ruiz and a few other engineers. "It comes from living
things—trees—which are unique, just like people are unique." In
Spain, paper is made from eucalyptus; in Kentucky, from southern
pine; in the Northwest, from Douglas fir. To transform these trees
into copy paper, you must first turn them into wood chips, which
are then mashed into pulp. The pulp is bleached, and run through

screens and chemical processes that remove biological gunk until only water and wood fiber remain. In building-size paper mills, the fiber is sprayed onto rollers turning 35 miles per hour, which press it into fat cylinders of paper 40 reams wide. It doesn't take much to reverse this process. When paper gets too wet, it liquefies; when it gets too dry, it crumbles to dust.

To a sheet of paper, a paper path is like a Tough Mudder—a multistage obstacle course that must be run in hostile conditions. With a hint of swagger, Warner walked me through the paper path of a hulking, truck-size iGen printing press (around $1 million and 150 pages per minute). "We start by sucking a sheet off the stack with vacuum feeders," she said. "Then it travels along thirty feet of path at one thousand three hundred and fifty millimeters per second, changing speed and direction at accelerations reaching 3g." In xerographic printers, she continued—she had to shout above the press's vacuum pumps, which sound like a copier's, but louder —"the sheets are charged with sixty-five hundred volts. In ink-jets, they're soaked in liquid. Then we have to keep the image from shaking or wiping off." Warner pointed to the back of the paper path, where the fuser was situated: a set of black rubber rollers heated to 385 degrees. "It's like wringing a shirt through an old washing machine," she said, miming the motion with her hands. Later, she gave me a flowchart of the printing process; it featured a cartoon of a paper sheet, its mouth agape in terror.

Ruiz gestured down the length of the iGen, which resembled many copiers daisy-chained together. "The straighter the path, the less probability of damaging the paper," he explained. For this reason, printing-press paper paths tend to sprawl horizontally. Office printers must be smaller, and so their paths must fold back on themselves, making a series of hairpin turns. "Think about being in a car," Ruiz said. "The more turns you take, the more likely you are to get into an accident." Contemplating the "tight radiuses" of office printers and their other daunting requirements—they must be quiet, cheap, and low-power, and "people without master's degrees" must be able to clear their jams—Ruiz shook his head with parental indulgence. "For us, the smaller ones are more challenging than the bigger ones."

The owners of printing presses have exotic tastes: they print on magnets, tinfoil, windshield decals. Xerox executives push the engineers to accommodate new kinds of stock, which might open new markets. But even plain office paper is full of hidden dangers. In the facility some engineers call the Paper Torture Lab —officially, it's the Media Technology Center—Bruce Katz, a soft-spoken paper technologist, examined some copy paper through a microscope. "The edge of a sheet of paper is really a third dimension," he said. Magnified, the edge resembles a snowy mountain range about four-thousandths of an inch thick; the snow is paper dust, ready to drift into a printer's jammable gears. More expensive paper is more cleanly split, and its straighter edges have less dust-generating surface area. (They are also more likely to cause paper cuts.)

"Papers are not created equally," John Viavattine, the head of the Torture Lab, said. Some stocks generate excessive friction; others swell in the humidity. (In general, winter jams are more common than summer jams.) Sheets cut from the same 40-ream roll can vary in quality. At the center of the roll, paper fibers tend to arrange themselves in an orderly matrix; nearer the edges, they become jumbled. ("Think of logs going down a river; the flow is different at the edges of the river from down the middle," Katz said.) When heated, wood fibers contract; neatly arranged fibers contract equally in both dimensions, but badly aligned fibers do so unevenly, creating curl. The team from the Paper Torture Lab travels around the world, helping paper mills improve their product, and raising the quality of printer paper has played a major role in increasing print speeds. Still, even the highest-quality paper can be ruined by poor "paper handling." A half-used package of paper left to sit will grow damp and curly or dry and "tight." Reams of paper that are thrown around or kept in stacks can develop hidden curls that lead to jams.

At a hip Rochester restaurant called Nosh, Viavattine held the menu up to the light to assess its "flocculation" (the degree to which its fibers had clumped infelicitously together). He launched into a fabulous paper-jam war story. "I was asked to go to Chicago to visit the Chicago children's court," he said. "This was the mid-nineties, and a sales rep had put our printers—I think they were 400 Series—all over the court system. What was happening was, lawyers had to deliver certain court documents to the defense attorneys within a certain amount of time. Otherwise, the defendant was let go. And they were losing two out of three cases because of paper jams." He paused. "*Two out of three* defendants were *gone* —walking out the door—because of paper jams!"

Ruiz looked both fascinated and skeptical. "So, just so I understand—the repeated jams were delaying the process so much that—?"

"That two out of three times they would be late, and the defendant would be released!" Viavattine said. "And the problem was that they were using some off-brand, really down-in-the-dumps paper."

Ruiz turned to me with a twinkle in his eye. "Paper jams!" he said. "Now you know why the crime rate in Chicago went down."

Paper jams are a species within a larger genus. Traffic jams, too; so do tape decks, guns, and sewing machines. On humid days, voting machines jam, leading to recounts; over the eons, tectonic plates jam, resulting in earthquakes. Ice floating down a river makes an ice jam; floating logs join up into logjams. (Before railroads transformed the transportation of lumber, logjams had to be addressed by "jam breakers"—experts who spotted and removed the "key logs" jamming up the river.) Jamming happens whenever something that's supposed to flow through a space fails to do so, perhaps because of overcrowding, or bending, or because its constant movement degrades the space through which it travels.

To some extent, jamming is what engineers call a "scheduling" problem. Picture a warehouse in which thousands of packages are traveling on intersecting conveyor belts. If the distance between the packages isn't carefully maintained, they will collide and pile up, creating jams. Printer designers solve this problem by making the paper path smart. In a typical office photocopier, a host of small optical sensors monitor the location, angle, and speed of individual sheets of paper; if one gets too close to its neighbor, the rollers slow it down. Similarly, if a sheet is subtly off-angle, rollers on the slow side accelerate to straighten it; if the sheet is duplex—that is, printed on both sides—they adjust on the fly to ensure that both sides are aligned. Printer engineers call this "agile registration."

"The tolerances are very tight," Ruiz said. "When you're moving a box from here to there, if you're off an inch it's probably fine. But our images cannot be off by more than eighty-five microns"— a third of a thousandth of an inch—"or else they'll be fuzzy." Dave Gurak, a software engineer who designs printer control systems

("It's his brain in there!" Ruiz said) thinks that the biggest jump in print speed happened in the 1990s, when cheaper microprocessors enabled paper-path designers to control scheduling at a minute level. Today, he said, "twenty-five thousand independent events happen per page." In some printers, if a sensor in a paper tray detects a curl in a sheet the tray tilts to make up for it.

In the largest sense, jamming is a problem in a field called tribology—the study of friction, lubrication, and wear between interacting surfaces. In the 1960s, the British government asked an engineer named H. Peter Jost to investigate this subject; the 1966 "Jost Report" found that poorly lubricated surfaces—sticky ball bearings, rusty train rails, and the like—cost Britain 1.4 percent of its GDP. (The term *tribology*, coined by Jost, comes from the Greek verb "to rub.") The smooth functioning of the world depends on invisible tribological improvements. We rely on axles and gears that don't grind, artificial joints that don't stick, and hard drives that spin smoothly. Everything in a printer, likewise, must slide quickly and smoothly over everything else. Paper-path engineers work to accelerate a system that wants to get stuck.

Tim Slattery, a recent graduate of RIT, stood in Erwin Ruiz's paper lab, inspecting a stacker—the final component of a large printing press. The owners of an identical machine hoped to print on thick, laminated plastic labels—the kind that might mark the price of an item in a big-box store. The problem was static electricity. "There's so much static between the sheets that they levitate in the stacker," Slattery said. I grabbed one, and had to make a concerted effort to push one sheet across another. "Our fluffers are constantly on, and we're alternating our vacuum and air knife," Slattery noted, but it wasn't working.

"Instead of sliding the sheet, we're going to corrugate it," Ruiz said.

"Corrugating is when we put an intentional wave in the sheet, like in a piece of corrugated cardboard," Vicki Warner explained. "It adds stiffness." The plan was to corrugate the sheet lengthwise by running it over a line of rollers turning at variable speeds before "flying it" into the stacker. If a physical fix was necessary, a part might be 3-D-printed and installed, on-site, by one of the engineers. (On some occasions, printer-jam fixes are propagated through software updates.)

For a little while, I watched the team at work. Then I asked whether it would ever be possible to build a jamless printer.

"Well, we have printers on submarines, and also in space," Ruiz said. "For the right amount of money you can build lots of redundant systems. So I think the answer is maybe yes."

"I think the answer is no," Warner said. "It's paper. There will always be something unpredictable about paper that will cause a jam."

Perfectly made synthetic paper might eliminate jams; it might also create unforeseen problems of its own. They stood, contemplating the problem, while the copiers whirred.

Xerox's engineering campus can be a spooky place. Over time, the workforce there has dwindled. Warehouses contain pyramids of unused office chairs, and groups of copiers lurk in utility corridors like robots preparing to take over. (If the machines ever do rise up, jams may be what save us.) While we walked through mazes of cubicles, Ruiz thought about his future. He and his team are very good at their jobs—printing speeds keep rising, and jam rates hold steady or decline—and his promotion seems inevitable. But he loves paper jams too much to move on.

"Once, a cell phone company tried to hire me," he recalled. "They said, 'You're going to be working on the frames of the cell phones.' I said, 'What else?' They said, 'No, that's it—the frames of the cell phones.' That's so boring! I don't think they sell this job well enough. It's, like, 'Printers—I used to have one, it used to break.' But if you really want to learn more about everything, this is what you should do." He grinned. "I like solving problems. Once you go to Toner Tower"—Xerox's coal-black skyscraper in downtown Rochester—"life starts passing you by." In the hallway, we walked past another engineer, who gave Ruiz a discreet fist pump. Ruiz turned to me: "Volleyball buddies!"

In one of the company's climate-controlled testing chambers, the team working on the Asian dog-ear problem had gathered around a printing-press component identical to the one the customer owned. Earlier, conditions within the chamber had been set to 80 degrees and 80 percent humidity, to match those at the customer's facility; now the room was cooler and crowded with engineers. There was an atmosphere of convivial fascination. Everyone took a turn bending down to squint at the area that Bruce Thompson had represented so clearly in wireframe, on the computer.

"Can you see it?" Ruiz asked, sinking, with athletic fluidity, into a deep squat. He pointed to the small upside-down conveyor belt. "It's tiny! Actually, maybe just one finger would do it."

"Can you see the stripper fingers going?" Gurak asked. "Or is it just the air knife? We used to have the stripper fingers down there."

"Yeah," Ruiz said. "In theory, we could have stripper fingers pick up the lead edge, but then they might touch the belt, and that would be super bad."

Someone turned the machine on, and paper began flowing through the path. The engineers drew closer, looking for the flower arrangement.

JORDAN MICHAEL SMITH

The Professor of Horrible Deeds

FROM *The Chronicle of Higher Education*

ON A SATURDAY afternoon last spring, Fred Berlin leads me into a room where he holds group therapy sessions in his private practice in midtown Baltimore. In this sparsely furnished space, with peeling wallpaper, Berlin and the team he oversees treat patients with sexual disorders and difficulties—everything from gender dysphoria to overpowering sadism to pedophilia.

On one wall are grotesque sculptures representing the seven deadly sins. Berlin, who is an associate professor of psychiatric and behavioral sciences and director of the Sex and Gender Clinic at the Johns Hopkins University, is fascinated by shifting perceptions of sin—gluttony, he observes, was once categorized with other cardinal vices, but is now understood as more of a medical issue than a personal failing. For Berlin, what a culture considers abhorrent says as much about the culture as it does about the vice. That's why, next to the sin sculptures, in this chilly room where people share their most personal impulses, hangs a baby photo of Adolf Hitler.

"I put it there *not* because I have respect for the horrors of Hitler," he says. It is unlikely that any patient recognizes the picture. But for Berlin, the photo raises eternal questions about the nature of evil. "I could just label Hitler as evil, and we could all say, 'OK, he was evil.' Where does that get us? But if the question is, what about his life experiences? What about his biology? Was there some sort of psychosis?"

Berlin possesses an ultrarational, almost detached way of contemplating horrible deeds. As much as any other academic in the past 40 years, he has made arguably the most reviled type of person in our culture—the pedophile—into a legitimate subject of study and even empathy. "Fred needs tremendous recognition for making it clear to the general public that people don't choose to become pedophiles, that it's a neurobiological condition," says John Bradford, a forensic psychiatrist at the University of Ottawa. While it's unclear that the general public has, in fact, recognized this, it's undeniable that, as Bradford says about Berlin, "he has been a pioneer."

As one might imagine, Berlin's views have made him enemies. He is a "pedophile apologist," Judith Reisman, a law professor at Liberty University, has written (with Geoffrey B. Strickland), in the *Ave Maria International Law Journal.* "Dr. Berlin protected the predators in his care while ignoring their acknowledged ongoing child victims."

Explosive charges like those are routine for Berlin. When he began his work, in the mid-1970s, few people paid attention to the scant research on child molestation and pedophilia. Since then, daycare abuse scandals in the mid-1980s, revelations of child rape in the Roman Catholic Church, and assaults on minors at Penn State and Michigan State, mean anyone doing his kind of work has to withstand public suspicion, even vilification. Now, as he nears retirement, he is having difficulty finding a successor to run his clinic. Will anyone else want to devote a career to studying and helping pedophiles?

A growing number of experts have arrived at something of a consensus on pedophilia and its origins. First, adults do not choose to be attracted to children. They are either born with these attractions or develop them through ways that are mysterious to us. Second, not all those with pedophilia act on their attractions—indeed, many such individuals are horrified by their impulses and suppress them successfully. Third, most people who abuse children are not pedophiles. They are not exclusively attracted to children but rather are sexual opportunists. Finally, for some people, pedophilia is something that can be treated with a combination of psychotherapy and medication. People with pedophilia usually cannot cease being attracted to children, but sometimes they can control their urges to the point of never acting on them.

None of this was known to Berlin at the outset of his career. He received his PhD in psychology and MD from Canada's Dalhousie University in the early 1970s. His dissertation examined the effects of hypnosis on patients trying to quit smoking. But when he arrived at Johns Hopkins as a resident in 1975, his focus shifted. He began working under John Money, a sexologist who pioneered research into gender fluidity and transgenderism. Money coined the now-common terms *gender identity* and *gender role*.

Money's Johns Hopkins was an experimental environment. Given the paucity of research into sexuality—this was but a few years after the *Mad Men* era—it had to be. In the mid-1960s he established the world's first clinic devoted to transgender surgeries, the Johns Hopkins Gender Identity Clinic. Money himself engaged in nudism and group sex and advocated open marriages and the virtues of the porn flick *Deep Throat*. He was described by one writer as "suavely charismatic . . . with the long, elegantly cut features of a matinee idol." He would regularly sprinkle expletives into conversations and make sexual remarks at unusual times. The *New York Times* called him "an agent provocateur of the sexual revolution."

Money's brilliance could morph into arrogance. Most notoriously, he persuaded the parents of a son who lost his penis during circumcision to undergo sex-reassignment surgery and be raised as a girl—but the girl later reverted to identifying as male, suffered from depression, and committed suicide. Money also studied pedophilia and child abuse. And, as with other sexual practices he studied, he was sometimes blasé about the harm that could result from child molestation. "If I were to see the case of a boy aged 10 or 11 who's intensely erotically attracted toward a man in his 20s or 30s, if the relationship is totally mutual, and the bonding is genuinely totally mutual, . . . then I would not call it pathological in any way," he once told *Paidika*, a Dutch-based pro-pedophilia journal.

This was the avant-garde setting Berlin entered. "I remember back then people asking me, Is it *pedophile* or *peedophile?*" Berlin recalls. "People weren't even familiar with the term. If you looked at the literature, there was very, very little out there."

In 1963 the FDA approved a drug called Depo-Provera to be tested as a female contraceptive, but Money began experimenting with it on sex offenders. Combining the drug with therapy, he used it to suppress testosterone and diminish the brain's erotic imagery, reducing sexual urges and functioning. Since Money wasn't a medical doctor, the department asked Berlin to help write prescriptions for Depo-Provera.

Beginning in the early 1980s, hundreds of Maryland sex offenders were offered probation if they agreed to be injected with the drug under the supervision of Money and Berlin. (Money died in 2006; Berlin no longer works with the prison system.) This course has proved enticing to medical experts, government officials, and inmates alike as an alternative to prison, but civil libertarians and others have lambasted the process as essentially forcing people to take mind-altering drugs. "To me, it's still kind of coercive because they are inmates. It's a coerced choice," says Daniel C. Tsang, librarian emeritus and an LGBT activist at the University of California, Irvine, who wrote about Money for the *Journal of Homosexuality.*

Others identify the limitations of Berlin's advocacy of drugs and therapy for most pedophiles. "From a public-safety standpoint, my understanding is that it is only effective so long as the offender takes it," says John Stinneford, a law professor and assistant director of the Criminal Justice Center at the University of Florida. Some patients may simply stop taking the medication and disappear.

Still others question the science behind Depo-Provera. "Dr. Berlin's studies and those conducted by other psychiatrists and clinical psychologists have not provided any scientifically valid basis for assessing the safety and effectiveness of using Depo-Provera to treat sexual deviates and sex offenders," writes William Green, a Morehead State University professor, in his 2017 book, *Contraceptive Risk: The FDA, Depo-Provera, and the Politics of Experimental Medicine* (NYU Press).

That hasn't prevented many states from mandating what is often called "chemical castration" of sex offenders. And patients who benefit from the medication see it as vital. Jason (not his real name) is a 30-year-old married man living in West Texas. In 2011 he was charged with indecency with a child. As part of his plea deal, he received sexual counseling. But because his counselors doubled as his probation officers, he felt inhibited from being honest about his attraction to children. In addition, he couldn't find anybody who specialized in sex-related treatment in his region. In 2014 the courts gave Jason permission to travel to Maryland to be treated by Berlin. He flies to Baltimore every other month. The drug he is on, Lupron, which reduces the amount of testosterone in men and is commonly used to treat symptoms of prostate cancer, has "absolutely, literally, been life-changing," he says. "My addiction really controlled me." The injections of the drug don't make his attraction to children disappear, but they reduce the urge to a controllable level. "I shudder to think where I'd be without it."

Berlin's office is large and has a fireplace. Framed posters announcing the clinic's early days hang on a wall. Beside them are stills of Berlin speaking on the *Today* show and on *20/20*. A statue of Yoda stands in the corner ("I have to get wisdom from somewhere," Berlin jokes).

He is friendly and candid. But he is also wary of talking with reporters. He asks me to explain my interest in him, although I've explained it to him previously, over the phone. He's googled me thoroughly.

Berlin is remarkably equanimous about the controversy he's attracted, however. "I see on the internet sometimes the nasty statements that people make, but I felt very supported in this over the years [by] the people I work with," he says. "I've just tried to do a good job, to do what I feel is right, and trust that things will turn out well if I continue to approach it that way."

When Hopkins created a clinic for people with sexual disorders and problems, in 1980, the university was a pioneer. "There was a tremendous need," he recalls. "If you had schizophrenia, substance-abuse problems, anorexia, you name it, there were places you could go to get help. But there was virtually no place you could go, even though pedophilia was listed as a mental-health condition, if you wanted to get help."

Discussing sexual problems was so taboo that in the early years of the clinic, there were days when no patients showed up. But eventually they began showing up in droves. Few individuals in America have studied as many people with paraphilias—sexual abnormalities—as Berlin. One of his patients was a sadist who was afraid he was going to act on his violent fantasies. He was preparing to attack his wife with a weapon he had devised by attaching a club to a chain (Berlin still has the device).

By the early 1980s, Berlin's clinic had gained national attention. "Sexual Deviancy: Clinic at Hopkins Fills a Need," read the headline of a 1984 profile in the *Baltimore Sun*. The article's language is revealing of the times: "They are child molesters, all of them. In the language of medicine, they are 'pedophiles.'" The next year, the *Sun* reported on a man under Berlin's care who stopped by the downtown arcade to meet young boys before his visits to the Hopkins clinic. This validated criticism from child-advocacy groups, which charged that Hopkins was protecting predators at the expense of children. The man had previously been convicted of child molestation, compounding the blunder. Berlin admitted to the *Sun* that he and the staff at Hopkins had erred in allowing the man to go off Depo-Provera, but otherwise they stood their ground.

Berlin points to the decreased recidivism rates of his patients as evidence that the therapy-and-medication approach works. "Most child molesters are treatable," says the sociologist David Finkelhor, director of the Crimes Against Children Research Center at the University of New Hampshire. Of course, much of the public thinks otherwise. Many believe any punishment short of execution is too lenient to pedophiles. "This clinic was founded by John Money to give judges 'leeway' in sentencing sex offenders—that is, a place where they could send child molesters other than jail," writes Liberty University's Reisman, a leading figure in conservative efforts to roll back the sexual revolution. "Dr. Berlin was his disciple."

In 1987 the Maryland legislature introduced a bill requiring physicians and therapists to notify authorities if an adult patient told them he had abused a child. But Berlin told lawmakers that such legislation would prevent sex criminals—or those struggling with urges to commit sex crimes—from voluntarily pursuing psychiatric treatment. "I didn't want to deter undetected people from coming forward and getting help that might, in effect, make the community safer," he says. "We weren't trying to give anybody a pass —if somebody was in trouble, we weren't saying they shouldn't be prosecuted, we weren't saying if children are seen, they shouldn't be reported. But we didn't want to enable anybody who might want to get help, to be safer, not to be able to get it."

Initially, Maryland policymakers compromised with Berlin. If abuse was happening and the clinic knew about it, the law said, it had to be reported. But if the abuse had been committed in the past, it didn't. Critics nicknamed it "the Berlin exemption."

It proved deeply unpopular—local critics worried Johns Hopkins would attract pedophiles from across the country—and soon the exemption was revoked.

But Berlin and Hopkins found something of a loophole. "The Sexual Disorders Clinic adopted a policy of advising prospective patients that offenses against children would not be reported if their cases were referred to the clinic by an attorney, since such information could be protected by the attorney-client privilege," the clinical psychologist Douglas Peddicord wrote in the *Sun*. He said the clinic was trying "to hide behind lawyers to shield knowledge that the legislature has specifically refused to exempt." A different, front-page article in the *Sun* accused the clinic of "skirting the law" when it came to reporting sexual abuse. "Johns Hopkins Hospital Has a Sex Problem" was the headline in *Baltimore* magazine. Only after the state's then–attorney general, J. Joseph Curran Jr., spoke out against Berlin did he grudgingly agree to cease the practice, though he said that "an attorney general's opinion is not legally binding."

Then, in March 1988, a sexual-assault victim being treated for depression at Johns Hopkins Hospital said she had been sexually assaulted for a second time—by a fellow patient who broke into her room. They were on the same floor of the hospital, although he was in the Sexual Disorders Clinic and had raped teenage girls. The woman filed a lawsuit against Berlin and the hospital; her lawyer said they were responsible for "housing sexual deviants and sexual offenders including pedophiles with unsuspecting general psychiatric patients." The woman eventually dropped the complaint against Berlin and sued only the hospital.

Hopkins settled with the woman for an undisclosed amount, but the incident permanently altered the relationship between the hospital and Berlin. In 1992 the university distanced itself from the Sexual Disorders Clinic. Berlin moved the clinic into a private office and changed its name to the National Institute for the Study, Prevention, and Treatment of Sexual Trauma. The *Sun* reported the news by saying, "Over the years, Dr. Berlin earned the scorn of many with his position that therapists specializing in the treatment of pedophiles—child molesters—should not have to report sexual offenses to law-enforcement authorities . . . Child advocates have frequently pointed to patients and former patients who returned to deviant behavior after brief periods of good behavior."

"I didn't by any means agree with the decision, but I understood it, and I've tried to work to find common ground," Berlin says. He remains director of the Johns Hopkins Sex and Gender Clinic, and the university has repeatedly been forced to defend its relationship with him. The most notorious incident came in August 2011 at a conference on pedophilia and the *Diagnostic and Statistical Manual of Mental Disorders,* where Berlin was to deliver the keynote speech. The symposium was hosted by B4U-ACT, a support group that aims to help people attracted to children receive treatment before they act on their impulses. In his remarks, Berlin argued that "we need to have both a criminal-justice involvement and a public health perspective. I would make an analogy to alcoholism. We have to have laws against drunk driving—society has a right to protect itself—but it would be naive to think that you are going to solve the problems of alcoholism simply by putting drunk drivers on a registry, and to think that we can punish and legislate the problem away." Cued up by a *Daily Caller* article before the event, the media response, particularly from the right, was immediate and vicious. The *Washington Times, First Things,* and Fox News accused Berlin of trying to normalize child rape.

Hopkins released a statement decrying the "significant misinformation circulating about Fred Berlin . . . and his recent participation at a conference in Baltimore." It continued: "As an international expert in sexual disorders, Dr. Berlin speaks routinely at conferences in his role as an individual and as an expert in these matters. In this role, Dr. Berlin speaks for himself, and not as a representative of Johns Hopkins."

Berlin doesn't seem to comprehend how unpopular his views are in the wider culture. "There are very few who have come and talked to me who have left, you know, wanting to say bad things about me," he says. With his thinning hair and grandfatherly demeanor, it is easy to forget how contentious his work can be. But the B4U-ACT incident illustrated why his institution has few peers. "I've been trying to reproduce what he's done: a university-based sexual-disorders clinic," says Renee Sorrentino, an assistant professor of psychiatry at Harvard. "Most institutions are concerned with the stigma of having this kind of patient population at their university." Though administrators understand that the work is important, "getting the money and the resources" is next to impossible, she says. "Because of its history with John Money, Berlin was able to go make Hopkins" into a rare place.

A clinic attached to a university has several advantages over a private practice. It prioritizes research. It also can treat the indigent. (Research shows that individuals on the sex-offender registry have difficulty getting jobs.) As a result of her difficulties with Boston-area universities, Sorrentino established a private practice, the Institute for Sexual Wellness. She treats her patients with Lupron and psychotherapy, but laments that not everyone can afford private therapy. "Everyone agrees that someone has to do it, but nobody wants it in their facility," she says.

Some of the most cutting-edge research into pedophilia is taking place in Canada. An Ottawa hospital is claiming that its Sexual Behaviours Clinic can actually cure pedophilia, turning a man who was attracted to children into someone attracted to adults. The clinic won a 2015 award from the American Psychiatric Association, but most experts are deeply skeptical of claims that pedophilia can be eliminated, any more than straight men can cease being attracted to women. However, all agree that more research is beneficial—it's just hard to come by.

"There's no obvious reason why money should be spent directed toward learning more about pedophilia," Berlin concedes. "It probably is harder to do research—to get the funding to do research—because there's not this sense that it's deserving of research." He continues: "If I had to say there's one big thing holding back more research, more availability in clinical health and so on, often people don't have the sense that these are deserving human beings."

Berlin's empathy is almost limitless. He worked with the first person to be executed in New England since 1960, the serial killer and rapist Michael Ross. He put Ross on medication while he was incarcerated, so he wouldn't assault guards or obsess about sexual violence. Berlin doesn't think people like Ross could be rehabilitated or safely released, but he does think their condition could be improved. Similarly, Berlin visited the serial killer and cannibal Jeffrey Dahmer in prison in Milwaukee, and testified for the defense, saying that Dahmer had no control over himself. "There's a lot there showing a guy who really was struggling not to do this, but who eventually lost the battle," he says. "I couldn't help wonder if we had had a society where we encouraged somebody who was having these thoughts to come forward in the beginning, might these people who have all died been alive?"

Through all of this, Hopkins has maintained ties with Berlin, who remains grateful to the university. But now in his mid-70s, he is slowing down his pace, and having trouble finding a successor for his clinic. He can't offer a big salary, and anyone carrying on his role will have to be comfortable being vilified. Matthew Taylor, a former student of Berlin's and now a psychiatrist at Hopkins, admits that many of his colleagues hated having to work at Berlin's Hopkins clinic. "He can do this kind of work that is so distasteful for others," Taylor says of Berlin. "Many find it too difficult."

Berlin is often asked why he'd want to spend his life studying and helping people who are attracted to children. To him, the answer is easy. "I don't think the general public today has the sense that, if a person is sexually attracted to children, and certainly if they've acted on it, that there's any chance at all this could be a decent person struggling, deserving of help," Berlin says.

"I believe that's the case. I believe that, in time, society will see it that way."

SHANNON STIRONE

Welcome to the Center of the Universe

FROM *Longreads*

THE POWER HAS just gone out in mission control. I look to Jim McClure, operations manager at the Space Flight Operations Facility, and he assures me that everything is fine. A power outage like this hasn't happened at NASA's Jet Propulsion Laboratory in nearly eight years, and while it's only been out for a few seconds, the Deep Space Network is disconnected and NASA has temporarily lost contact with *Cassini*, the nearly 20-year-old space probe in orbit around Saturn, as well as all spacecraft beyond the moon.

We're standing in JPL's mission control, known simply as the Dark Room to those who work here. Five men and women are glued to their screens, the artificial pink-and-white glow highlighting their faces. I've been here twice before, but I have never seen this many people running the consoles. The operators are calm and hyperfocused despite the unexpected hiccup, both hands typing, eyes darting at one another's screens.

While the quiet panic plays out, I walk over to a sunken plaque in the middle of the room that glows with blue neon lights: THE CENTER OF THE UNIVERSE. Above it is a large metal coin embossed with the images of three spacecraft and a DSN antenna; below is JPL's motto, DARE MIGHTY THINGS. Teddy Roosevelt offered these words during an 1899 speech in approbation of the virtues of a "strenuous life" and they are now synonymous with the risks taken when it comes to spaceflight. "Far better is it to dare mighty things," he said, "to win glorious triumphs, even though checkered by failure . . . than to rank with those poor spirits who neither enjoy nor suffer much because they live in a gray twilight that knows not victory nor defeat."

I catch a bit of conversation. "Are you having any luck over there?" the data controller asks the person sitting at the Tracking Support Specialist desk. "Not yet." Above the consoles near the ceiling are six large television screens that curve around the room. Usually these screens stream real-time telemetry from dishes around the world and are labeled with the name of the spacecraft they're talking to. Right now, most of them are blacked out. The only active monitors display images of celebrities who've visited JPL: Matt Damon in the Mars Yard, William Shatner giving the Vulcan salute.

McClure is nervously tapping a stack of round CENTER OF THE UNIVERSE stickers on a table. "The data is always stored, so it's fine," he says, trying to reassure me. "Once it hits the ground it's stored." The staff speaks to one another like doctors in an emergency room moments before attempting to jump-start a quiet heart. "OK, trying to reconnect now." The data controller grabs the paddles. "Not getting anything. Nothing. Trying again." The *Cassini* mission ACE, the liaison between Earth and the spacecraft, rushes in, his messenger bag slung over his shoulder, and mumbles something to McClure. He hurries to his station, lit up in neon blue, past the barricade with a homemade sign that reads DO NOT FEED THE ACE — TO THE WOLVES. He plops his bag onto the floor, hunches over his desk, taps the keyboard, and begins trying to talk to Saturn.

When a mission launches into space, whether it is to Venus, Mars, or as far out as Pluto, we have to be able to track it, send commands, and receive data—all over a signal about as powerful as the wattage of a refrigerator light bulb. These faint whispers are hard to hear, and losing track of them for any length of time can be a harrowing experience. If the Deep Space Network goes down, if we permanently lose our connection to *Cassini*, it would not only be a loss of billions of dollars but also two decades of work.

The heart of the Deep Space Network started beating on Christmas Eve 1963, when JPL confirmed their long-term intentions of sending missions into deep space. It hasn't been turned off since. Its dishes, operators, and radio astronomers around the world have worked 24 hours a day, seven days a week for the past 54 years. The DSN has many vital roles, but one of its biggest is to serve as the communication link between Earth and its robotic emissaries in deep space—anything from the moon and beyond. Every image we've ever received from deep space, every relay of scientific data, even those famous words "The *Eagle* has landed" was collected by the dishes of the Deep Space Network. (In addition to being a vital communication link, the DSN also tracks potentially hazardous near-Earth asteroids, monitoring around 100 each year.)

Like so many of NASA's projects, the first generation of the Deep Space Network was born out of a military order. In 1955 the Department of Defense approved a JPL mission that would come to be known as Explorer 1; it was the first from JPL to use radio waves to track a satellite as it orbited the Earth. As the satellite curved around the planet, antennas placed around the globe would lock onto its location and send back telemetry to the people on the ground. This nascent tracking system was called Microlock, and it was the very beginning of the Deep Space Network.

By 1958, the leaders of JPL wanted out of military projects. They didn't see their special center being used to build rockets alongside the army; they wanted to go to space. After much negotiation, JPL wiggled loose from the army and were officially free to explore the solar system robotically with NASA. Their center would not send humans to the moon or Mars. They would conceive, design, and build robots to do the exploring in our stead. With as much foresight as one can have, the director of JPL at the time, William Pickering, knew this meant that they would always need a way to talk to their spacecraft. The functional but small Microlock system needed to expand.

Today the DSN has three permanent stations around the globe: in Goldstone, California; Madrid, Spain; and Canberra, Australia. Each location has one 70-meter dish standing 20 stories high and weighing a staggering 6 million pounds. These dishes are the DSN's workhorses, built between 1966 and 1974. They are the antennas with the biggest ears, most suited to collecting high-resolution data like color images or a lot of science.

The rest of the network is made up of dishes with 34-meter antennas that can either talk to spacecraft on their own or can "array together" to form a more powerful signal. In total, there are 13 dishes positioned 120 degrees apart around the world talking to

35 different spacecraft. In 2014 the DSN clocked around 18,000 passes and averaged over 100,000 hours of tracking. As the Earth spins, there is always a dish able to receive or place a call, creating a level of assurance for the science teams on the ground. The dishes are so powerful that their aim only works if you're at least 30,000 miles away from the Earth. The trouble is, the farther we venture out, the harder it is to hear the call.

Years ago, an intern at JPL created a website that allowed anyone to watch the antennas talking to the spacecraft in real time. DSN NOW, as it's called, shows the downlinking and uplinking of data in real time. A solid squiggly line means we have found the spacecraft and are "holding hands" with it. A jagged squiggly line means we're passing "envelopes" with information back and forth. Sometimes those envelopes contain the uplink, or commands. *Curiosity: Drive to the left 30 degrees for five minutes to that weird rock and stop, shoot your laser, and send back the data when you're done.* The downlink from the spacecraft then comes back. *Here's the chemical composition of that weird rock you made me shoot at.*

These conversations happen every day, and have for every spacecraft since NASA began exploring the solar system. This morning, before the blackout, the team was scheduled to uplink commands to *Cassini.* It is spring 2017, and there are only a few months left on the mission, so it is vital to get as much information back from the spacecraft as possible. McClure has to leave to check on the systems and make sure everything is OK with the network. The power was only out for a minute, but the fact that William Shatner is staring down from overhead is not a good sign. I open the DSN NOW page on my phone—still silent.

When you put out a call and get nothing back, it can feel lonely and scary. It's a reminder that we really do rely on these robotic explorers to get us the answers that we need. While NASA sends half a dozen people every year to live in the International Space Station, they spend the majority of their time operating a fleet of aluminum explorers. For the most part, our space program isn't human, it's robotic. And when these robots don't respond, it's a reminder of how alone we are on this little blue dot.

After 54 years of constant operation, the Deep Space Network is worn out.

The problem is not with the speed of the system itself, but

rather the hard blows of budget cuts, cyberthreats, and the decay of decades-old hardware. Like much of NASA, which receives less than 1 percent of the total federal budget every year, the DSN is underfunded. Every year is a constant fight just to keep the money they have. From the NASA budget, Space Communications and Navigation (SCaN), which the DSN is a part of, gets less than 1 percent of that. The budget for the DSN is only around $200 million a year, and it covers everything from maintaining the dishes to the ongoing upgrades to the antennas to paying the 300-plus people who work at the dishes around the world. For comparison, the total value of all the robotic missions currently in deep space is around $25 billion and growing.

This insufficient infrastructure support has left the predominantly robotic space program gasping for air. During the launches to Mars around 2012, the system became overloaded, causing glitches in the uplinking and downlinking of dozens of spacecraft already in the network. Many scientists began to worry that the DSN wouldn't be able to support the onslaught of missions vying for antenna time. With the Mars 2020 rover, Europa Clipper, Exo-Mars, and the James Webb Space Telescope all coming online in the next few years, the DSN will once again run into what NASA calls a time of "contention." When multiple spacecraft need to talk to Earth at once, the phone lines can glitch and cause a sudden disconnection, creating gaps in the data sent and received, and depending on how crucial the commands are, these crowded times could mean the difference between a healthy spacecraft or one doomed to wander the cosmos alone.

NASA alone fights every year for proper funding from the government, receiving only one half of 1 percent of the federal budget since people walked on the moon. Since the early 1970s, NASA's only major boost in funding has gone toward the SLS, NASA's long-delayed super rocket. Funding for operations like the Deep Space Network has been slashed. In 2013 SCaN ordered the DSN to cut its funding even further, and as a result it has lost $100 million in five years, delaying upgrades to antennas and the installation of new dishes in all three centers. There is no sign that their budget will increase anytime soon.

It's not just funding that the DSN is battling with. A 2015 NASA Office of Investigation report found that the network was extremely vulnerable to break-ins. Any large network is susceptible to hacking, but no other network is flying $25 billion worth of irreplaceable spacecraft.

According to a 2015 report by NASA's Office of Investigation, the Deep Space Network has been hacked several times. The worst break-in on record originated in China in 2011, when intruders managed to infiltrate the highest security-level clearance on JPL's computers, gaining the ability to copy, delete, and modify user accounts and sensitive files. The investigation is still ongoing, even seven years after the attack.

There are many, varied security protocols in place at each NASA center, but the 2015 report found several areas where JPL and NASA were in violation of the standards meant to keep them secure. NASA isn't lax about the state of security at their centers, but any vulnerability could lead to a disastrous outcome.

Multiple break-ins in 2009, 2011, and 2012 resulted in several arrests and jeopardized classified NASA information. These attacks have JPL and NASA officials on high alert, and while they are actively working on fixing the problems, the system as a whole is still extremely vulnerable. One significant software glitch or one really successful hack could shut down the entire system, potentially causing a multibillion-dollar disaster. And while some of these issues, as the report states, are fixable by simply being more careful —using a 12-digit password rather than 8, for example—others are due to the age of the system. Many of the components for the DSN are as old as the network itself, 54 years old and counting.

When Richard Stephenson interviewed for a job at the Deep Space Network, he was warned up front about the kangaroos. The road to the center in Canberra winds through the Tidbinbilla Valley, a quiet alcove surrounded by velvety green hills. It has the charms of a picturesque Australian landscape, but hidden within this verdant oasis are monoliths of steel. White round antennas litter the landscape, while the fauna of Australia meander around the 20-story-high pedestals. A white metal sign hangs at the entrance gate asking all visitors to turn off their cell phones, laptops, and anything else that emits radio waves, so that they can "help listen to whispers from space."

Just out of college, Stephenson planned to be a radio operator for the British Merchant Navy, but after a lecturer in one of his classes asked if anyone wanted to work for NASA in Australia, he jumped at the chance and ended up working for the Canberra Deep Space Communication Complex as a 20-year-old. Little did he know that he would be navigating ships of a different kind. It was early 1988, and he was hired specifically to help *Voyager 2* during its final and most challenging planetary encounter yet: the Neptune flyby. It would take two years for Stephenson to train to be a DSN link controller, leading up to the five most intense days of his life to date—the final hurrah for the *Voyager* mission.

For events like this, distance is the enemy of a DSN operator. *Voyager 2* would be the farthest from Earth it had ever been, and it was becoming almost impossible to talk to. For a planetary encounter, everything has to work perfectly. The spacecraft has to fly by just the right spot. It has to have its antenna turned exactly toward Earth, and its cameras and scientific instruments have to work. The only way to make sure all of this is happening is by talking through the antennas of the DSN. For the Neptune flyby, the stakes were so high that the Deep Space Network alone wasn't going to cut it. They had only one chance to visit Neptune, so scientists and mission engineers wanted to get as much data down as possible. Because of the extreme 2.7-billion-mile distance between Earth and the eighth planet, the bandwidth had to be increased. Not only were all the DSN dishes around the world on duty those five days, but the Parkes radio telescope in Australia, all 27 of the Very Large Array dishes in New Mexico, and the 64-meter Usuda dish in Japan all pointed their antennas at once toward *Voyager 2*. As it approached Neptune, the entire world was trying desperately to talk to the spacecraft—and to listen for an answer.

When the day finally arrived, Stephenson's job was to turn the knob to marry the signal from the Parkes radio telescope and the dishes at the DSN. "I can still remember standing there munching on a sandwich because we couldn't leave our consoles . . . It was like an adrenaline rush." And their efforts worked. For the first time in human history, the world saw images of Neptune. It was an ultramarine blue sphere featuring a large, dark, oval-shaped storm, appearing almost like a passageway into the planet. At the end of those five days, *Voyager 2* would begin its journey out of the solar system. Today it continues to travel quietly into deep space, constantly sending back tones from 11 billion miles away—tones we only hear when we turn a dish to listen.

Stephenson's hands have been on every deep space mission

launched from Earth since 1988. He has seen rovers try and fail to land on Mars, and other spacecraft try to enter into orbit, never to be heard from again. He's also seen the DSN blossom from an array of antennas built for 1970s exploration, grow to become the most powerful deep space communication system in the world. Dishes have been upgraded. Some have been decommissioned, scrapped, and replaced.

Stephenson has also seen the DSN fight for funding year after year. "There's never enough money," he says. "But over the last ten years, when you have a budget and halfway through the year it's cut, and then it's cut again before the end of that financial year. Then you're told that the budget you have for next year is going to be cut as well. It makes it very hard."

The people who work around the world in Madrid, Canberra, and Goldstone do not work nine-to-five shifts or take lunches promptly at noon. Spacecraft call in at all times of the day and night. Sometimes the alignment between Earth and Mars is best for a 4:00 a.m. uplink. Having teams of people working odd hours and employees stationed at three centers around the world can be expensive. To account for the mandatory budget cuts, changes in DSN operations were initiated in November 2017 with a program called Follow the Sun. Instead of each DSN center running their own dishes whenever the spacecraft aligned with their antennas, one team of operators will run all DSN dishes during its daytime hours. "As we go home at night, we'll be handing over to Madrid, and then Madrid will be handing over to Goldstone. So for the first time ever, we'll all be dabbling with each other's antennas," Stephenson explains. "That's quite a paradigm shift for the DSN."

This new shift system cuts down on how many staff members are working at once, and should account for millions in dollars in savings for the DSN portion of the budget. But this is not a long-term solution, and this system only works when there are no big events happening.

Eventually, as we stretch our arms out even farther into space, sending humans back to the moon or to Mars, the ability to send video will be required, as well as higher-resolution data. As it stands now, the bandwidth on the dishes will max out for those requirements. As an alternative, NASA engineers are looking to optical communication using lasers to carry more bits of data at a higher rate. And they need to do it quickly. Each spacecraft that is selected and funded is designed in part based on the DSN's ability to send and receive certain kinds of information. Antennas and scientific components are chosen in tandem with the rates and availability of the dishes. Just next year, the DSN is slated to begin communication for NASA's Orion EM (Exploration Mission) test flight around the moon. Soon the DSN will begin splitting its time between communication with human missions and the armada of robots already in flight.

You can talk to anyone from any NASA department anywhere, and they will tell you the same thing: there is not enough money. Badri Younes, the deputy associate administrator of SCaN at NASA Headquarters, is one of the people who decide where the money goes. "When you are operating with a budget that's flat, no one can get everything they want. You have to reconcile based on priority . . . In terms of maintenance, we may take some risk instead of replacing a piece."

But the centers aren't fighting over $100 million. "Sometimes we're fighting over two or three million dollars, or ten thousand dollars," says Stephenson. The budget cuts are having a real effect on daily operations. The department even bought antenna parts on eBay, as some of them are now obsolete. "We're starting to lose expertise," Stephenson added. "There are whole networks of people who are starting to retire. Our recovery times now are being impacted as well. Missions are seeing a direct link between the underfunded resources of the network and the data that's being delivered to them"—the data that's required for the success of any space mission.

Data is just another word for "information," but when I think about what the data *really* is in this case, it takes on a new meaning. Each bit of information that is sent back to Earth helps us understand more about other planets or their moons, which in turn informs us about our own existence. It's hard to imagine how data can be romantic, but when transmitted by the Deep Space Network, it represents answers to some of life's biggest questions. Why is one planet solid and another gaseous? Why is Earth the way it is and not like Venus? How do we fit into the universe's grander scheme?

After the power is restored in Mission Control, and the operators are getting everything back online, I walk down the dark hallway to a freight elevator. Jim Chu, the Data Center manager at

JPL, is taking me to the first of several new server rooms, the place where the DSN's data lives.

The doors open to braided ropes of caution tape and construction workers. "Pardon our mess," someone says, tipping their hard hat like a cowboy. As we walk through the dusty halls, there are rooms as big as houses with missing walls. Workers are putting up drywall, and dozens of people are hammering and drilling, building the entire floor from scratch.

Chu warns me it's going to be loud inside the Data Center, thanks to the cooling system. He grabs the security badge tethered to his belt and swipes it across the sensor. He yanks open the thick black fire door, and we're hit with an almost-deafening white noise from the machines. We instantly have to yell if we want to talk to each other. The walls are a soothing white to match the rows of white server racks, which look a bit like lockers. It even smells new, or what I imagine a newly minted server room to smell like, barely used plastic and fresh paint. Each section is about 10 feet tall and 20 feet long, separated into areas that each represent a group of spacecraft.

Chu walks me straight to a rack and gently opens the door. Red, blue, and white wires are neatly wound around each other, connections from the top trail down to below. It looks like a stomach sliced open with innards spilling out. These lockers are the guts of the Deep Space Network. We walk around the room, and he opens locker after locker and points: "This entire row is all Earth science." One row is devoted just to an Earth satellite called SMAP that measures moisture in the soil.

I really want to see *Voyager*, but Chu informs me the *Voyager* data has always been held off-site, just a mile from JPL at a place called Woodbury. When this floor is completed, all the *Voyager* data will be moved here and the entire DSN system will finally be housed under one roof. I ask Chu about the other spacecraft's data sets as if they were celebrities. Where's *Cassini*? Where's *Curiosity*? Where's *Juno*? He walks to a row toward the very back of the room, opens a single locker, and points to a dozen neatly stacked 3-terabyte hard drives at the bottom. "That's part of Mars and *Juno*"—a spacecraft currently in orbit around Jupiter. Two DVD player–sized stacks of hard drives house data for some of the biggest space missions we've ever launched. Unlike the Earth missions, these deep space missions don't require a lot of room for storage because it takes so long to get the data in the first place. Their bit rates are small. They trickle in slowly because of their distance from Earth, and as a result they require less square footage.

Before we leave, we stop and stare at the rows of new white lockers and listen to the soothing hum of the cooling system. All I can see in my head are the shiny, grooved rings of Saturn, the swirls of Jupiter's red spot, grains of dust on Mars, the heart of Pluto, the haze of Venus. It was all here, in the noisy confines of a light, bright room so unlike the dark quietude of space.

The Deep Space Network sees the beginnings of things, and keeps an ear to the long, hushed middle; only rarely does it see the end. It's the listeners whose time is fleeting.

When the twin *Voyager* spacecraft departed Earth in the summer of 1977, they were destined to embark on the most epic journey any mission had ever attempted, a distinction they still hold 41 years later. Not only would they visit every planet and take the first pictures of Uranus and Neptune, but they would also both pierce through the boundary of plasma from the sun that envelopes our solar system. Their goal was to leave us forever.

Suzy Dodd was just 23 when she joined the *Voyager* team, two years before the Uranus flyby in 1984, three years before Richard started at Canberra, and four years before *Voyager* would skim past Neptune. She's currently both the project manager on *Voyager* and also the director for the Interplanetary Network, a directorate at JPL that manages the DSN. Images of the Spitzer Space Telescope, *Voyager*, and Neptune cover the walls of her office, along with awards of just about every kind NASA offers. While she began her work on the team for the Uranus encounter, she's quick to say that it was the final *Voyager* flyby that was the most special to her.

The Neptune visit was the denouement of *Voyager*'s long story. Sure it would continue out into the depths of the solar system, sending back science data, but after Neptune there would be no more photos, just darkness. "It was really special," says Dodd, with a warm expression. "It was also the realization that wow, we're done."

September 14, 2017, was the day before *Cassini* was scheduled to deorbit, enter Saturn's atmosphere, and burn up—20 years after its launch. I was among the few members of the press allowed into the Center of the Universe for a tour. We didn't know we'd soon bear witness to a real-time downlink from the spacecraft. The red DSN TRACK IN PROGRESS sign was placed atop the ACE's desk. *Cassini* was calling home.

This conversation happened suddenly, and the human chatter became hushed as soon as the *Cassini* ACE's phone rang. It was a call from the center in Goldstone. They were checking the connection. After all, this was close to the end. For a moment, the spacecraft dropped out of lock. The clasped fingers slipped, and they were attempting to reconnect. I stood behind the ACE's chair, just inches from him, and watched his computer screens quickly fill with lines of jumbled numbers and letters. Hidden within this code were images and updates from the spacecraft—it was OK, *Cassini* was still on a collision course with the planet.

I looked up to the monitors that months earlier were blank. Now they were filled with working antennas and the spacecraft they were talking to. The countdown clock to *Cassini* indicated the end of the mission was near: T-0 14:06:20.x.

The next morning, hundreds of people awoke before dawn to head back to JPL, where together we would witness the end of *Cassini*. The only way we could know it was over was by listening via the Deep Space Network. We sat in silence in those last few minutes. We knew it would be sometime around 5:00 a.m. but there was no telling how long the spacecraft would survive before we lost it. The center in Canberra would be the last to talk to *Cassini*; they had their biggest ears listening with Dish #43. The screens at JPL displayed the radio downlink information. This is what they were watching in Canberra. We were connected to *Cassini* via two radio connections on an X band and an S band. These movements look like any EKG or heart monitor. You can almost hear the rhythms coming all the way from Saturn: ba-bump-ba-bump-ba-bump. There were peaks and drops. We still had it! Ba-bump-ba-bump-ba-bump. No one blinked, and then suddenly at 4:55 a.m. PST, the heartbeats from the S band fell flat.

> Mission Control: I call loss of signal at 11:55:46 for the S band. So that would be the end of the spacecraft.
> *Cassini* Project Manager: Copy that. There may be a trickle of telemetry left, but you just heard the signal from the spacecraft is gone and in the next 45 seconds so will be the spacecraft.
> I hope you're all deeply proud of this amazing accomplishment.

> Congratulations to you all. This has been an incredible mission, an incredible spacecraft, and you're all an incredible team. I'm going to call the End of Mission. Project manager, off the net.

In that moment, I thought of something Richard Stephenson had said, speaking with a colleague about the upcoming death of *Cassini*. "He said, 'We'll be with you till your last bit,' and I felt myself getting all teary. I suppose we always stamp human emotion on these things."

In 1899 Nikola Tesla wrote a treatise called "Talking with the Planets," and his visions are indiscernible from what's come to pass.

> At the present stage of progress, there would be no insurmountable obstacle in construction of a machine capable of conveying a message to Mars, nor would there be any great difficulty in recording signals transmitted to us by the inhabitants of that planet, if they be skilled electricians, communication once established, even in the simplest way, by a mere interchange of numbers, the progress towards more intelligible communication would be rapid. Absolute certitude as to the receipt and interchange of messages would be reached soon as we could respond with the number *four*, say in reply to the signal *one two* and *three*. The Martians or the inhabitants of whatever planet had signaled to us would understand at once that we had caught their message across the gulf of space and had sent back a response. What a tremendous stir this would make in the world! How soon will it come?

While we have yet to find intelligent civilizations wandering around our cosmic neighborhood, there is in fact intelligent life in the universe—we put it there.

Voyager 1 has already departed the solar system. For the past five years it's been sailing between our star and another, and every day it still calls home. One day it will stop calling. For years the team has been slowly turning off instruments on both *Voyager 1* and *Voyager 2* in order to preserve the most important feature—the communication link. Suzy Dodd thinks the spacecraft have several years left. There's no way to know for sure what *Voyager*'s final call will be. "You don't exactly know when you get to say goodbye." she tells me. "So every day you should say goodbye."

Before leaving JPL, I stop in the quad for a moment to take in the Pasadena sunshine and open my phone to bring up the DSN

NOW website. It takes a moment to load, and then suddenly there was *Voyager* sending us whispers, in squiggly lines and solid lines, telemetry, data, tones, heartbeats. We were talking again.

As you read this, there are men and women at the DSN stations checking links, turning dishes, and talking to space. Some might be eating a sandwich at their console, watching the jagged lines disappear as a connection between Earth and deep space temporarily severs and is filled again with the black silence of the cosmos. It is likely the Deep Space Network will forever remain the silent partner of the space program, but no doubt its heart will continue to beat, its dishes will sway to meet the rise of a spacecraft over the horizon, its operators will bring its radio waves together, and its explorers will turn their faces to Earth and say hello.

LINDA VILLAROSA

The Hidden Toll: Why Are Black Mothers and Babies in the United States Dying at More Than Double the Rate of White Mothers and Babies? The Answer Has Everything to Do with the Lived Experience of Being a Black Woman in America

FROM *The New York Times Magazine*

WHEN SIMONE LANDRUM felt tired and both nauseated and ravenous at the same time in the spring of 2016, she recognized the signs of pregnancy. Her beloved grandmother died earlier that year, and Landrum felt a sense of divine order when her doctor confirmed on Muma's birthday that she was carrying a girl. She decided she would name her daughter Harmony. "I pictured myself teaching my daughter to sing," says Landrum, now 23, who lives in New Orleans. "It was something I thought we could do together."

But Landrum, who was the mother of two young sons, noticed something different about this pregnancy as it progressed. The trouble began with constant headaches and sensitivity to light; Landrum described the pain as "shocking." It would have been reasonable to guess that the crippling headaches had something to do with stress: Her relationship with her boyfriend, the baby's

father, had become increasingly contentious and eventually physically violent. Three months into her pregnancy, he became angry
at her for wanting to hang out with friends and threw her to the
ground outside their apartment. She scrambled to her feet, ran
inside, and called the police. He continued to pursue her, so she
grabbed a knife. "Back up—I have a baby," she screamed. After the
police arrived, he was arrested and charged with multiple offenses,
including battery. He was released on bond pending a trial that
would not be held until the next year. Though she had broken up
with him several times, Landrum took him back, out of love and
also out of fear that she couldn't support herself, her sons, and the
child she was carrying on the paycheck from her waitress gig at a
restaurant in the French Quarter.

As her January due date grew closer, Landrum noticed that her
hands, her feet, and even her face were swollen, and she had to
quit her job because she felt so ill. But her doctor, whom several
friends had recommended and who accepted Medicaid, brushed
aside her complaints. He recommended Tylenol for the headaches. "I am not a person who likes to take medicine, but I was always popping Tylenol," Landrum says. "When I told him my head
still hurt, he said to take more."

At a prenatal appointment a few days before her baby shower in
November, Landrum reported that the headache had intensified
and that she felt achy and tired. A handwritten note from the appointment, sandwiched into a printed file of Landrum's electronic
medical records that she later obtained, shows an elevated blood
pressure reading of 143/86. A top number of 140 or more or a
bottom number higher than 90, especially combined with headaches, swelling, and fatigue, points to the possibility of preeclampsia: dangerously high blood pressure during pregnancy.

High blood pressure and cardiovascular disease are two of
the leading causes of maternal death, according to the Centers
for Disease Control and Prevention, and hypertensive disorders
in pregnancy, including preeclampsia, have been on the rise over
the past two decades, increasing 72 percent from 1993 to 2014. A
Department of Health and Human Services report last year found
that preeclampsia and eclampsia (seizures that develop after preeclampsia) are 60 percent more common in African American
women and also more severe. Landrum's medical records note
that she received printed educational material about preeclampsia

during a prenatal visit. But Landrum would comprehend the details about the disorder only months later, doing online research on her own.

When Landrum complained about how she was feeling more forcefully at the appointment, she recalls, her doctor told her to lie down—and calm down. She says that he also warned her that he was planning to go out of town and told her that he could deliver the baby by C-section that day if she wished, six weeks before her early-January due date. Landrum says it seemed like an ultimatum, centered on his schedule and convenience. So she took a deep breath and lay on her back for 40 minutes until her blood pressure dropped within normal range. Aside from the handwritten note, Landrum's medical records don't mention the hypertensive episode, the headaches or the swelling, and she says that was the last time the doctor or anyone from his office spoke to her. "It was like he threw me away," Landrum says angrily.

Four days later, Landrum could no longer deny that something was very wrong. She was suffering from severe back pain and felt bone-tired, unable to get out of bed. That evening, she packed a bag and asked her boyfriend to take her sons to her stepfather's house and then drive her to the hospital. In the car on the way to drop off the boys, she felt wetness between her legs and assumed her water had broken. But when she looked at the seat, she saw blood. At her stepfather's house, she called 911. Before she got into the ambulance, Landrum pulled her sons close. "Mommy loves you," she told them, willing them to stay calm. "I have to go away, but when I come back I will have your sister."

By the time she was lying on a gurney in the emergency room of Touro Infirmary, a hospital in the Uptown section of New Orleans, the splash of blood had turned into a steady stream. "I could feel it draining out of me, like if you get a jug of milk and pour it onto the floor," she recalls. Elevated blood pressure—Landrum's medical records show a reading of 160/100 that day—had caused an abruption: the separation of the placenta from her uterine wall.

With doctors and nurses hovering over her, everything became both hazy and chaotic. When a nurse moved a monitor across her belly, Landrum couldn't hear a heartbeat. "I kept saying: 'Is she OK? Is she all right?'" Landrum recalls. "Nobody said a word. I have never heard a room so silent in my life." She remembers that the emergency room doctor dropped his head. Then he looked into her eyes. "He told me my baby was dead inside of me. I was like: What just happened? Is this a dream? And then I turned my head to the side and threw up."

Sedated but conscious, Landrum felt her mind growing foggy. "I was just so tired," she says. "I felt like giving up." Then she pictured the faces of her two young sons. "I thought, Who's going to take care of them if I'm gone?" That's the last thing she recalls clearly. When she became more alert sometime later, a nurse told her that she had almost bled to death and had required a half dozen units of transfused blood and platelets to survive. "The nurse told me: 'You know, you been sick. You are very lucky to be alive,'" Landrum remembers. "She said it more than once."

A few hours later, a nurse brought Harmony, who had been delivered stillborn via C-section, to her. Wrapped in a hospital blanket, her hair thick and black, the baby looked peaceful, as if she were dozing. "She was so beautiful—she reminded me of a doll," Landrum says. "I know I was still sedated, but as I held her, I kept looking at her, thinking, Why doesn't she wake up? I tried to feel love, but after a while I got more and more angry. I thought, Why is God doing this to me?"

The hardest part was going to pick up her sons empty-handed and telling them that their sister had died. "I felt like I failed them," Landrum says, choking up. "I felt like someone had taken something from me, but also from them."

In 1850, when the death of a baby was simply a fact of life, and babies died so often that parents avoided naming their children before their first birthdays, the United States began keeping records of infant mortality by race. That year, the reported black infant-mortality rate was 340 per 1,000; the white rate was 217 per 1,000. This black/white divide in infant mortality has been a source of both concern and debate for over a century. In his 1899 book, *The Philadelphia Negro*, the first sociological case study of black Americans, W. E. B. Du Bois pointed to the tragedy of black infant death and persistent racial disparities. He also shared his own "sorrow song," the death of his baby son, Burghardt, in his 1903 masterwork, *The Souls of Black Folk*.

From 1915 through the 1990s, amid vast improvements in hygiene, nutrition, living conditions, and healthcare, the number of babies of all races who died in the first year of life dropped by over 90 percent—a decrease unparalleled by reductions in other causes of death. But that national decline in infant mortality has since slowed. In 1960 the United States was ranked 12th among developed countries in infant mortality. Since then, with its rate largely driven by the deaths of black babies, the United States has fallen behind and now ranks 32nd out of the 35 wealthiest nations. Low birth weight is a key factor in infant death, and a new report released in March by the Robert Wood Johnson Foundation and the University of Wisconsin suggests that the number of low-birth-weight babies born in the United States—also driven by the data for black babies—has inched up for the first time in a decade.

Black infants in America are now more than twice as likely to die as white infants—11.3 per 1,000 black babies, compared with 4.9 per 1,000 white babies, according to the most recent government data—a racial disparity that is actually wider than in 1850, 15 years before the end of slavery, when most black women were considered chattel. In one year, that racial gap adds up to more than 4,000 lost black babies. Education and income offer little protection. In fact, a black woman with an advanced degree is more likely to lose her baby than a white woman with less than an eighth-grade education.

This tragedy of black infant mortality is intimately intertwined with another tragedy: a crisis of death and near-death in black mothers themselves. The United States is one of only 13 countries in the world where the rate of maternal mortality—the death of a woman related to pregnancy or childbirth up to a year after the end of pregnancy—is now worse than it was 25 years ago. Each year, an estimated 700 to 900 maternal deaths occur in the United States. In addition, the CDC reports more than 50,000 potentially preventable near-deaths, like Landrum's, per year—a number that rose nearly 200 percent from 1993 to 2014, the last year for which statistics are available. Black women are three to four times as likely to die from pregnancy-related causes as their white counterparts, according to the CDC—a disproportionate rate that is higher than that of Mexico, where nearly half the population lives in poverty —and as with infants, the high numbers for black women drive the national numbers.

Monica Simpson is the executive director of SisterSong, the country's largest organization dedicated to reproductive justice for women of color, and a member of the Black Mamas Matter

Alliance, an advocacy group. In 2014 she testified in Geneva before the United Nations Committee on the Elimination of Racial Discrimination, saying that the United States, by failing to address the crisis in black maternal mortality, was violating an international human rights treaty. After her testimony, the committee called on the United States to "eliminate racial disparities in the field of sexual and reproductive health and standardize the data-collection system on maternal and infant deaths in all states to effectively identify and address the causes of disparities in maternal- and infant-mortality rates." No such measures have been forthcoming. Only about half the states and a few cities maintain maternal-mortality review boards to analyze individual cases of pregnancy-related deaths. There has not been an official federal count of deaths related to pregnancy in more than 10 years. An effort to standardize the national count has been financed in part by contributions from Merck for Mothers, a program of the pharmaceutical company, to the CDC Foundation.

The crisis of maternal death and near-death also persists for black women across class lines. This year, the tennis star Serena Williams shared in *Vogue* the story of the birth of her first child and in further detail in a Facebook post. The day after delivering her daughter, Alexis Olympia, via C-section in September, Williams experienced a pulmonary embolism, the sudden blockage of an artery in the lung by a blood clot. Though she had a history of this disorder and was gasping for breath, she says medical personnel initially ignored her concerns. Though Williams should have been able to count on the most attentive healthcare in the world, her medical team seems to have been unprepared to monitor her for complications after her cesarean, including blood clots, one of the most common side effects of C-sections. Even after she received treatment, her problems continued; coughing, triggered by the embolism, caused her C-section wound to rupture. When she returned to surgery, physicians discovered a large hematoma, or collection of blood, in her abdomen, which required more surgery. Williams, 36, spent the first six weeks of her baby's life bedridden.

The reasons for the black/white divide in both infant and maternal mortality have been debated by researchers and doctors for more than two decades. But recently there has been growing acceptance of what has largely been, for the medical establishment, a shocking idea: For black women in America, an inescapable atmosphere of societal and systemic racism can create a kind of toxic physiological stress, resulting in conditions—including hypertension and preeclampsia—that lead directly to higher rates of infant and maternal death. And that societal racism is further expressed in a pervasive, long-standing racial bias in healthcare—including the dismissal of legitimate concerns and symptoms—that can help explain poor birth outcomes even in the case of black women with the most advantages.

"Actual institutional and structural racism has a big bearing on our patients' lives, and it's our responsibility to talk about that more than just saying that it's a problem," says Dr. Sanithia L. Williams, an African American OB-GYN in the Bay Area and a fellow with the nonprofit organization Physicians for Reproductive Health. "That has been the missing piece, I think, for a long time in medicine."

After Harmony's death, Landrum's life grew more chaotic. Her boyfriend blamed her for what happened to their baby and grew more abusive. Around Christmas 2016, in a rage, he attacked her, choking her so hard that she urinated on herself. "He said to me, 'Do you want to die in front of your kids?'" Landrum said, her hands shaking with the memory.

Then he tore off her clothes and sexually assaulted her. She called the police, who arrested him and charged him with second-degree rape. Landrum got a restraining order, but the district attorney eventually declined to prosecute. She also sought the assistance of the New Orleans Family Justice Center, an organization that provides advocacy and support for survivors of domestic violence and sexual assault. Counselors secreted her and her sons to a safe house, before moving them to a more permanent home early last year.

Landrum had a brief relationship with another man and found out in March 2017 that she was pregnant again and due in December. "I'm not going to lie; though I had a lot going on, I wanted to give my boys back the sister they had lost," Landrum said, looking down at her lap. "They don't forget. Every night they always say their prayers, like: 'Goodnight, Harmony. Goodnight, God. We love you, Sister.'" She paused and took a breath. "But I was also afraid, because of what happened to me before."

Early last fall, Landrum's case manager at the Family Justice

Center, Mary Ann Bartkowicz, attended a workshop conducted by Latona Giwa, the 31-year-old cofounder of the Birthmark Doula Collective. The group's 12 racially diverse birth doulas, ages 26 to 46, work as professional companions during pregnancy and childbirth and for six weeks after the baby is born, serving about 400 clients across New Orleans each year, from wealthy women who live in the upscale Garden District to women from the Katrina-ravaged Lower Ninth Ward and other communities of color who are referred through clinics, school counselors, and social-service organizations. Birthmark offers pro bono services to these women in need.

Right away, the case manager thought of her young, pregnant client. Losing her baby, nearly bleeding to death, and fleeing an abusive partner were only the latest in a cascade of harrowing life events that Landrum had lived through since childhood. She was 10 when Hurricane Katrina devastated New Orleans in 2005. She and her family first fled to a hotel and then walked more than a mile through the rising water to the Superdome, where thousands of evacuees were already packed in with little food, water, or space. She remembers passing Charity Hospital, where she was born. "The water was getting deeper and deeper, and by the end, I was on my tippy-toes, and the water was starting to go right by my mouth," Landrum recalls. "When I saw the hospital, honestly I thought, I'm going to die where I was born." Landrum wasn't sure what doulas were, but once Bartkowicz explained their role as a source of support and information, she requested the service. Latona Giwa would be her doula.

Giwa, the daughter of a white mother and a Nigerian immigrant father, took her first doula training while she was still a student at Grinnell College in Iowa. She moved to New Orleans for a fellowship in community organizing before getting a degree in nursing. After working as a labor and delivery nurse and then as a visiting nurse for Medicaid clients in St. Bernard Parish, an area of southeast New Orleans where every structure was damaged by Katrina floodwaters, she devoted herself to doula work and childbirth education. She founded Birthmark in 2011 with Dana Keren, another doula who was motivated to provide services for women in New Orleans who most needed support during pregnancy but couldn't afford it.

"Being a labor and delivery nurse in the United States means seeing patients come in acute medical need, because we haven't been practicing preventive and supportive care all along," Giwa says. Louisiana ranks 44th out of all 50 states in maternal mortality; black mothers in the state die at 3.5 times the rate of white mothers. Among the 1,500 clients the Birthmark doulas have served since the collective's founding seven years ago, 10 infant deaths have occurred, including late-term miscarriage and stillbirth, which is lower than the overall rate for both Louisiana and the United States, as well as the rates for black infants. No mothers have died.

A scientific examination of 26 studies of nearly 16,000 subjects first conducted in 2003 and updated last year by Cochrane, a nonprofit network of independent researchers, found that pregnant women who received the continuous support that doulas provide were 39 percent less likely to have C-sections. In general, women with continuous support tended to have babies who were healthier at birth. Though empirical research has not yet linked doula support with decreased maternal and infant mortality, there are promising anecdotal reports. Last year, the American College of Obstetricians and Gynecologists released a statement noting that "evidence suggests that, in addition to regular nursing care, continuous one-to-one emotional support provided by support personnel, such as a doula, is associated with improved outcomes for women in labor."

In early November, the air was thick with humidity as Giwa pulled up to Landrum's house, half of a wood-frame duplex, for their second meeting. Landrum opened the door, happy to see the smiling, fresh-faced Giwa, who at first glance looked younger than her 23-year-old client. Giwa would continue to meet with Landrum weekly until her December 22 due date, would be with her during labor and delivery, and would make six postpartum home visits to assure that both mother and baby son remained healthy. Landrum led Giwa through her living room, which was empty except for a tangle of disconnected cable cords. She had left most of her belongings behind—including her dog and the children's new Christmas toys—when she fled from her abusive boyfriend, and she still couldn't afford to replace all her furniture.

They sat at the kitchen table, where Giwa asked about Landrum's last doctor visit, prodding her for details. Landrum reassured her that her blood pressure and weight, as well as the baby's size and position, were all on target.

"Have you been getting rid of things that are stressful?" Giwa asked, handing her a tin of lavender balm, homemade from herbs in her garden.

"I'm trying not to be worried, but sometimes. . . ." Landrum said haltingly, looking down at the table as her hair, tipped orange at the ends, brushed her shoulders. "I feel like my heart is so anxious."

Taking crayons from her bag, Giwa suggested they write affirmations on sheets of white paper for Landrum to post around her home, to see and remind her of the good in her life. Landrum took a purple crayon, her favorite color, and scribbled in tight, tiny letters. But even as she wrote the affirmations, she began to recite a litany of fears: bleeding again when she goes into labor, coming home empty-handed, dying and leaving her sons motherless. Giwa leaned across the table, speaking evenly. "I know that it was a tragedy and a huge loss with Harmony, but don't forget that you survived, you made it, you came home to your sons," she said. Landrum stopped writing and looked at Giwa.

"If it's OK, why don't I write down something you told me when we talked last time?" Giwa asked. Landrum nodded. "I know God has his arms wrapped around me and my son," Giwa wrote in large purple letters, outlining *God* and *arms* in red, as Landrum watched. She took out another sheet of paper and wrote, "Harmony is here with us, protecting us." After the period, she drew two purple butterflies.

Landrum's eyes locked on the butterflies. "Every day, I see a butterfly, and I think that's her. I really do," she said, finally smiling, her large, dark eyes crinkling into half moons. "I like that a lot, because I think that's something that I can look at and be like, Girl, you going to be OK."

With this pregnancy, Landrum was focused on making sure everything went right. She had switched to a new doctor, a woman who specialized in high-risk pregnancies and accepted Medicaid, and she would deliver this baby at a different hospital. Now she asked Giwa to review the birth plan one more time.

"On November 30, I go on call, and that means this phone is always on me," Giwa said, holding up her iPhone.

"What if . . ." Landrum began tentatively.

"I'm keeping a backup doula informed of everything," Giwa said. "Just in case."

"I think everything's going to be OK this time," Landrum said. But it sounded like a question.

When the black/white disparity in infant mortality first became the subject of study, discussion, and media attention more than two decades ago, the high rate of infant death for black women was widely believed by almost everyone, including doctors and public health experts, to affect only poor, less educated women—who do experience the highest numbers of infant deaths. This led inevitably to blaming the mother. Was she eating badly, smoking, drinking, using drugs, overweight, not taking prenatal vitamins or getting enough rest, afraid to be proactive during prenatal visits, skipping them altogether, too young, unmarried?

At *Essence* magazine, where I was the health editor from the late 1980s to the mid-'90s, we covered the issue of infant mortality by encouraging our largely middle-class black female readers to avoid unwanted pregnancy and by reminding them to pay attention to their health habits during pregnancy and make sure newborns slept on their backs. Because the future of the race depended on it, we also promoted a kind of each-one-teach-one mentality: encourage teenagers in your orbit to just say no to sex and educate all the "sisters" in your life (read: your less educated and less privileged friends and family) about the importance of prenatal care and healthful habits during pregnancy.

In 1992 I was a journalism fellow at the Harvard T. H. Chan School of Public Health. One day a professor of health policy, Dr. Robert Blendon, who knew I was the health editor of *Essence*, said, "I thought you'd be interested in this." He handed me the latest issue of the *New England Journal of Medicine*, which contained what is now considered the watershed study on race, class, and infant mortality. The study, conducted by four researchers at the CDC—Kenneth Schoendorf, Carol Hogue, Joel Kleinman, and Diane Rowley—mined a database of close to a million previously unavailable linked birth and death certificates and found that infants born to college-educated black parents were twice as likely to die as infants born to similarly educated white parents. In 72 percent of the cases, low birth weight was to blame. I was so surprised and skeptical that I peppered him with the kinds of questions about medical research that he encouraged us to ask in his course. Mainly I wanted to know *why.* "No one knows," he told me, "but this might have something to do with stress."

Though I wouldn't learn of her work until years later, Dr. Arline Geronimus, a professor in the department of health behavior and health education at the University of Michigan School of Public Health, first linked stress and black infant mortality with her theory of "weathering." She believed that a kind of toxic stress triggered the premature deterioration of the bodies of African American women as a consequence of repeated exposure to a climate of discrimination and insults. The weathering of the mother's body, she theorized, could lead to poor pregnancy outcomes, including the death of her infant.

After graduating from the Harvard School of Public Health, Geronimus landed at Michigan in 1987, where she continued her research. That year, in a report published in the journal *Population and Development Review,* she noted that black women in their mid-20s had higher rates of infant death than teenage girls did —presumably because they were older and stress had more time to affect their bodies. For white mothers, the opposite proved true: teenagers had the highest risk of infant mortality, and women in their mid-20s the lowest.

Geronimus's work contradicted the widely accepted belief that black teenage girls (assumed to be careless, poor, and uneducated) were to blame for the high rate of black infant mortality. The backlash was swift. Politicians, media commentators, and even other scientists accused her of promoting teenage pregnancy. She was attacked by colleagues and even received anonymous death threats at her office in Ann Arbor and at home. "At that time, which is now twenty-five or so years ago, there were more calls to complain about me to the University of Michigan, to say I should be fired, than had happened to anybody in the history of the university," recalls Geronimus, who went on to publish in 1992 what is now considered her seminal study on weathering and black women and infants in the journal *Ethnicity and Disease.*

By the late 1990s, other researchers were trying to chip away at the mystery of the black/white gap in infant mortality. Poverty on its own had been disproved to explain infant mortality, and a study of more than 1,000 women in New York and Chicago, published in the *American Journal of Public Health* in 1997, found that black women were less likely to drink and smoke during pregnancy, and that even when they had access to prenatal care, their babies were often born small.

Experts wondered if the high rates of infant death in black women, understood to be related to small, preterm babies, had a genetic component. Were black women passing along a defect that was affecting their offspring? But science has refuted that theory too: a 1997 study published by two Chicago neonatologists, Richard David and James Collins, in the *New England Journal of Medicine,* found that babies born to new immigrants from impoverished West African nations weighed more than their black American-born counterparts and were similar in size to white babies. In other words, they were more likely to be born full term, which lowers the risk of death. In 2002 the same researchers made a further discovery: the daughters of African and Caribbean immigrants who grew up in the United States went on to have babies who were smaller than their mothers had been at birth, while the grandchildren of white European women actually weighed more than their mothers had at birth. It took just one generation for the American black/white disparity to manifest.

When I became pregnant in 1996, this research became suddenly real for me. When my Park Avenue OB-GYN, a female friend I trusted implicitly, discovered that my baby was far smaller than her gestational age would predict, even though I was in excellent health, she put me on bed rest and sent me to a specialist. I was found to have a condition called intrauterine growth restriction (IUGR), generally associated with mothers who have diabetes, high blood pressure, malnutrition, or infections including syphilis, none of which applied to me. During an appointment with a perinatologist—covered by my excellent health insurance—I was hounded with questions about my "lifestyle" and whether I drank, smoked, or used a vast assortment of illegal drugs. I wondered, Do these people think I'm sucking on a crack pipe the second I leave the office? I eventually learned that in the absence of a medical condition, IUGR is almost exclusively linked with mothers who smoke or abuse drugs and alcohol. As my pregnancy progressed but my baby didn't grow, my doctor decided to induce labor one month before my due date, believing that the baby would be healthier outside my body. My daughter was born at four pounds 13 ounces, classified as low birth weight. Though she is now a bright, healthy, athletic college student, I have always wondered: Was this somehow related to the experience of being a black woman in America?

Though it seemed radical 25 years ago, few in the field now dispute that the black/white disparity in the deaths of babies is related not to the genetics of race but to the lived experience of race in this country. In 2007 David and Collins published an even more thorough examination of race and infant mortality in the *American Journal of Public Health*, again dispelling the notion of some sort of gene that would predispose black women to preterm birth or low birth weight. To make sure the message of the research was crystal clear, David, a professor of pediatrics at the University of Illinois at Chicago, stated his hypothesis in media-friendly but blunt-force terms in interviews: "For black women," he said, "something about growing up in America seems to be bad for your baby's birth weight."

On a December morning three days before her due date, Landrum went to the hospital for her last ultrasound before the birth. Because of the stillbirth the previous year, her doctor did not want to let the pregnancy go past 40 weeks, to avoid the complications that can come with post-term delivery, so an induction had been scheduled in 48 hours.

During Giwa's last prenatal visit, the day before, she explained to Landrum that she would be given Pitocin, a synthetic version of the natural hormone that makes the uterus contract during labor, to start her contractions. "Will inducing stress out the baby?" Landrum asked. "I can't lie; I used to wake up and scream, when I'd be dreaming about getting cut open again. I know my body is fine, and I'm healthy, but I don't want to die."

"I respect how honest you are, and your trauma is real," Giwa told her, slowing down her words. "But my hope for you is, this birth can be a part of your healing. Your uterus is injured and has been scarred, but you've pushed out two babies, so your body knows what it's doing."

Now, lying on the table, Landrum looked out the window, smiling as the sound of her baby's heartbeat filled the room. A few minutes later, the technician returned and looked at the monitor. The baby's heart rate appeared less like little mountains than chicken scratching. He was also either not moving consistently or not breathing properly. A nurse left the room to call Landrum's doctor to get her opinion. The nurse returned in 20 minutes and gave Landrum the news that the baby would be induced not in two days but now. "We don't want to wait; we're going to get him out today," she said to Landrum.

"I'm very anxious," Landrum told Giwa on the phone as she walked to labor and delivery, a few floors up in the same hospital, "but I'm ready." An hour later, Giwa arrived, wearing purple scrubs, her cloth bag filled with snacks, lavender lotion, and clary sage oil. She made sure the crayon-drawn affirmations were taped on the wall within Landrum's line of vision, then settled into a chair next to the bed, low-key but watchful. Though some doctors resent or even forbid the presence of a doula during labor and delivery—and some doulas overstep their roles and create conflict with doctors and nurses—Giwa says she and the other Birthmark doulas try to be unobtrusive and focused on what's best for the mother.

A medical resident, who was white, like all of the staff who would attend Landrum throughout her labor and delivery, walked into the room with paperwork. Right away, she asked Landrum briskly, "Have you had any children before?"

She hadn't read the chart.

"Yes, I've had three babies, but one died," Landrum explained warily, for the third time since she had arrived at the hospital that day. Her voice was flat. "I had a stillbirth."

"The demise was last year?" the resident asked without looking up to see Landrum stiffen at the word *demise.*

"May I speak to you outside," Giwa said to the nurse caring for Landrum. In the hall, she asked her to please make a note in Landrum's chart about the stillbirth. "Each time she has to go over what happened, it brings her mind back to a place of fear and anxiety and loss," Giwa said later. "This is really serious. She's having a high-risk delivery, and I would hope that her care team would thoroughly review her chart before walking into her room."

One of the most important roles that doulas play is as an advocate in the medical system for their clients. "At the point a woman is most vulnerable, she has another set of ears and another voice to help get through some of the potentially traumatic decisions that have to be made," says Dána-Ain Davis, the director of the Center for the Study of Women and Society at the City University of

New York, the author of a forthcoming book on pregnancy, race, and premature birth and a black woman who is a doula herself. Doulas, she adds, "are a critical piece of the puzzle in the crisis of premature birth, infant and maternal mortality in black women."

Over the next 10 hours, Giwa left Landrum's side only briefly. About 5 hours in, Landrum requested an epidural. The anesthesiologist required all visitors to leave the room while it was administered. When Giwa returned about a half hour later, Landrum was angry and agitated, clenching her fists and talking much faster than usual. She had mistakenly been given a spinal dose of anesthesia—generally reserved for C-sections performed in the operating room—rather than the epidural dose usually used in vaginal childbirth. Now she had no feeling at all in her legs and a splitting headache. When she questioned the incorrect dose of anesthesia, Landrum told Giwa, one nurse said, "You ask a lot of questions, don't you?" and winked at another nurse in the room and then rolled her eyes.

As Landrum loudly complained about what occurred, her blood pressure shot up, while the baby's heart rate dropped. Giwa glanced nervously at the monitor, the blinking lights reflecting off her face. "What happened was wrong," she said to Landrum, lowering her voice to a whisper. "But for the sake of the baby, it's time to let it go."

She asked Landrum to close her eyes and imagine the color of her stress.

"Red," Landrum snapped, before finally laying her head onto the pillow.

"What color is really soothing and relaxing?" Giwa asked, massaging her hand with lotion.

"Lavender," Landrum replied, taking a deep breath. Over the next 10 minutes, Landrum's blood pressure dropped within normal range as the baby's heart rate stabilized.

At 1:00 a.m., a team of three young female residents bustled into the room; the labor and delivery nurse followed them, flipping on the overhead light. They were accompanied by an older man Landrum had never seen. He briefly introduced himself as the attending physician before plunging his hand between Landrum's legs to feel for the baby. Landrum had been told that her OB-GYN might not deliver her infant, but a nurse had reassured her earlier in the day that if her doctor was not available, her doctor's husband, also an OB-GYN, would cover for her. This doctor, however, was not the husband, and no one explained the switch. Giwa raised an eyebrow. The Listening to Mothers Survey III, a national sampling of 2,400 women who gave birth in 2011 and 2012, found that more than a quarter of black women meet their birth attendants for the first time during childbirth, compared with 18 percent of white women.

"He's ready," the doctor said, snapping off his gloves. "It's time to push."

One resident stepped forward and took his place, putting her hand into Landrum's vagina, feeling for the baby. Landrum gripped the side of the bed and closed her eyes, grimacing. "You're a rock star," Giwa said. The nurse, standing at her side, told Landrum: "Push! Now. You can do it." After about 20 minutes of pushing, the baby's head appeared. "This is it," the nurse told her. "You can do this," Giwa whispered on her other side.

Landrum bore down and pushed again. "You're doing amazing," Giwa said, not taking her eyes away from Landrum. The attending physician left the room to put on a clean gown. Landrum breathed in, closed her eyes, and pushed. More of the infant's head appeared, a slick cluster of black curls. The senior resident motioned to the third and most junior of the women, standing at her shoulder, and told her, "Here's your chance." The young resident took the baby's head and eased the slippery infant out. Landrum was oblivious to the procession of young residents taking turns between her legs or the fact that the attending physician wasn't in the room at all. She was sobbing, shaking, laughing —all at the same time—flooded with the kind of hysterical relief a woman feels when a baby leaves her body and emerges into the world.

The resident lay the infant, purple, wrinkled, and still as a stone, on Landrum's bare chest. "Is he all right? Is he OK?" Landrum asked, panicking as she looked down at the motionless baby. A second later, his tiny arms and legs tensed, and he opened his mouth and let out a definitive cry.

"He's perfect," Giwa told her, touching her shoulder.

"I did it," Landrum said, looking up at Giwa and laying her hands on the baby's back, still coated with blood and amniotic fluid. She had decided to name him Kingston Blessed Landrum.

"Yes," Giwa said, finally allowing herself a wide smile. "You did."

*

In 1995 a pregnant African American doctoral student had a preterm birth after her water broke unexpectedly at 34 weeks. Her baby was on a ventilator for 48 hours and a feeding tube for 6 days during his 10-day stay in the neonatal intensive-care unit.

The woman was part of a team of female researchers from Boston and Howard Universities working on the Black Women's Health Study, an ongoing examination, funded by the National Institutes of Health, of conditions like preterm birth that affect black women disproportionately. The team had started the study after they noticed that most large, long-term medical investigations of women overwhelmingly comprised white women. The Black Women's Health Study researchers, except for two black women, were also all white.

What happened to the doctoral student altered the course of the study. "We're thinking, Here's a middle-class, well-educated black woman having a preterm birth when no one else in our group had a preterm birth," says Dr. Julie Palmer, associate director of the Slone Epidemiology Center at Boston University and a principal investigator of the continuing study of 59,000 subjects. "That's when I became aware that the race difference in preterm birth has got to be something different, that it really cuts across class. People had already done some studies showing health effects of racism, so we wanted to ask about that as soon as possible."

In 1997 the study investigators added several yes-or-no questions about everyday race-related insults: I receive poorer service than others; people act as if I am not intelligent; people act as if I am dishonest; people act as if they are better than me; people act as if they are afraid of me. They also included a set of questions about more significant discrimination: I have been treated unfairly because of my race at my job, in housing, or by the police. The findings showed higher levels of preterm birth among women who reported the greatest experiences of racism.

The bone-deep accumulation of traumatizing life experiences and persistent insults that the study pinpointed is not the sort of "lean in" stress relieved by meditation and "me time." When a person is faced with a threat, the brain responds to the stress by releasing a flood of hormones, which allow the body to adapt and respond to the challenge. When stress is sustained, long-term exposure to stress hormones can lead to wear and tear on the cardiovascular, metabolic, and immune systems, making the body vulnerable to illness and even early death.

Though Arline Geronimus's early research had focused on birth outcomes mainly in disadvantaged teenagers and young women, she went on to apply her weathering theory across class lines. In 2006 she and her colleagues used government data, blood tests, and questionnaires to measure the effects of stress associated with weathering on the systems of the body. Even when controlling for income and education, African American women had the highest allostatic load scores—an algorithmic measurement of stress-associated body chemicals and their cumulative effect on the body's systems—higher than white women and black men. Writing in the *American Journal of Public Health,* Geronimus and her colleagues concluded that "persistent racial differences in health may be influenced by the stress of living in a race-conscious society. These effects may be felt particularly by black women because of [the] double jeopardy of gender and racial discrimination."

People of color, particularly black people, are treated differently the moment they enter the healthcare system. In 2002 the groundbreaking report *Unequal Treatment: Confronting Racial and Ethnic Disparities in Health Care,* published by a division of the National Academy of Sciences, took an exhaustive plunge into 100 previous studies, careful to decouple class from race, by comparing subjects with similar income and insurance coverage. The researchers found that people of color were less likely to be given appropriate medications for heart disease, or to undergo coronary bypass surgery, and received kidney dialysis and transplants less frequently than white people, which resulted in higher death rates. Black people were 3.6 times as likely as white people to have their legs and feet amputated as a result of diabetes, even when all other factors were equal. One study analyzed in the report found that cesarean sections were 40 percent more likely among black women compared with white women. "Some of us on the committee were surprised and shocked at the extent of the evidence," noted the chairman of the panel of physicians and scientists who compiled the research.

In 2016 a study by researchers at the University of Virginia examined why African American patients receive inadequate treatment for pain not only compared with white patients but also relative to World Health Organization guidelines. The study found that white medical students and residents often believed incorrect and sometimes "fantastical" biological fallacies about racial differences in patients. For example, many thought, falsely, that blacks have less-sensitive nerve endings than whites, that black people's blood coagulates more quickly and that black skin is thicker than white. For these assumptions, researchers blamed not individual prejudice but deeply ingrained unconscious stereotypes about people of color, as well as physicians' difficulty in empathizing with patients whose experiences differ from their own. In specific research regarding childbirth, the Listening to Mothers Survey III found that one in five black and Hispanic women reported poor treatment from hospital staff because of race, ethnicity, cultural background, or language, compared with 8 percent of white mothers.

Researchers have worked to connect the dots between racial bias and unequal treatment in the healthcare system and maternal and infant mortality. Carol Hogue, an epidemiologist and the Jules & Uldeen Terry Chair in Maternal and Child Health at the Rollins School of Public Health at Emory University and one of the original authors of the 1992 *New England Journal of Medicine* study on infant mortality that opened my own eyes, was a coauthor of a 2009 epidemiological review of research about the association between racial disparities in preterm birth and interpersonal and institutional racism. Her study, published by the Johns Hopkins School of Public Health, contains an extraordinary list of 174 citations from previous work. "You can't convince people of something like discrimination unless you really have evidence behind it," Hogue says. "You can't just say this—you have to prove it."

Lynn Freedman, director of the Averting Maternal Death and Disability Program at Columbia University's Mailman School of Public Health, decided to take the lessons she and her colleagues learned while studying disrespect and abuse in maternal care in Tanzania—where problems in pregnancy and childbirth lead to nearly 20 percent of all deaths in women ages 15 to 49—and apply them to New York City and Atlanta. Though the study is still in its preliminary phase, early focus groups of some 50 women who recently delivered babies in Washington Heights and Inwood, as well as with doulas who work in both those areas and in central Brooklyn, revealed a range of grievances—from having to wait one to two months before an initial prenatal appointment to being ignored, scolded, and demeaned, even feeling bullied or pushed into having C-sections. "Disrespect and abuse means more than just somebody wasn't nice to another individual person," Freedman says. "There is something structural and much deeper going on in the health system that then expresses itself in poor outcomes and sometimes deaths."

Two days after the birth of Landrum's baby, she had moved out of labor and delivery and into a hospital room, with the butterfly-decorated, crayon-drawn affirmations taped above her bed. She'd had a few hours of sleep and felt rested and cheerful in a peach-colored jumpsuit she brought from home, with baby Kingston, who had weighed in at a healthy six pounds 13 ounces, napping in a plastic crib next to her bed. But over the next hours, Landrum's mood worsened. When Giwa walked into her room after leaving for a few hours to change and nap, Landrum once again angrily recounted the mishap with the epidural and complained about the nurses and even the hospital food. Finally, Giwa put her hand on Landrum's arm and asked, "Simone, where are the boys?"

Landrum stopped, and her entire body sagged. She told Giwa that her sons were staying on the other side of town with her god-mother, whom she called Nanny. But with children of her own, Nanny was unable to make the 40-minute drive to bring Landrum's sons to the hospital to see their mama and meet their brother. "After they lost their sister, it's really important that they see Kingston," Landrum said.

"I understand," Giwa said, stroking her shoulder. "You need the boys to see their brother, to know that he is alive, that this is all real." Landrum nodded. She made several phone calls from her hospital bed but could find no one to get the boys, so I left to drive across town and pick them up. It took Giwa's attentive eyes, and the months of building trust and a relationship with Landrum, to recognize a problem that couldn't be addressed medically but one that could have emotional and physical consequences.

The doula consumer market has been largely driven by and tailored for white women, but the kind of support Giwa was providing to Landrum was actually originated by black women, the granny midwives of the South. Inspired by that historic legacy and by increasingly visible reproductive-justice activism, dozens of doula groups like Birthmark in New Orleans have emerged or expanded in the past several years in Brooklyn, Los Angeles, Atlanta, Dallas, Memphis, Miami, Washington, and many other cities, providing services to women of color, often free or on a sliding scale.

The By My Side Birth Support Program in New York City, administered by the city's Department of Health, offers free doula services during pregnancy, labor and delivery, and postpartum for mothers in central and eastern Brooklyn's predominantly black and brown neighborhoods where maternal and infant mortality are highest. A team of 12 doulas has served more than 800 families since 2010, and an analysis of the program showed that from 2010 to 2015, mothers receiving doula support had half as many preterm births and low-birth-weight babies as other women in the same community.

Interventions that have worked to bring down maternal- and infant-mortality rates in other parts of the world have been brought back to the United States. Rachel Zaslow, a midwife and doula based in Charlottesville, Virginia, runs a program in northern Uganda, where a woman has a 1 in 25 lifetime chance of dying in childbirth, through her nonprofit organization, Mother Health International. In Zaslow's program, community health workers—individuals selected by the community and given medical training—link local pregnant women to trained midwives and nurse-midwives. Since 2008, a mother has never died in Zaslow's program, and the infant-mortality rate is 11 per 1,000, compared with 64 per 1,000 for the country at large.

Three years ago, when she became aware of high rates of infant and maternal mortality in pockets of Virginia, Zaslow decided to take her Ugandan model there: a collective of 45 black and Latina doulas in Charlottesville, called Sisters Keeper, that offers birthing services free to women of color. "The doula model is very similar to the community health worker model that's being used a lot, and successfully, throughout the global South," Zaslow says. "For me, when it comes to maternal health, the answer is almost always some form of community health worker." Since 2015, the Sisters Keeper doulas have attended about 300 births—with no maternal deaths and only one infant death among them.

"It is really hard for American healthcare professionals to get their heads around that when you have an organized community-based team that connects technical clinical issues with a deep, embedded set of relationships, you can make real breakthroughs," says Dr. Prabhjot Singh, the director of the Arnhold Institute for Global Health at the Icahn School of Medicine at Mount Sinai, who studies community health worker models and how they can be used in the United States. "In the US, doulas can't do it by themselves, but based on work that's taken place globally, they can help reduce infant and maternal deaths using what is essentially a very simple solution."

An hour and a half after Giwa noticed that Landrum needed to have her sons with her, Caden and Dillon burst through the door of the hospital room. Holding Kingston in her lap, Landrum lit up at the sight of the boys. Caden, who is four, ran to his new brother, gleefully grabbing at the infant. "Calm down," Landrum said, smiling and patting the side of the bed. "Put out your arms, strong, like this," she told him, arranging his small arms with her free hand. Gently, she lay Kingston in his brother's outstretched arms. "It's my baby," he said excitedly, leaning down to kiss the infant all over his cheeks and forehead. "I luh you, brother."

Dillon, seven, was more cautious. He stood near the door, watchful. "Don't you want to meet your brother, Dillon?" Landrum asked. He inched closer, looking at the floor. "Come on, boy, don't be shy. This is Kingston." He sat on the other side of his mother, and she took the baby from Caden and placed him in Dillon's arms. He looked down at the newborn, nervous and still hesitant. "It's a real baby," he said, looking up at his mother and finally smiling. "Mommy, you did it."

"At that moment, I felt complete," Landrum said later, tearing up, "seeing them all together."

On a cool, sunny afternoon in March, Landrum led me into her living room, which now held a used couch—a gift from a congregant of her church, where she is an active member. A white plastic Christmas tree strewn with multicolored Mardi Gras beads, left up after the holidays, added a festive touch. Landrum handed me Kingston, now three months old, dressed in a clean onesie with a little blue giraffe on the front. Plump and rosy, with cheeks chunky from breast milk and meaty, dimpled thighs, he smiled when I sang him a snippet of a Stevie Wonder song.

Landrum had lost the baby weight and looked strong and healthy in an oversize T-shirt and leggings, wearing her hair in pink braids that hung down her back. There was a lightness to her that wasn't apparent during her pregnancy. One word tumbling over the next, she told me that the new baby had motivated her to put her life in order. She had been doing hair and makeup for church members and friends out of her house to earn money to buy a car. She had applied to Delgado Community College to study to be an ultrasound technician. "I love babies," she said. "When I look at ultrasound pictures, I imagine I see the babies smiling at me."

Latona Giwa had continued to care for Landrum for two months after Kingston's birth. The CDC measures American maternal mortality not just by deaths that occur in pregnancy or childbirth, or in the immediate days afterward, but rather all deaths during pregnancy and the year after the end of pregnancy—suggesting the need for continued care and monitoring, especially for women who are most at risk of complications.

It was Giwa who drove Landrum and the baby home from the hospital, moving her own two-year-old daughter's car seat from the back of her Honda and replacing it with a backward-facing infant seat, when Landrum had no other ride. It was Giwa who ushered the new mother into her home and then surprised her by taking a bag of groceries and a tray of homemade lasagna, still warm, from the back of the car. And it was Giwa who asked her, six weeks after childbirth, if she had talked to her doctor about getting a contraceptive implant to avoid pregnancy. When Landrum told her that her doctor had never called her about a checkup, Giwa was livid. "High-risk patients with complicated maternal histories often have an appointment two weeks after they've been discharged," she said later, after insisting that Landrum call to make an appointment. "Her life is hectic; she's at home with three children. Luckily she's fine, but at minimum someone should've called to check on her."

For Giwa's work with Landrum, from October to February, she earned just $600. Like the other Birthmark doulas, Giwa can't make ends meet just doing doula work; she is employed as a lactation consultant for new mothers both privately and at a "latch clinic" in a New Orleans office of the federal Women, Infants and Children food and nutrition service that supports low-income pregnant and postpartum women.

"We need to recognize that there is actual medical benefit to having doula support—and make the argument that insurance should pay for it," says Williams, the Bay Area OB-GYN. "It is a job. People do have to be paid for that work." Insurance would mean some standardization; Williams notes that many programs securing public funding or grants to provide doula support to lower-income women can't match the kind of money that private doulas can command. These programs often have "all black women who are doulas," she says. "Yes, it's fantastic that these women are training to be doulas and supporting other black women—but they're not making as much as these other doulas." If, she asks, "doula support is important and can have this beneficial outcome for women, especially black women, how can we actually move forward to make that more accessible to everybody?"

In her home on that March afternoon, Landrum put Kingston into a baby carrier. He fell asleep as we walked five blocks to meet Dillon and Caden, who were due home from school at two different bus stops. The boys jumped off their buses, dressed in identical red polo shirts, their hair freshly cut, each dragging a large backpack, and ran to their mother. Dillon could hardly wait to pull out his report card and show his mother his grades; he had received four out of six "exceptional" marks. "He's smart," Landrum said, and he gave her a huge, gaptoothed smile.

Then he raced ahead, his backpack lurching as he leapt over bumps in the sidewalk, full of pent-up little-boy energy; Caden was right behind him, doing his best to keep up with his brother's longer strides. "Hey, y'all, you be careful!" Landrum called, keeping her eyes trained on them. "You hear me?!"

Kingston stirred when he heard his mother's voice. He lifted his head briefly and looked into Landrum's face. Their eyes met, his still slightly crossed with new-baby nearsightedness. Landrum paused long enough to stroke his head and kiss his damp cheek. The baby sighed. Then he burrowed his head back into the warmth and safety of his mother's chest.

ED YONG

When the Next Plague Hits

FROM *The Atlantic*

AT SIX O'CLOCK in the morning, shortly after the sun spills over
the horizon, the city of Kikwit doesn't so much wake up as ignite.
Loud music blares from car radios. Shops fly open along the main
street. Dust-sprayed jeeps and motorcycles zoom eastward toward
the town's bustling markets or westward toward Kinshasa, the Dem-
ocratic Republic of the Congo's capital city. The air starts to heat
up, its molecules vibrating with absorbed energy. So, too, the city.

By late morning I am away from the bustle, on a quiet, exposed
hilltop some five miles down a pothole-ridden road. As I walk,
desiccated shrubs crunch underfoot and butterflies flit past. The
only shade is cast by two lines of trees, which mark the edges of
a site where more than 200 people are buried, their bodies piled
into three mass graves, each about 15 feet wide and 70 feet long.
Nearby, a large blue sign says IN MEMORY OF THE VICTIMS OF
THE EBOLA EPIDEMIC IN MAY 1995. The sign is partly obscured
by overgrown grass, just as the memory itself has been occluded
by time. The ordeal that Kikwit suffered has been crowded out by
the continual eruption of deadly diseases elsewhere in the Congo,
and around the globe.

Emery Mikolo, a 55-year-old Congolese man with a wide, angular face, walks with me. Mikolo survived his own encounter with
Ebola in 1995. As he looks at the resting place of those who didn't,
his solemn demeanor cracks a bit. In the Congo, when people
die, their bodies are meant to be cleaned by their families. They
should be dressed, caressed, kissed, and embraced. These intense
rituals of love and community were corrupted by Ebola, which

harnessed them to spread through entire families. Eventually, of necessity, they were eliminated entirely. Until Ebola, "no one had ever taken bodies and thrown them together like sacks of manioc," Mikolo tells me.

The Congo—and the world—first learned about Ebola in 1976, when a mystery illness emerged in the northern village of Yambuku. Jean-Jacques Muyembe, then the country's only virologist, collected blood samples from some of the first patients and carried them back to Kinshasa in delicate test tubes, which bounced on his lap as he trundled down undulating roads. From those samples, which were shipped to the Centers for Disease Control and Prevention in Atlanta, scientists identified the virus. It took the name Ebola from a river near Yambuku. And, having been discovered, it largely vanished for almost 20 years.

In 1995 it reemerged in Kikwit, about 500 miles to the southwest. The first victim was 35-year-old Gaspard Menga, who worked in the surrounding forest raising crops and making charcoal. In Kikongo, the predominant local dialect, his surname means "blood." He checked into Kikwit General Hospital in January and died from what doctors took to be shigellosis—a diarrheal disease caused by bacteria. It was only in May, after the simmering outbreak had flared into something disastrous, after wards had filled with screams and vomit, after graves had filled with bodies, after Muyembe had arrived on the scene and again sent samples abroad for testing, that everyone realized Ebola was back. By the time the epidemic abated, 317 people had been infected and 245 had died. The horrors of Kikwit, documented by foreign journalists, catapulted Ebola into international infamy. Since then, Ebola has returned to the Congo on six more occasions; the most recent outbreak, which began in Bikoro and then spread to Mbandaka, a provincial capital, is still ongoing at the time of this writing.

Unlike airborne viruses such as influenza, Ebola spreads only through contact with infected bodily fluids. Even so, it is capable of incredible devastation, as West Africa learned in 2014, when, in the largest outbreak to date, more than 28,000 people were infected and upwards of 11,000 died. Despite the relative difficulty of transmission, Ebola still shut down health systems, crushed economies, and fomented fear. With each outbreak, it reveals the vulnerabilities in our infrastructure and our psyches that a more contagious pathogen might one day exploit.

These include forgetfulness. In the 23 years since 1995, new generations who have never experienced the horrors of Ebola have been born in Kikwit. Protective equipment to shield doctors and nurses from contaminated blood has vanished, even as the virus has continued to emerge in other corners of the country. The city's population has tripled. New neighborhoods have sprung up. In one of them, I walk through a market, gazing at delectable displays of peppers, eggplants, avocados, and goat meat. Pieces of salted fish sell for 300 Congolese francs—about the equivalent of an American quarter. Juicy white grubs go for 1,000. And the biggest delicacy of all goes for 13,000—a roasted monkey, its charred face preserved in a deathly grimace.

The monkey surprises me. Mikolo is surprised to see only one. Usually, he says, these stalls are heaving with monkeys, bats, and other bushmeat, but rains the night before must have stranded any hunters in the eastern forests. As I look around the market, I picture it as an ecological magnet, drawing in all the varied animals that dwell within the forest—and all the viruses that dwell within them.

The Congo is one of the most biodiverse countries in the world. It was here that HIV bubbled into a pandemic, eventually detected half a world away, in California. It was here that monkeypox was first documented in people. The country has seen outbreaks of Marburg virus, Crimean-Congo hemorrhagic fever, chikungunya virus, yellow fever. These are all zoonotic diseases, which originate in animals and spill over into humans. Wherever people push into wildlife-rich habitats, the potential for such spillover is high. Sub-Saharan Africa's population will more than double during the next three decades, and urban centers will extend farther into wilderness, bringing large groups of immunologically naive people into contact with the pathogens that skulk in animal reservoirs —Lassa fever from rats, monkeypox from primates and rodents, Ebola from God-knows-what in who-knows-where.

On average, in one corner of the world or another, a new infectious disease has emerged every year for the past 30 years: MERS, Nipah, Hendra, and many more. Researchers estimate that birds and mammals harbor anywhere from 631,000 to 827,000 unknown viruses that could potentially leap into humans. Valiant efforts are underway to identify them all, and scan for them in places like poultry farms and bushmeat markets, where animals and people are most likely to encounter each other. Still, we likely won't ever be able to predict which will spill over next; even long-known viruses like Zika, which was discovered in 1947, can suddenly develop into unforeseen epidemics.

The Congo, ironically, has a good history of containing its diseases, partly because travel is so challenging. Most of the country is covered by thick forest, crisscrossed by just 1,700 miles of road. Large distances and poor travel infrastructure limited the spread of Ebola outbreaks in years past.

But that is changing. A 340-mile road, flanked by deep valleys, connects Kikwit to Kinshasa. In 1995 that road was so badly maintained that the journey took more than a week. "You'd have to dig yourself out every couple of minutes," Mikolo says. Now the road is beautifully paved for most of its length, and can be traversed in just eight hours. Twelve million people live in Kinshasa—three times the combined population of the capitals affected by the 2014 West African outbreak. About eight international flights depart daily from the city's airport.

If Ebola hit Kikwit today, "it would arrive here easily," Muyembe tells me in his office at the National Institute for Biomedical Research, in Kinshasa. "Patients will leave Kikwit to seek better treatment, and Kinshasa will be contaminated immediately. And then from here to Belgium? Or the US?" He laughs, morbidly.

"What can you do to stop that?" I ask.

"Nothing."

One hundred years ago, in 1918, a strain of H1N1 flu swept the world. It might have originated in Haskell County, Kansas, or in France or China—but soon it was everywhere. In two years, it killed as many as 100 million people—5 percent of the world's population, and far more than the number who died in World War I. It killed not just the very young, old, and sick, but also the strong and fit, bringing them down through their own violent immune responses. It killed so quickly that hospitals ran out of beds, cities ran out of coffins, and coroners could not meet the demand for death certificates. It lowered Americans' life expectancy by more than a decade. "The flu resculpted human populations more radically than anything since the Black Death," Laura Spinney wrote in *Pale Rider,* her 2017 book about the pandemic. It was one of the deadliest natural disasters in history—a potent reminder of the threat posed by disease.

Humanity seems to need such reminders often. In 1948, shortly after the first flu vaccine was created and penicillin became the first mass-produced antibiotic, US Secretary of State George Marshall reportedly claimed that the conquest of infectious disease was imminent. In 1962, after the second polio vaccine was formulated, the Nobel Prize–winning virologist Sir Frank Macfarlane Burnet asserted, "To write about infectious diseases is almost to write of something that has passed into history."

Hindsight has not been kind to these proclamations. Despite advances in antibiotics and vaccines, and the successful eradication of smallpox, *Homo sapiens* is still locked in the same epic battle with viruses and other pathogens that we've been fighting since the beginning of our history. When cities first arose, diseases laid them low, a process repeated over and over for millennia. When Europeans colonized the Americas, smallpox followed. When soldiers fought in the first global war, influenza hitched a ride, and found new opportunities in the unprecedented scale of the conflict. Down through the centuries, diseases have always excelled at exploiting flux.

Humanity is now in the midst of its fastest-ever period of change. There were almost 2 billion people alive in 1918; there are now 7.6 billion, and they have migrated rapidly into cities, which since 2008 have been home to more than half of all human beings. In these dense throngs, pathogens can more easily spread and more quickly evolve resistance to drugs. Not coincidentally, the total number of outbreaks per decade has more than tripled since the 1980s.

Globalization compounds the risk: airplanes now carry almost 10 times as many passengers around the world as they did four decades ago. In the 1980s, HIV showed how potent new diseases can be, by launching a slow-moving pandemic that has since claimed about 35 million lives. In 2003, another newly discovered virus, SARS, spread decidedly more quickly. A Chinese seafood seller hospitalized in Guangzhou passed it to dozens of doctors and nurses, one of whom traveled to Hong Kong for a wedding. In a single night, he infected at least 16 others, who then carried the virus to Canada, Singapore, and Vietnam. Within six months, SARS had reached 29 countries and infected more than 8,000 people. This is a new epoch of disease, when geographic barriers disappear and threats that once would have been local go global.

Last year, with the centennial of the 1918 flu looming, I started looking into whether America is prepared for the next pandemic. I fully expected that the answer would be no. What I found, after talking with dozens of experts, was more complicated—reassuring in some ways, but even more worrying than I'd imagined in others. Certainly, medicine has advanced considerably during the past century. The United States has nationwide vaccination programs, advanced hospitals, the latest diagnostic tests. In the National Institutes of Health, it has the world's largest biomedical-research establishment, and in the CDC, arguably the world's strongest public health agency. America is as ready to face down new diseases as any country in the world.

Yet even the US is disturbingly vulnerable—and in some respects is becoming quickly more so. It depends on a just-in-time medical economy, in which stockpiles are limited and even key items are made to order. Most of the intravenous bags used in the country are manufactured in Puerto Rico, so when Hurricane Maria devastated the island last September, the bags fell in short supply. Some hospitals were forced to inject saline with syringes—and so syringe supplies started running low too. The most common lifesaving drugs all depend on long supply chains that include India and China—chains that would likely break in a severe pandemic. "Each year, the system gets leaner and leaner," says Michael Osterholm, the director of the Center for Infectious Disease Research and Policy at the University of Minnesota. "It doesn't take much of a hiccup anymore to challenge it."

Perhaps most important, the United States is prone to the same forgetfulness and shortsightedness that befall all nations, rich and poor—and the myopia has worsened considerably in recent years. Public health programs are low on money; hospitals are stretched perilously thin; crucial funding is being slashed. And while we tend to think of science when we think of pandemic response, the worse the situation, the more the defense depends on political leadership.

When Ebola flared in 2014, the science-minded president Barack Obama calmly and quickly took the reins. The White House is now home to a president who is neither calm nor science-minded. We should not underestimate what that may mean if risk becomes reality.

Bill Gates, whose foundation has studied pandemic risks closely, is not a man given to alarmism. But when I spoke with him upon my return from Kikwit, he described simulations showing that a severe flu pandemic, for instance, could kill more than 33 million people worldwide in just 250 days. That possibility, and the world's continued inability to adequately prepare for it, is one of the few things that shake Gates's trademark optimism and challenge his narrative of global progress. "This is a rare case of me being the bearer of bad news," he told me. "Boy, do we not have our act together."

Preparing for a pandemic ultimately boils down to real people and tangible things: A busy doctor who raises an eyebrow when a patient presents with an unfamiliar fever. A nurse who takes a travel history. A hospital wing in which patients can be isolated. A warehouse where protective masks are stockpiled. A factory that churns out vaccines. A line on a budget. A vote in Congress. "It's like a chain—one weak link and the whole thing falls apart," says Anthony Fauci, the director of the National Institute of Allergy and Infectious Diseases. "You need no weak links."

Among all known pandemic threats, influenza is widely regarded as the most dangerous. Its various strains are constantly changing, sometimes through subtle mutations in their genes, and sometimes through dramatic reshuffles. Even in nonpandemic years, when new viruses aren't sweeping the world, the more familiar strains kill up to 500,000 people around the globe. Their ever-changing nature explains why the flu vaccine needs to be updated annually. It's why a disease that is sometimes little worse than a bad cold can transform into a mass-murdering monster. And it's why flu is the disease the United States has invested the most in tracking. An expansive surveillance network constantly scans for new flu viruses, collating alerts raised by doctors and results from lab tests, and channeling it all to the CDC, the spider at the center of a thrumming worldwide web.

Yet just 10 years ago, the virus that the world is most prepared for caught almost everyone off guard. In the early 2000s the CDC was focused mostly on Asia, where H5N1—the type of flu deemed most likely to cause the next pandemic—was running wild among poultry and waterfowl. But while experts fretted about H5N1 in birds in the East, new strains of H1N1 were evolving within pigs in the West. One of those swine strains jumped into humans in

Mexico, launching outbreaks there and in the US in early 2009. The surveillance web picked it up only in mid-April of that year, when the CDC tested samples from two California children who had recently fallen ill.

One of the most sophisticated disease-detecting networks in the world had been blindsided by a virus that had sprung up in its backyard, circulated for months, and snuck into the country unnoticed. "We joked that the influenza virus is listening in on our conference calls," says Daniel Jernigan, who directs the CDC's Influenza Division. "It tends to do whatever we're least expecting."

The pandemic caused problems for vaccine manufacturers, too. Most flu vaccines are made by growing viruses in chicken eggs —the same archaic method that's been used for 70 years. Every strain grows differently, so manufacturers must constantly adjust to each new peculiarity. Creating flu vaccines is an artisanal affair, more like cultivating a crop than making a pharmaceutical. The process works reasonably well for seasonal flu, which arrives on a predictable schedule. It fails miserably for pandemic strains, which do not.

In 2009 the vaccine for the new pandemic strain of H1N1 flu arrived slowly. (Then–CDC director Tom Frieden told the press, "Even if you yell at the eggs, it won't grow any faster.") Once the pandemic was officially declared, it took four months before the doses even *began* to roll out in earnest. By then the disaster was already near its peak. Those doses prevented no more than 500 deaths—the fewest of any flu season in the surrounding 10-year period. Some 12,500 Americans died.

The egg-based system depends on chickens, which are themselves vulnerable to flu. And since viruses can mutate within the eggs, the resulting vaccines don't always match the strains that are circulating. But vaccine makers have few incentives to use anything else. Switching to a different process would cost billions, and why bother? Flu vaccines are low-margin products, which only about 45 percent of Americans get in a normal year. So when demand soars during a pandemic, the supply is not set to cope.

American hospitals, which often operate unnervingly close to full capacity, likewise struggled with the surge of patients. Pediatric units were hit especially hard by H1N1, and staff became exhausted from continuously caring for sick children. Hospitals almost ran out of the life-support units that sustain people whose lungs and hearts start to fail. The healthcare system didn't break, but it came too close for comfort—especially for what turned out to be a training-wheels pandemic. The 2009 H1N1 strain killed merely 0.03 percent of those it infected; by contrast, the 1918 strain had killed 1 to 3 percent, and the H7N9 strain currently circulating in China has a fatality rate of 40 percent.

"A lot of people said that we dodged a bullet in 2009, but nature just shot us with a BB gun," says Richard Hatchett, the CEO of the Coalition for Epidemic Preparedness Innovations. Tom Inglesby, a biosecurity expert at the Johns Hopkins Bloomberg School of Public Health, told me that if a 1918-style pandemic hit, his hospital "would need in the realm of seven times as many critical-care beds and four times as many ventilators as we have on hand."

That the United States could be so ill-prepared for flu, of all things, should be deeply concerning. The country has a dedicated surveillance web, antiviral drugs, and an infrastructure for making and deploying flu vaccines. None of that exists for the majority of other emerging infectious diseases.

As I walk down a seventh-floor hallway of the University of Nebraska Medical Center, Kate Boulter, a nurse manager, points out that the carpet beneath my feet has disappeared, exposing bare floors that are more easily cleaned. In an otherwise unmarked corridor, this, she says, is the first sign that I am approaching the biocontainment unit—a special facility designed to treat the victims of bioterror attacks, or patients with a deadly infectious disease such as Ebola or SARS.

There is nothing obviously special about the 4,100 square feet, but every detail has been carefully designed to give patients maximal access to the best care, and viruses minimal access to anything. A supply room is stocked with scrubs, underwear, and socks, so that no piece of clothing staff members wear at work will make its way home. There are two large autoclaves—pressure cookers that use steam to sterilize equipment—so that soiled linens and clothes can be immediately decontaminated. The space is under negative air pressure: when doctors enter the hallway, or any of the five patient rooms, air flows in with them, preventing viruses from drifting out. This also dries the air. Working here, I'm told, is murder on the skin.

Almost everything in the unit is a barrier of some form. Floor seams are welded. Light and plumbing fixtures are sealed. The ventilation and air-conditioning systems are separate from those for the rest of the hospital, and rigorously filtered. Patients can be wheeled in on a tented gurney with built-in glove ports; it looks like a translucent caterpillar whose legs have been pushed inward. A separate storage room is stocked with full-body suits, tape for sealing the edges of gloves, and space-suit-like hoods with their own air filter. A videoconferencing system allows team members —and family—to monitor what happens in the patient rooms without having to suit up themselves. A roll of heavy-duty metallic wrapping paper can be used to seal the body of anyone who dies.

The unit is currently empty, as it has been for most of its existence. The beds are occupied only by four hyperrealistic mannequins, upon which nurses can practice medical procedures while wearing cumbersome protective layers. "We've named all the mannequins," Boulter tells me. Pointing to the largest one: "That one's Phil, after Dr. Smith."

Phil Smith began pushing the hospital to build the biocontainment unit in 2003, back when he was a professor of infectious diseases. SARS had emerged from nowhere, and monkeypox had broken out in the Midwest; Smith realized the United States had no facilities that could handle such diseases, beyond a few high-security research labs. With support from the state health department, he opened the unit in 2005.

And then, nothing happened.

For nine years, the facility was dormant, acting mostly as an overflow ward. "We didn't know if it would be needed, but we planned and prepared as if it would," says Shelly Schwedhelm, the head of the hospital's emergency-preparedness program, who for years kept the unit afloat on a shoestring budget. Her efforts paid off in September 2014, when the State Department called, telling Schwedhelm and her team to prepare for possible Ebola patients. Over 10 weeks, the unit's 40 staff members took care of three infected Americans who had been evacuated from West Africa. They worked around the clock in teams of six, some staffers treating the patients directly, others helping their colleagues put on and take off their gear, and still others supervising from the nurses' station. Two of the patients—Rick Sacra, a physician, and Ashoka Mukpo, a journalist—were cured and discharged. The third—a surgeon named Martin Salia—was already suffering from organ failure by the time he arrived, and died two days later. A green-marble plaque now hangs in the unit to honor him.

The University of Nebraska Medical Center is one of the best in the country at handling dangerous and unusual diseases, Ron Klain, who was in charge of the Obama administration's Ebola response, tells me. Only the NIH and Emory University Hospital have biocontainment units of a similar standard, he says, but both are smaller. Those three hospitals were the only ones ready to take patients when Ebola struck in 2014, but within two months, Klain's team had raised the number to 50 facilities. It was "a lot of hard work," he says. "But ultimately, we had 144 beds." A more contagious and widespread disease would have overwhelmed them all.

Preparing hospitals for new epidemics is challenging in the United States, Klain says, because healthcare is so decentralized: "You and I could decide that every hospital should have three beds capable of isolating people with a dangerous disease, and Trump could agree with us, and there's no way of making that happen." Hospitals are independent entities; in this fractured environment, preparedness is less the result of governmental mandate and more the product of individual will. It comes from dedicated visionaries like Smith and skilled managers like Schwedhelm, who can keep things going when there's no immediate need.

The trio of Ebola patients in 2014 produced 3,700 pounds of contaminated linens, gloves, and other waste among them, all of which demanded careful handling. Treating them cost more than $1 million. That kind of care quickly reaches its limits as an epidemic spreads. In June 2015 the Samsung Medical Center, in Seoul—one of the most advanced medical centers in the world —was forced to suspend most of its services after a single man with MERS arrived in its overcrowded emergency room. American hospitals wouldn't fare much better. But at the very least, they can plan for the worst.

Schwedhelm, with a 100-person team, has been creating plans for how every aspect of hospital operation would need to work during a pandemic. How much should hospitals stockpile? How would they provide psychological support during a weeks-long crisis? How could they feed people working longer-than-usual shifts? When would they cancel elective surgeries? Where could they get extra disinfectant, mop heads, and other cleaning supplies?

At a single meeting, I hear two dozen people discuss how they would care for the 400 or so patients on the hospital's organ-transplant list. How would they get such patients into the facility safely? At what point would it become too risky to pump them with immunosuppressants? If ICUs are full, where could they create clean spaces for post-transplant recovery? It matters that the hospital has considered these questions. It matters just as much that the people in charge have met, talked, and established a bond.

The members of the team running the biocontainment unit all work in different parts of the hospital, as pediatricians, critical-care specialists, obstetricians. But even during the unit's long dormancy, Schwedhelm would gather them for quarterly training sessions. That's why, when the moment came, they were ready. When they escorted the Ebola patients off their respective planes, the staff members recalled what they had learned during practice drills.

"We do a lot of team building," Boulter says, showing me a photo of the group at a ropes course.

"It was the scariest thing I've ever done," Schwedhelm says. They followed that up with something more sedate—a movie night in the hospital auditorium. They watched *Contagion*.

Kikwit General Hospital has no biocontainment unit. Instead, it has Pavilion 3.

Emery Mikolo, who works at the hospital as a nurse supervisor, takes me into the blue-walled, open-windowed building that is now the pediatrics ward. In one room, mosquito nets are suspended hammocklike over 16 closely packed beds, on which mothers care for young children and newborn babies. This is a place of new life. But in 1995 it was the infamous "death ward," where Ebola patients were treated. Exhausted doctors struggled to control the outbreak; outside the hospital, the military established a perimeter to turn back fleeing patients. The dead were laid in a row on the pavement.

We walk into another room, which is largely empty except for a poster of a cartoonish giraffe, a few worn mattresses, and some old bed frames. Mikolo touches one of them. It was his, he says. He looks around quietly and shakes his head. Many of the people who shared this room with him were his colleagues who had become infected while they cared for patients. Ebola's symptoms are sometimes mythologized: organs don't liquefy; blood seldom pours from orifices.

But the reality is no less gruesome. "It was like a horror movie," he says. "All these people I worked with—my friends—throwing up, screaming, dying, falling out of bed." At one point, delirious with fever, he too rolled off his mattress. "There was vomit and piss and shit on the ground, but at least it was cool."

Many of the people who worked at the hospital during the outbreak are still there. Jacqui, a nurse, worked in Pavilion 3 and returned there only three years ago. She was terrified at first, but she soon habituated. I ask whether she's worried that Ebola might return. "I'm not afraid," she says. "It's never coming back."

If it does, is there any protective equipment at the hospital? "No," she tells me.

Mikolo laughs. "Article Fifteen," he says.

Article 15 is something of a Congolese catchphrase, referring to a fictional but universally recognized 15th article of the country's constitution: *Débrouillez-vous*—"Figure it out yourself." I hear it everywhere. It is simultaneously a testament to the Congolese love for droll humor, a weary acknowledgment of hardship, a screw-you to the establishment, and a motivational mantra. *No one's going to fix your problems. You must make do with what you've got.*

In a nearby room, dried blood dots the floor around an old operating table, where a sick lab technician once passed Ebola to five other medical staff members, starting a chain of transmission that eventually enveloped Mikolo and many of his friends. The phlebotomist who drew the blood samples that were used to confirm Ebola also still works at the hospital. I watch as he handles a rack of samples with his bare hands. "Ask someone here, 'Where are the kits that protect you from Ebola?'" Donat Kuma-Kuma Kenge, the hospital's chief coordinator, tells me. "There aren't any. I know *exactly* what I'm meant to do, but there are no materials—here, in the place where there was Ebola.

"*Débrouillez-vous*," he adds.

The hospital's challenges are considerable, but as I walk around, I realize that they are familiar. Even though the United States is 500 times as wealthy as the Congo, the laments I heard from people in both countries were uncannily similar—different in degree, but not in kind. Protective equipment is scarce in the Congo, but even America's stockpiles would quickly be depleted in a serious epidemic. Unfamiliarity with Ebola allowed the virus to spread among the staff of Kikwit's hospital, just as it did among nurses in Dallas, where an infected patient landed in September 2014. In Kikwit, a lack of running water makes hygiene a luxury, but even in the United States, getting medical professionals to wash their hands or follow other best practices is surprisingly hard; every year, at least 70,000 Americans die after picking up infections in hospitals. And most of all, the people in both countries worry that brief spates of foresight and preparedness will always give way to negligence and entropy.

In the United States, attention and money have crested and then crashed with each new crisis: anthrax in 2001, SARS in 2003. Resources, hurriedly assembled, dwindle. Research into countermeasures fizzles. "We fund this thing like Minnesota snow," Michael Osterholm says. "There's a lot in January, but in July it's all melted."

Take the Hospital Preparedness Program. It's a funding plan that was created in the wake of 9/11 to help hospitals ready themselves for disasters, run training drills, and build their surge capacity—everything that Shelly Schwedhelm's team does so well in Nebraska. It transformed emergency planning from an after-hours avocation into an actual profession, carried out by skilled specialists. But since 2003, its $514 million budget has been halved.

Another fund—the Public Health Emergency Preparedness program—was created at the same time to help state and local health departments keep an eye on infectious diseases, improve their labs, and train epidemiologists. Its budget has been pruned to 70 percent of its $940 million peak. Small wonder, then, that in the past decade, local health departments have cut more than 55,000 jobs. That's 55,000 people who won't be there to answer the call when the next epidemic hits.

These sums of money are paltry compared with what another pandemic might cost the country. Diseases are exorbitantly expensive. In response to just 10 cases of Ebola in 2014, the United States spent $1.1 billion on domestic preparations, including $119 million on screening and quarantine. A severe 1918-style flu pandemic would drain an estimated $683 billion from American coffers, according to the nonprofit Trust for America's Health. The World Bank estimates that global output would fall by almost 5 percent—totaling some $4 trillion.

The US is not unfamiliar with the concept of preparedness. It currently spends roughly half a trillion dollars on its military —the highest defense budget in the world, equal to the combined budgets of the next seven top countries. But against viruses—more likely to kill millions than any rogue state is—such consistent investments are nowhere to be found.

At a modern building in Holly Springs, on the outskirts of Raleigh, North Carolina, I walk down a wide corridor where the words IT REALLY IS A MATTER OF LIFE AND DEATH have been stenciled on a yellow wall. The walkway leads to a refrigerator-cool warehouse, where several white containers sit on a blue pallet. The containers are full of flu vaccine, and each holds enough to immunize more than 1 million Americans. When their contents are ready to be used, they head toward a long, Rube Goldberg–esque machine that dispenses the vaccine into syringes—more than 400,000 a day.

Instead of eggs, the facility grows flu viruses in lab-grown dog cells, which fill 5,000-liter steel vats one floor above. The cells are infected with flu viruses, which quickly propagate. The technique is faster than using eggs, and produces vaccines that are a closer match to circulating strains.

This facility is the result of a partnership between the pharmaceutical company Seqirus and a government agency called the Biomedical Advanced Research and Development Authority. Established in 2006, BARDA acts more or less as a venture-capital firm, funding the development of vaccines, drugs, and other epidemic countermeasures that would otherwise be unprofitable. In 2007 it entered into a $1 billion partnership to create the Holly Springs plant, which started making vaccines in 2011. "No one would have taken the risk of disposing of egg manufacturing unless they could reach the scale we have here," says Marie Mazur, Seqirus's vice president of pandemic response.

The facility will soon be able to make 200 million doses of vaccine within the first six months of a new pandemic—enough to immunize more than one in every three Americans. Six months is still a long time, though, and there are limits to how quick the process can be. To vaccinate people during that window, Seqirus also prepares vaccines against the flu strains that BARDA deems most likely to cause a pandemic. Those doses are stockpiled, and can be used to immunize healthcare workers, government employees, and the military while the Holly Springs plant churns out more.

Yet even this strategy is imperfect. When H7N9 first appeared in China, in 2013, the plant did its job, creating a vaccine that was then stockpiled. Since then, H7N9 has mutated, and the hoarded doses may be ineffective against the current strains. "We occasionally have to chase a pre-pandemic," says Anthony Fauci, the National Institute of Allergy and Infectious Diseases (NIAID) director. "We have to do it," but the strategy remains wasteful and reactive.

What society really needs, Fauci tells me, is a *universal* flu vaccine — one that protects against every variant of the virus and provides long-term protection, just as the vaccines against measles and mumps do. One vaccine to bind them all: it's hard to overstate what a win that would be. No more worrying about strain mismatches or annual injections. "It would be the epitome of preparedness," Fauci says, and he has committed his institute to developing one.

Flu viruses are studded with a molecule called hemagglutinin (the *H* in H1N1 and other such names), which looks like a stumpy Pez dispenser. Vaccines target the head, but that's the part that varies most among strains, and evolves most quickly. Targeting the stem, which is more uniform and stable, might yield better results. The stem, however, is usually ignored by the immune system. To draw attention to it, Fauci's team decapitates the molecule and sticks the stem onto a nanoparticle. The result looks like a flu virus, but encourages the immune system to go after the stable stem instead of the adaptable head. In a preliminary study, his team used this approach to build a vaccine using an H1 virus, which then protected ferrets against a very different H5N1 strain.

This type of work is promising, but flu is such an adaptive adversary that the quest for a universal vaccine might take years, even decades, to fulfill. Progress will be incremental, but each increment will have value in itself. A universal-ish vaccine that, say, protected against all H1N1 strains would have prevented the 2009 pandemic. And reducing flu's menace, even in some of its variants, would free up resources and intellectual capacity for dealing with other deadly diseases for which no vaccines exist at all.

Many of those diseases strike poor countries first and are — for now — rare. Creating vaccines for them is painstaking and often unprofitable, and therefore little gets done. Last year, to help change that, the Coalition for Epidemic Preparedness Innovations was created, and now has $630 million pledged by governments and nonprofits. It will focus first on Lassa fever, Nipah, and MERS, and its ambition is to yank promising vaccines out of developmental purgatory, push them through trials, and stockpile them by the hundreds of thousands. (One goal is to avoid a repeat of 2014, when Ebola ravaged West Africa while an experimental vaccine that could potentially have stopped it was languishing in a freezer, where it had been for a decade.)

More important, the coalition is looking to fund so-called platform technologies that could create a vaccine against any new virus far more quickly than can be done today: within 16 weeks of its discovery. Most current vaccines work by presenting the immune system with dead, weakened, or fragmented microbes. Every microbe is unique, so every vaccine must be unique, which is one reason they're so time-consuming to create. But by loading key parts of a given microbe onto a standard molecular chassis, scientists could build plug-and-play vaccines that could be swiftly customized.

In the same way that movable type revolutionized printing by allowing people to rapidly set up new pages without carving bespoke woodblocks, such vaccines could greatly accelerate the defense against emerging infections. In 2016 a team of researchers used the concept to create a vaccine against Zika that is now being tested in clinical trials across the Americas. The process took four months — the shortest development time in vaccinology's 222-year history.

The possibilities of vaccine science—a universal flu vaccine, plug-and-play platforms—are exciting. But they are only possibilities. No matter how brilliant and dedicated the people involved, they face a long and uncertain road. Missteps and failures are assured along the way; dogged effort and consistent support are essential to sustain the journey. These latter necessities, unavoidably, bring us to politics—where they are, predictably, in short supply.

Anthony Fauci's office walls are plastered with certificates, magazine articles, and other mementos from his 34-year career as NIAID director, including photos of him with various presidents. In one picture, he stands in the Oval Office with Bill Clinton and Al Gore, pointing to a photo of HIV latching onto a white blood cell. In another, George W. Bush fastens the Presidential Medal of Freedom around his neck. Fauci has counseled every president from Ronald Reagan through Barack Obama about the problem of epidemics, because each of them has needed that counsel. "This transcends administrations," he tells me.

Reagan and the elder Bush had to face the emergence and proliferation of HIV. Clinton had to deal with the arrival of West Nile virus. Bush the younger had to contend with anthrax and SARS. Barack Obama saw a flu pandemic in his third month in office, MERS and Ebola at the start of his second term, and Zika at the dusk of his presidency. The responses of the presidents varied, Fauci told me: Clinton went on autopilot; the younger Bush made public health part of his legacy, funding an astonishingly successful anti-HIV program; Obama had the keenest intellectual interest in the subject.

And Donald Trump? "I haven't had any interaction with him yet," Fauci says. "But in fairness, there hasn't been a situation."

There surely will be, though. At some point, a new virus will emerge to test Trump's mettle. What happens then? He has no background in science or health, and has surrounded himself with little such expertise. The President's Council of Advisers on Science and Technology, a group of leading scientists who consult on policy matters, is dormant. The Office of Science and Technology Policy, which has advised presidents on everything from epidemics to nuclear disasters since 1976, is diminished. The head of that office typically acts as the president's chief scientific consigliere, but to date no one has been appointed.

Other parts of Trump's administration that will prove crucial during an epidemic have operated like an Etch A Sketch. During the nine months I spent working on this story, Tom Price resigned as secretary of health and human services after using taxpayer money to fund charter flights (although his replacement, Alex Azar, is arguably better prepared, having dealt with anthrax, flu, and SARS during the Bush years). Brenda Fitzgerald stepped down as CDC director after it became known that she had bought stock in tobacco companies; her replacement, Robert Redfield, has a long track record studying HIV, but relatively little public health experience.

Rear Admiral Tim Ziemer, a veteran malaria fighter, was appointed to the National Security Council, in part to oversee the development of the White House's forthcoming biosecurity strategy. When I met Ziemer at the White House in February, he hadn't spoken with the president, but said pandemic preparedness was a priority for the administration. He left in May.

Organizing a federal response to an emerging pandemic is harder than one might think. The largely successful US response to Ebola in 2014 benefited from the special appointment of an

"Ebola czar"—Klain—to help coordinate the many agencies that face unclear responsibilities. In 2016, when Obama asked for $1.9 billion to fight Zika, Congress devolved into partisan squabbling. Republicans wanted to keep the funds away from clinics that worked with Planned Parenthood, and Democrats opposed the restriction. It took more than seven months to appropriate $1.1 billion; by then, the CDC and NIH had been forced to divert funds meant to deal with flu, HIV, and the next Ebola.

How will Trump manage such a situation? Back in 2014, he called Obama a "psycho" for not banning flights from Ebola-afflicted countries, even though no direct flights existed, and even though health experts noted that travel restrictions hadn't helped control SARS or H1N1. Counterintuitively, flight bans increase the odds that outbreaks will spread by driving fearful patients underground, forcing them to seek alternative and even illegal transport routes. They also discourage health workers from helping to contain foreign outbreaks, for fear that they'll be denied reentry into their home country. Trump clearly felt that such Americans *should* be denied reentry. "KEEP THEM OUT OF HERE!" he tweeted, before questioning the evidence that Ebola is not as contagious as is commonly believed.

Trump called Obama "dumb" for deploying the military to countries suffering from the Ebola outbreak, and he now commands that same military. His dislike of outsiders and disdain for diplomacy could lead him to spurn the cooperative, outward-facing strategies that work best to contain emergent pandemics.

Perhaps the two most important things a leader can personally provide in the midst of an epidemic are reliable information and a unifying spirit. In the absence of strong countermeasures, severe outbreaks tear communities apart, forcing people to fear their neighbors; the longest-lasting damage can be psychosocial. Trump's tendency to tweet rashly, delegitimize legitimate sources of information, and readily buy into conspiracy theories could be disastrous.

Emery Mikolo greets me warmly, with one outstretched hand. We shake, do a little ankle tap, and say, "Nous sommes ensemble"— *We are together*. This is the greeting of the Kikwit Ebola Survivors' Association, of which Mikolo is a cofounder and the vice president. Fifteen of the 42 members file into the breakfast room of Hotel

Kwilu, the men in simple shirts and the women in glorious kaleidoscopic dresses. The youngest are in their mid-30s, the oldest in their late 70s. They speak softly as they reconnect over plates of bread, cheese, and Nutella.

There is still no definitive treatment for Ebola. In 1995, like most of the survivors, Mikolo fought the virus off on his own, over three grueling weeks. After he recovered, he donated his blood —and the virus-fighting antibodies within it—to others, saving the lives of Shimene Mukungu and Emilienne Luzolo, who are also here today. Blood spreads Ebola. Sometimes, blood cures it.

The outbreak destroyed entire families. Afterward, some of the survivors found themselves the sole providers for several children. Others were orphans. Worst of all, they became pariahs. "Here, for we who live in communities, it is solitude that kills us," Mikolo says. He rolls up his trouser leg and shows me the scars inflicted by fearful neighbors, who hurled stones at him when he tried to return home. Like others, he discovered that his house and belongings had been burned.

The survivors banded together. "We had to take care of ourselves," Norbert Mabanza, the association's president, tells me. "Those with a little bit of strength could support those who were weaker. *Débrouillez-vous.*"

I listen to their stories in the company of Anne Rimoin, an epidemiologist from UCLA. During her 16 years working in the Congo, Rimoin has shown that monkeypox is on the rise, helped discover a new virus, and worked to create the first truly accurate maps of the country, down to the most-isolated villages. The Congo is a second home for her. When Rimoin's father died shortly before her wedding, Muyembe, the virologist who first encountered Ebola, flew to Los Angeles to walk her down the aisle.

Rimoin emphasized to me the social rupture that disease outbreaks wreak on unprepared communities, and the difficulty of repair. She also said that until the Congo and other developing countries can control the diseases at their doorsteps, it is imperative for richer nations like the United States to help them. That was a truth acknowledged by every expert I spoke with: the best way to prevent pandemics is to contain outbreaks at their source. The United States cannot possibly consider itself protected if other nations are not.

America's prior investments in global health preparedness

—the largest of any nation's—have already made a tangible difference. In 2010 the CDC helped Uganda set up a new surveillance system for viral hemorrhagic fevers like Ebola and Marburg. Health workers there are now trained to recognize these diseases, and have tools for collecting samples safely. Labs have diagnostic equipment. Response teams are ready to go. "It's been incredible to watch," says Inger Damon, who oversaw the CDC's 2014 Ebola response. "It used to take two weeks to respond to an outbreak. By the time you understood what was going on, you'd have twenty to thirty cases, and eventually hundreds. Now they can respond in two days." Sixteen outbreaks have been detected since 2010, but they were typically much smaller and shorter than before. Half of them involved just one case.

And in July 2014, in the midst of the West African Ebola outbreak, those investments very likely prevented a horrific catastrophe that might otherwise still be unfolding today. A Liberian American man brought the virus into Lagos, Nigeria, home to 21 million people and one of Africa's busiest airports. "If it had gone out of control in Lagos, it would have gone all over Africa for years," Tom Frieden, the former CDC director, says. "We were right on the edge of the abyss."

But Nigeria responded quickly. For years, it had used investments from the United States and other countries to build infrastructure for eradicating polio. It had a command center and a crack team of CDC-trained epidemiologists. When Ebola hit Lagos, the team dropped its polio work. It found every person who'd contracted Ebola, and every person with whom those infected had had contact. In only three months, after just 19 cases and eight deaths, it brought Ebola to heel and stopped it from spreading to any other country.

With patience and money—not even very much money compared with the vastness of rich-country spending—this kind of victory could be commonplace. An international partnership called the Global Health Security Agenda has already laid out a road map for nations to plug their vulnerabilities against infectious threats. Back in 2014, the United States committed $1 billion to the effort over five years. With it came a clear, if implicit, statement: pandemic threats should be a global priority. *Nous sommes ensemble.*

Given that sense of commitment, and with the related funding in hand, the CDC made a large bet: it began helping 49 countries improve their epidemic preparedness, on the assumption that demonstrating success would assure a continued flow of money. But that bet now looks uncertain. Trump's budget for 2019 would cut 67 percent from current annual spending.

If investments start receding, the CDC will have to wind down its activity in several countries, and its field officers will look for other jobs. Their local knowledge will disappear, and the relationships they have built will crumble. Trust is essential for controlling outbreaks; it is hard won, and not easily replaced. "In an outbreak, there's so little time to learn things, make connections, learn how to not offend people," Rimoin tells me. "We're here in the Congo all the time. People know us."

Until Rimoin arrived in Kikwit last summer, the Ebola survivors had for decades refused to collaborate with outsiders. "Others see us as people to study," Mikolo tells her. "But you came to us with friendship and humanity. You haven't abandoned us." Indeed, while Rimoin is studying the blood of the survivors, she is also trying to set up a clinic where survivors, half of whom are medically trained, can provide primary care to one another and to their communities. She has used donations and some of her own money to help Mabanza, the association's president, get a master's degree in public health.

Rimoin and I take the same flight out of Kinshasa; she will likely be back in a few months. I think about her ties to the Congo as our plane soars over one of the most biodiverse rain forests in the world, on the first of three legs that will put me back within a stone's throw of the White House in 28 hours. Below my flight path, the sparks of a new Ebola outbreak are flickering, unbeknownst to me or any of the scientists with whom I'd spoken. (It would be discovered in the weeks that followed.)

I think about the survivors of Kikwit, and how our connectedness is both the source of our greatest vulnerability and the potential means of our salvation. I think about whether it is possible to break the old cycle of panic and neglect, to fully transition from *Débrouillez-vous* to *Nous sommes ensemble*. I think about this amid bouts of restless sleep, as the plane flies westward across the Atlantic, stuck in the shadow of the world, until finally, dawn catches up.

ILANA YURKIEWICZ

Paper Trails: Living and Dying with Fragmented Medical Records

FROM *Undark*

IT WAS A cool April day in 2016 when Michael Champion's wife, Leah, noticed that her husband's forehead was drenched in sweat. She took his temperature and couldn't believe the number: 102.4 degrees Fahrenheit. He had been lying in bed more often, and as she watched him over the previous weekend, she'd hoped he would improve. But when Michael became so weak he couldn't even hold up his head, she knew this couldn't wait any longer and she called an ambulance.

I was a first-year resident working at the Palo Alto veterans' hospital when I got a call to evaluate an elderly vet with fevers. Michael was not able to tell me his symptoms. From his chart, I gathered that he was a long-standing diabetic. He'd also suffered a stroke several years earlier that left him with a weak right side, slurred speech, and trouble swallowing. He had a permanent feeding tube and required bladder catheter insertions several times every day. Stripped of the ability to interact the way he could before his stroke, he spent most of his time in bed or in his wheelchair. Leah was his primary caregiver.

Now she sat by his bedside, coloring in the details of a medical history he was unable to voice himself. She told me how tired and disoriented he seemed. He was pale. It wasn't right.

I ordered blood and urine tests that traced his fevers to a multi-drug-resistant infection in his bladder. We treated the infection with antibiotics and worked on techniques for hygienic catheterization. Because of the infection, his blood sugar ran consistently high, so we also added extra insulin to his diabetes regimen.

He was doing better within a few days and his mental status had perked up. Every morning, when I asked how he was feeling, he was able to provide one- or two-word answers. Several times he gave me a thumbs-up.

But his hospitalization had taken a toll—especially on Leah, who now realized Michael would need to regain his strength for her to care for him at home. He could hardly move from his bed to a chair without two people assisting him, and even that left him drained. Our case manager identified a skilled nursing facility nearby, just south of San Francisco, with continued physical therapy and around-the-clock nursing care. I remember Leah expressing relief about the choice. She would be able to visit him every day but still rely on a dedicated team of professionals to help him until he fully recovered.

I performed the rituals of hospital discharge that had become second nature to me as a resident. I typed out a discharge summary outlining each of his medical problems. I spelled out his antibiotics plan. I wrote his new insulin regimen—an additional injection every six hours and extra doses with his tube feeds, on top of his usual morning dose. I summarized what we were thinking, what we had done, and what needed to be done next. Whenever possible, I had learned to bolster that sheet. I used simple and straightforward language. I bolded. I double-checked my medication list. I knew this sheet was often the only guidance a nursing facility would receive. If I didn't write something here, it very often didn't exist.

But I also knew that even if I did write it down, it might not be read by the caregivers and healthcare professionals who would treat him next. And I knew that only fragments of *their* notes and charts were likely to get passed down the line of Michael's care, too. And that's precisely what happened. Over the next few weeks, Michael would return to my hospital more than once, in bad shape thanks to unconnected records that were not easy to transfer. In one instance, he didn't receive the insulin doses that I had so carefully marked in his discharge form, which left him in a near-comatose state.

Michael isn't alone. Every year, an untold number of patients undergo duplicate procedures—or fail to get them in the first place—because key pieces of their medical history go missing. Countless others suffer from medication errors. Hospitals, nursing homes, and other medical facilities use a patchwork of methods to track records, relying on proprietary technology or old-fashioned communications such as faxes and paper notes. These systems don't always sync, and the collective costs to patients, hospitals, and the economy as a whole are impossible to quantify—although some experts say consistent and cohesive health-information technology could save billions of dollars. An initiative from the US Department of Health and Human Services aims to unify these disparate systems, but we remain far from a universal electronic medical record that would solve the problem.

Meanwhile, we have stories like this one, which is a story about gaps. Michael would slip through these gaps. But as I filled out his record for the first time that day in 2016, neither Leah nor I knew just how short our best efforts would fall. Back at the hospital, she and I shook hands and she thanked me for my care.

More than a year and a half later I would ask how Leah had felt at this point. She said: "I had my concerns about him leaving." But we had faith in the system and she was relieved her husband would be in a skilled nursing facility, receiving help for medical problems that had become too overwhelming for her to manage alone.

He would get stronger soon, she thought. Then, they'd be able to go home.

The American healthcare system is dynamic by design. Patients move from one hospital to another; from a hospital to a rehab facility; from the wards to the intensive care unit (ICU); from the hospital to a primary care setting. These transfers are inevitable, as a person's health either improves or declines, or if that person simply desires a second opinion. In nonemergency situations, it's somewhat of a medical free market. Patients have every right to take advantage of it. But when the patients move, their histories often lag behind.

When I meet a new patient, I have to gather slips of these histories from various sources—electronic records, paper documentation, outside faxes, notes in wallets, family members—to piece together a meaningful narrative. Why is this person on a steroid medication and is there a plan for tapering? Is this poor kidney function new? How had Michael's fluctuating blood sugar been managed when he'd been sick before?

While most hospitals in the United States today use electronic health records, they remain disparate, with hundreds of different interfaces and minimal data sharing from one care facility to the next. Today, less than half of hospitals electronically integrate data from other hospitals outside of their system, and only 30 percent of skilled nursing facilities share data outside their walls, according to the Office of the National Coordinator for Health Information Technology (ONC), a division of the US Department of Health and Human Services.

Without an easy way to get a patient's full medical files, I must ask where their prior doctors were, have the patients sign a release form, fax it to the other hospitals, and receive stacks of papers in return. Then I dig in. Unable to use "control-find" on a stack of paper, I sift through several to hundreds of pages to find the few values of importance. If I'd like to see images, such as MRI or CT scans, it's more involved: I request CDs, wait for them to be mailed over, walk them to our radiology department, fill out another form, and then wait another day or so until I can see them in the computer system. The more places the patients have gone, the more there is to unravel.

When a patient with a complex medical history like Michael arrives under my care, it's like opening a book to page 200 and being asked to write page 201. That can be challenging enough. But on top of that, maybe the middle is mysteriously ripped out, pages 75 to 95 are shuffled, and several chapters don't even seem to be part of the same story.

Meanwhile, everyone around me is urging: write now.

Mere days after the Champions left, my senior resident asked me to evaluate a new patient. He had a history of diabetes and stroke, and he was so tired he couldn't keep his eyes open. "Admit to medicine," the note from the emergency room had read. When I pulled back the curtain, my heart sank. The patient was Michael Champion. His frail body lay still in the hospital bed, eyes closed, unable to communicate. When the emergency doctor gave Michael's sternum a brisk rub, he awoke only to instantly fall back asleep.

Leah was next to him, her face contorted. She told me that things went wrong the moment they had arrived at the nursing facility. The scheduling of his tube feeds was disrupted by the antibiotics regimen. The extra insulin I had written in his chart had also been withheld.

A blood test revealed one likely source of Michael's lethargy: his blood sugar was nearly four times higher than normal. A reading just slightly higher can tip a diabetic into a condition called hyperosmolar hyperglycemic state, with risks of extreme dehydration, electrolyte imbalances, and in the worst cases, brain swelling. Though the latter is rare, its mortality rate is high. I spoke to Leah at the bedside. Incredibly, she wasn't angry. She knew it was a fixable problem and she relied on us to fix it, just as we had his infection. But she was confused. The entire time he was in the hospital, he was given extra doses of insulin every morning, afternoon, and evening. At the nursing facility, he had been given none. How could this have happened?

Leah's question gets at the million-dollar one: *Why?*

The answer lies in the tangled evolution of e-health technology. In 2004 President George W. Bush created the ONC within the US Department of Health and Human Services. In 2009 Congress then authorized and funded related legislation known as the Health Information Technology for Economic and Clinical Health Act (HITECH) to stimulate the conversion of paper medical records into electronic charts. And indeed, many hospitals and doctor's offices did this successfully, says Karen DeSalvo, the national coordinator from 2014 to 2016, by "digitizing the care experience of every American." But each electronic health vendor made proprietary systems that weren't always compatible with one another, which made it hard for records to transfer between medical facilities. Now, says DeSalvo, "we need to get them blended."

In 2014 DeSalvo's office had about $100 million to hit this goal. But even with this level of support "that kind of transformation is difficult," she says. There are hundreds of different vendors using diverse technological platforms, each with a unique way to organize patient data. Finding a universal way to blend the systems has been a bureaucratic and engineering nightmare. Sometimes the barriers are intentional, DeSalvo says, as vendors don't necessarily want to make it easy to share data with hospitals using competitors' systems and may charge them a fee to do so. Blocking also occurs because of a misunderstanding of patient privacy rights. As efforts to exchange data go public, some critics worry about security leaks.

The handwringing over privacy may not be necessary. According to Lucia Savage, a lawyer and the former chief privacy officer at the ONC, the same privacy rules apply no matter how patient files are shared—whether by handwritten notes, faxes, or electronic files. "Doctors, nurses, physician assistants—they have ethical rules," she says. "They're not supposed to be snooping around someone's data because he's Steve Jobs . . . We have to assume people act professionally and ethically in this space."

For patients, the trust often already exists. When anyone comes into any healthcare facility, we review all the records we have. Together, we openly discuss and dissect deeply personal information. In fact, patients may be the biggest advocates for sharing medical information, says Mark Savage, director of health policy at the University of California at San Francisco's Center for Digital Health Innovation. (He's also married to Lucia Savage.) Patients tend to take it for granted that their doctors are talking to one another, he adds, and "they're frustrated when information is not exchanged."

There are data backing up these ideas. A 2014 survey of more than 2,000 patients by Mark Savage, then with the National Partnership for Women and Families, and colleagues showed that 95 percent felt electronic records "were useful in assuring timely access to relevant information by all of their health care providers." And more than three-quarters said they already share information with their healthcare providers all or most of the time. Those who used electronic records compared to paper records reported more trust in their providers to protect privacy.

These results resonate for me. Over my last few years as a doctor, I can't think of a time when a patient complained that a doctor knew too much of their medical history. Yet many times, I've heard frustration that we don't know enough.

In the absence of a system that reliably transmits information, we get creative. When the system fails to fill in the cracks, we hope the patients can help. I ask them: What did the doctors say? What did the testing show? Was it a loud machine where you lay flat or did someone use a probe coated with cold gel?

When patients don't know, the burden of creativity falls on motivated bystanders—sometimes other healthcare providers or family—who, through grit, persistence, and clever workarounds, find ways to cobble information together and pass it forward.

Michael Champion was fortunate to have a living, breathing medical record in Leah. Early on when we met, I noticed how she pulled from a purse a yellow legal pad. Later, I learned that the pad accompanied her wherever Michael went, and she used it to jot down the minutiae of his care. She spoke with a deep knowledge of his medical issues and hospital staff would often ask her if she was a doctor or nurse. No, she would say. This was a role she had acquired out of necessity.

By the time we lowered Michael's blood sugar and he was ready to be discharged for the second time, I learned my lesson. I had performed my rituals of discharge diligently. But I needed more safeguards. When in doubt, duplicate. I had a plan, and it had three parts. One: I would print my discharge instructions, as before. Two: I would call the nursing facility and verbally pass off Michael's care plan. Three: I would rely on Leah. As the only source of continuity in Michael's many transitions, Leah would be the glue. If everything else fell through, she could advocate.

Many patients don't have a Leah, though I'd seen a lucky handful who had people go above and beyond to fill in the gaps. There was the oncologist who needed the biopsy results for one of her new cancer patients. The biopsies had been done long ago, at another hospital, and the patient couldn't remember who had done them or what they showed. So the oncologist cold-called all the physicians in the department; when none of them knew the patient, she asked for the names of all the recently retired doctors and called them, too, until she got what she needed.

I also remember the Italian man whose clever physician spared him an invasive procedure into his heart from 6,000 miles away. The monitors capturing the man's heart rhythm showed ST elevations—the scary sign of a serious heart attack colloquially known as the "widowmaker." The patient saw the concerned looks on our faces as we watched his monitor and then calmly reached into his pocket and pulled out a folded paper that had been cut to wallet size. It was an old rhythm strip, labeled 10 years ago, and it had the same ST elevations. His primary care physician had told him to keep it on him at all times, so that any new doctor would learn he had an irregular baseline and not wheel him straight to the cardiology lab whenever he walked through a hospital door.

I thought of these stories on the morning of Michael's discharge. He was headed to a different nursing facility, about 10 miles south of the first one. I spoke to Leah candidly. I told her disappointing truths: Discharges are the most dangerous times of a hospital stay and I had seen plans break down even at the best institutions. Because of that, I said, I am going to tell you the details of his medical care. The nursing facility should already know these things. But if things go wrong, you can fill in the gaps.

It was unfair. Beyond dealing with the emotional weight of Michael's illness, I was asking her to carry his nuanced medical care as well. I was asking her to perform the jobs that the medical system around her was supposed to do. But no one was incentivized to care as much as Leah and she took on the challenge with enthusiasm.

We spent the next 30 minutes talking shop. She asked questions. I answered them. She took notes on her yellow sheets. We brainstormed, together.

"I want to empower you to advocate for him," I said.

She put down her legal pad, reached out to shake my hand, and then, reconsidering for a moment, instead softly wrapped her arms around me for a long hug. She was about to head into the medical unknown.

The next time I saw Leah, she sat by her husband's side as he lay in a hospital bed again. It was one week later. We were in the ICU. She rose from her seat when she saw me. With every visit, I swore I could see new wrinkle lines. It was as though she was in a time warp, aging a year through every week of her husband's medical decline.

"It was like you were prescient," I recall her saying. "I did everything you said, and it still didn't work."

This time, the gap was in Michael's tube feeds. Through his feeding tube, he was meant to receive both nutritional formula and water. Water is critical to balancing electrolytes in the bloodstream; without it, the relative sodium levels in the blood can rise. The cascade of side effects increases as the sodium does: confusion, seizures, coma, death. But the second nursing facility hadn't given Michael the right amount of water. His sodium soared and, once again, he was nearly comatose.

Leah had alerted the physician. This is completely different, she pointed out. He's not speaking. To which she said she was told: This is just progression of his illness. He's a sick person, you know.

But Leah was right—though she didn't know just how sick Michael was. She came back from running errands shortly thereafter and found Michael being loaded into an ambulance. In the emergency room, his labs showed a sodium level nearly 20 points higher than the upper limit of normal. It was infuriating. Michael was now in the ICU, less than three weeks after he was admitted and discharged and admitted and discharged again. It was preventable.

Leah's question was potent: "How could this have happened?" At the time, I imagined many possible answers. Maybe my printed discharge summary had gotten lost. Maybe his initial nurse left it out during a verbal handoff—where the responsibility for care is transferred from one provider to another—to be forever erased from his history.

More likely, a nurse or doctor had held a handful of pages in their left hand and a handful of pages in their right, trying to re-create a medical story from scratch. And the details of his sodium management were lost in the shuffle.

In 1999 the Institute of Medicine, now called the National Academy of Medicine, published one of the most famous reports in medical history. *To Err Is Human* noted that between 44,000 and 98,000 people die in hospitals each year from preventable medical errors—a figure that was later compared to the equivalent of a jumbo jet crashing every day. More recent studies suggested this number was too low, raising it to more than 400,000 deaths per year. If they were a disease, medical errors would now rank as the third leading cause of death in the United States, a 2016 analysis published in the *BMJ* found, right behind heart disease and cancer.

I wanted to find data on errors that occur as patients move across systems. What is the cost of these transitions? I imagined both errors of commission—of repeat lab tests, scans, and even procedures—along with errors of omission—delayed or missed diagnoses because of an incomplete medical context.

The data we do have is unsettling to read. The Joint Commission, a large accreditor of healthcare organizations with a focus on patient safety, studied transitions of care and concluded that they are largely ineffective, leading to adverse events, hospital readmissions, and soaring costs. Their conclusions were supported by several key studies. One reported that nearly 20 percent of patients experienced adverse events—like Michael Champion's sugar and sodium spikes—within three weeks of discharge. Almost half were deemed preventable. And when patients are admitted to hospitals, more than half have at least one discrepancy in their medications.

How much is faulty record sharing to blame? My colleague, Marta Almli, an internal medicine doctor, surveyed the resident physicians at my hospital and others and found widespread dissatisfaction with how we obtain medical records: among 58 physicians surveyed, 81 percent said it was "somewhat difficult" or "extremely difficult" to get information about patients who transferred from another healthcare facility. This was, notably, in spite of a majority of the same physicians saying they had a "good sense" of how to get transfer materials and reporting that they regularly evaluate a new patient's file when that person arrives under their care.

When the system fails, leaders in medicine suggest some tricks for improvement: better documentation, better verbal handoffs, and double-checking it all. The advice usually amounts to this: When the system fails you, be more careful. Work harder.

And so we do. But on a large scale, it breaks down. We can do every one of these steps and more, but stories like Michael's will persist because humans are fallible, memories are fickle, and it's an intellectually gargantuan task to distill decades of history in a few sentences.

We can cold-call retired physicians from distant hospitals and we can print rhythm strips to stuff in wallets. Something relevant will eventually go missing. Maybe not this time, and maybe not the next. But as an aggregate, this can't be counted on as a backbone system of safety. It's also a waste of resources to rewrite a new chart every time a patient enters a new building. I've seen doctors go above and beyond in every possible way and yet I've seen how hard it is to always get it right. It's as engineer W. Edwards Deming said: "A bad system will beat a good person every time."

When hospitals share information, care for everyone improves. Researchers at the University of Michigan found that when emergency rooms shared files, it was far less likely for patients to have repeat CT scans, ultrasound, or chest X-rays. Another study, from Israel, found that sharing health data compared to looking at only internal data cut down on redundant hospital admissions.

Slowly but surely, hospitals are venturing into data sharing. For instance, one of the most widely used electronic health vendors, called Epic, has a "Care Everywhere" platform, which helps facilities share electronic patient records quickly and efficiently. More recently, the company upped the ante with a "Share Everywhere" tool, giving patients control to share their health data with doctors anywhere in the world. Care for veterans improved when the electronic health record system VistA (Veterans Information Systems and Technology Architecture) included a tab for sharing data from one veterans hospital to the next anywhere in the country.

These are powerful steps. But many experts see the last frontier as a universal electronic health record accessible across all hospitals and patients. Such a health record would transcend institutional borders. It would mean no more creative workarounds. No more paper faxes. No gaps.

Michael's medical team eventually corrected his high sodium and, after two months of rehab, Leah finally took him home—with hospice services. Generally, in order to qualify for hospice, a doctor must certify that someone has six months or less to live. But sometimes when we cut out the aggressive medical care, amazing things happen. Some people live longer.

Such was the case with Michael.

On a windy afternoon in January 2018, nearly two years after I first met the Champions, I saw them once more, this time at their home. Michael was doing well, considering what he'd been through. Leah had decided to do Michael's insulin and water herself, and she also had nursing aides come visit the home four days a week. In the end, she trusted herself most. She was the pillar of continuity Michael needed—that many others do not have.

During my visit, I told Leah that I never found out exactly where the breakdown in communication happened at the nursing facilities nearly two years ago. I retraced their steps and paid visits to both, trying to re-create their story. Along the way, I found no electronic trail of Michael's stay. There were some records, but for discharge patients they were stored in a paper binder in a separate storage facility.

I sat with the admissions coordinator at one of the nursing facilities as she ran me through the process. How they obtain records depends on where the patient is coming from, she explained. Some places send electronic forms and some go through fax. What arrives is a standard set of papers: the physician's discharge summary and the most recent medication list. The nursing facility doesn't receive a full set of lab values unless the physician copies them into the discharge summary, or any notes prior to the hospitalization, such as primary care records.

I wish finding the hole was simple: a broken fax machine; a paper chart that ended up on the wrong desk. But when the process is roundabout, requiring multiple steps and multiple workarounds, the breakdown points multiply. Sometimes it's remarkable that things turn out well so much of the time, given all the places they could go wrong.

Michael had more fortitude than almost anyone I've met. He recovered, again and again—sometimes because of, and sometimes in spite of, his medical care. But six months after my visit, just this past July, Michael died—on his own terms. He'd had pneumonia and it showed signs of coming back. Leah asked if he wanted to treat it or have nature take its course. "His gift to me was clarity," Leah told me later. "I didn't have to make the decision for him." Michael's last words to Leah had been: "Let it go." She brought in family and friends to say their goodbyes. Then they stopped his tube feeds.

Since then, I've thought back to my last visit with Michael, when Leah asked if I'd like to talk to him privately. When we were alone, I asked him: "Are you happy?"

Michael looked out the window and then turned back to me.

"Yes," he nodded.

I nodded back, reflecting on what it was like to reconstruct the entire human narrative of Michael Champion from a handful of scattered, disconnected fragments. As I looked at him, I saw all my patients, and I thought of how to get them the informed medical care they so deeply deserve. I eagerly await the day a universal electronic health record connects the dots. But it's not that day, so we in the healthcare system must do everything in our power to deliver good care to those who trust us with it. We will continue to push papers through fax machines, to wait on hold as we cold-call those who may provide answers, and to repeat tests from scratch when we're stalled.

We know it's not a perfect system. We know there will be gaps. But what choice do we have?

Contributors' Notes

*Other Notable Science and Nature
Writing of 2018*

Contributors' Notes

Philip Ball is a freelance writer and broadcaster, and worked previously for over 20 years as an editor for *Nature*. He writes regularly in the scientific and popular media, and has authored many books on the interactions of the sciences, the arts, and the wider culture, including *H2O: A Biography of Water*. He trained as a chemist at the University of Oxford, and as a physicist at the University of Bristol. His latest book is *Beyond Weird* (2018), a survey of what quantum mechanics means.

Rebecca Boyle is an award-winning freelance journalist. Her magazine writing encompasses astronomy and physics, geoscience and climate change, and the history of science. She is at work on a book about humanity's relationship with the moon. She is a contributing writer for *The Atlantic*, and her work regularly appears in *FiveThirtyEight*, *Scientific American*, *Quanta*, and many other publications. Rebecca grew up in Denver, Colorado, and lives in St. Louis, Missouri, with her family.

Peter Brannen is a science journalist whose work has appeared in *The Atlantic*, the *New York Times*, and elsewhere. He is the author of *The Ends of the World: Volcanic Apocalypses, Lethal Oceans, and Our Quest to Understand Earth's Past Mass Extinctions*, published in 2017. He lives in Boulder, Colorado.

Chris Colin has written about endangered noodles, chimp filmmakers, ethnic cleansing, Obama's Irish roots, Japan's rent-a-friend industry, George Bush's pool boy, and more for the NewYorker.com, the *New York Times Magazine*, *Saveur*, *Atavist*, *Pop-Up Magazine*, and *Wired*. He's a contributing writer for *The California Sunday Magazine* and *Afar*, and his most recent book was *What to Talk About*.

Douglas Fox (www.douglasfox.org) is a freelance journalist who writes about biological, earth, and Antarctic sciences. His stories have appeared in *Scientific American, National Geographic, Esquire, Virginia Quarterly Review, High Country News, Discover, Nature, Slate,* the *Christian Science Monitor,* and other publications. Stories by Doug have garnered awards from the American Society of Journalists and Authors (2011), the National Association of Science Writers (2013), the American Geophysical Union (2015 and 2018), the Society of Environmental Journalists (2016), and the American Association for the Advancement of Science (2009 and 2017). Doug is a contributing author to *The Science Writers' Handbook* (2013).

Conor Gearin is a writer from St. Louis, Missouri. After studying grassland bird conservation in Nebraska, he has focused on understanding wildness and ecology in developed landscapes. His work has appeared in *The Atlantic* online, *The Millions, New Scientist, NOVA Next,* and *The New Territory,* where he is the features editor. He and his wife live in Massachusetts.

Ben Goldfarb is a conservation journalist whose writing has appeared in *Science, Mother Jones, Orion,* the *Guardian,* the *Washington Post, Pacific Standard, High Country News,* and many other publications. He is the author of *Eager: The Surprising, Secret Life of Beavers and Why They Matter,* for which he won the 2019 PEN/E. O. Wilson Literary Science Writing Award. He is a 2019 Alicia Patterson Foundation Fellow.

Gary Greenberg is a practicing psychotherapist and a contributing editor at *Harper's.* He is the author of four books, most recently *The Book of Woe: The DSM and the Unmaking of Psychiatry.*

Jeremy Hance writes a monthly column called "Saving Life on Earth: Words on the Wild" for *Mongabay.* He's a freelance journalist focusing on wildlife, forests, indigenous people, and climate change, with work appearing in the *Guardian, HuffPost,* the *Sydney Morning Herald, Ensia, Yale E360,* the *New Straits Times,* and *Alert: Conservation.* He lives in Minnesota with his wife, daughter, and two dogs. He is very busy working on his first book. He'd like to thank Isabel Esterman for her extraordinary editorial skills.

Holly Haworth was a recipient of the Middlebury Fellowship in Environmental Journalism. Her work appears in *Orion, Oxford American,* Terrain. org, and *Lapham's Quarterly,* among other places. It has been listed as notable in *The Best American Travel Writing* and nominated for a Pushcart Prize. She earned an MFA at Hollins University, where she was a Jackson Fellow. She lives in Floyd, Virginia.

Eva Holland is a freelance writer based in Canada's Yukon Territory. She is a correspondent for *Outside*, and her work has also appeared in *Wired*, *Pacific Standard*, *Hakai*, and many other publications in print and online. Her first book, *Nerve*, on the science of fear, is forthcoming in 2020.

Apricot Irving (www.apricotirving.com) is the author of *The Gospel of Trees*, an exploration of ecology, loss, and the tangled history of missions in Haiti. She is the recipient of a Rona Jaffe Foundation Writers' Award and a Literary Arts Fellowship, and reported from Haiti on post-earthquake rebuilding efforts for the radio program *This American Life*. She is the founder and director of the Boise Voices Oral History Project, a collaboration between youth and elders to record the stories of a changing neighborhood in North/Northeast Portland, and her writing has appeared in *Granta*, *Oregon Humanities*, and *On Being*.

Rowan Jacobsen is the author of *A Geography of Oysters*, *Fruitless Fall*, *The Living Shore*, *Shadows on the Gulf*, and other books. He has written for *Harper's*, *Outside*, *Mother Jones*, *Vice*, *Scientific American*, *Orion*, *Pacific Standard*, the *New York Times*, and other publications, and he frequently appears in *The Best American Science and Nature Writing* and *Best Food Writing* collections. He has received awards from the James Beard Foundation, the Society of American Travel Writers, the Overseas Press Club, the Alicia Patterson Foundation, and the McGraw Center for Business Journalism, and he was a 2017–18 Knight Science Journalism Fellow at MIT, focusing on the environmental and societal impacts of synthetic biology.

Brooke Jarvis is a contributing writer to *The New York Times Magazine* and *The California Sunday Magazine*. She won the Livingston Award for National Reporting and the NYU Reporting Award, and has been a finalist for the PEN Literary Award and the Livingston Award in International Reporting. Her work has been anthologized in *The Best American Science and Nature Writing* (2015 and 2019), *The Best American Travel Writing* (2019), *Love and Ruin: Tales of Obsession, Danger, and Heartbreak from the Atavist Magazine*, and *New Stories We Tell: True Tales by America's Next Generation of Great Women Journalists*. She lives in Seattle, Washington.

Matt Jones has an MFA in creative writing from the University of Alabama. He has published fiction and nonfiction in *The Southern Review*, *New England Review*, *The Atlantic*, *Michigan Quarterly Review*, the *Chicago Tribune*, and various other journals. He has received support from the Ohio Arts Council, Sitka Center for Art and Ecology, Vermont Studio Center, and The Leopold Writing Program. He is currently working on a book of nonfiction.

Kevin Krajick has reported on subjects ranging from criminal justice and immigration to science and nature from all 50 US states and 30-some countries. His work has appeared in *National Geographic, Newsweek, The New Yorker, Science, Smithsonian*, and many other publications. Among other honors, he is a two-time winner of the American Geophysical Union's Walter Sullivan Award for Excellence in Science Journalism. His 2001 book *Barren Lands* is the true account of how prospectors discovered diamond mines in Canada's remote Far North.

J. B. MacKinnon is a freelance journalist and author, most recently of *The Once and Future World: Nature As It Was, As It Is, As It Could Be*. His next book, which examines the consumer economy, will be published in 2020. He lives in Vancouver, Canada.

Bill McKibben is an author, environmentalist, and activist. In 1988 he wrote *The End of Nature*, the first book for a common audience about global warming. He is a cofounder and senior adviser at 350.org, an international climate campaign that works in 188 countries around the world.

Rebecca Mead is a staff writer at *The New Yorker*, and is the author of *My Life in Middlemarch*. She lives in London with her family.

Molly Osberg is a journalist who covers labor, crime, and the American healthcare system. She is a reporter with the Special Projects Desk, Gizmodo Media Group's investigative team. She lives on the East Coast.

Joshua Rothman is an editor and writer at *The New Yorker*.

Jordan Michael Smith has written for *The New York Times Magazine, The Atlantic*, the *Washington Post*, and many other publications. Formerly a speechwriter for New York City mayor Bill de Blasio, he is the author of the best-selling Kindle Single *Humanity*.

Shannon Stirone is a freelance writer in the San Francisco Bay Area. Her work focuses on exploration, policy, and how space and society intersect. Her work can be found in the *New York Times, The Atlantic, Longreads, National Geographic, Rolling Stone*, and other publications.

Linda Villarosa runs the journalism program at The City College of New York in Harlem and is a contributing writer for *The New York Times Magazine*. She is writing a book on race, inequality, and health.

Ed Yong is an award-winning science journalist who reports for *The Atlantic* and is based in Washington, DC. *I Contain Multitudes,* his first book, was a *New York Times* bestseller and a clue on *Jeopardy!* He has a Chatham Island black robin named after him.

Ilana Yurkiewicz is a physician at Stanford University and a medical journalist. She is a former Scientific American Blog Network columnist and AAAS Mass Media Fellow. Her writing has also appeared in the *New England Journal of Medicine, Aeon, Health Affairs,* and STAT News, and has been featured in *The Best Science Writing Online* and on CBC/Radio-Canada.

Other Notable Science and Nature Writing of 2018

RACHEL PEARSON
 The Challenge of "Chronic Lyme." *The New York Review of Books.* July 25, 2018.
CLINTON CROCKETT PETERS
 Snow Monkeys. *Proximity.* September 2018.
ELAINA PLOTT
 Living Water. *Pacific Standard.* September/October 2018.

DOUGLAS RUSHKOFF
 Survival of the Richest. *Medium.* July 5, 2018.

CHARLES SCHMIDT
 America's Misguided War on Childhood Lead Exposures. *Undark.* March 21, 2018.
MELODY SCHREIBER
 Solving the Suicide Crisis in the Arctic Circle. *Pacific Standard.* March 23, 2018.
KATHRYN SCHULZ
 Food Fight. *The New Yorker.* October 1, 2018.
ASHLEY SMART
 The War over Supercooled Water. *Physics Today.* August 22, 2018.
SEAN P. SMITH
 Ghost Tigers: Climate Change and the Escalation of Extinction. *Guernica.* March 21, 2018.
JOSHUA SOKOL
 Cracking the Cambrian. *Science.* November 23, 2018.
 Visiting the Mysterious Fairy Circles of the Namib Desert. *The Atlantic.* May 23, 2018.
CHRISTOPHER SOLOMON
 Who's Afraid of the Big Bad Wolf Scientist? *The New York Times Magazine.* July 8, 2018.

EDITH THURMAN
 Maltese for Beginners. *The New Yorker.* September 3, 2018.
JIA TOLENTINO
 Is There a Smarter Way to Think About Sexual Assault on Campus? *The New Yorker.* February 12 & 19, 2018.
REBECCA TUHUS-DUBROW
 "Do I Really Want to Hurt My Baby?" *The Cut.* December 5, 2018.

BOYCE UPHOLT
 A Killing Season. *The New Republic.* December 2018.

MOISES VELASQUEZ-MANOFF
 Can Dirt Save the Earth? *The New York Times Magazine.* April 22, 2018.
 The Great Meat Mystery. *The New York Times Magazine.* July 29, 2018.

THE BEST AMERICAN SERIES®

FIRST, BEST, AND BEST-SELLING

The Best American Comics

The Best American Essays

The Best American Food Writing

The Best American Mystery Stories

The Best American Nonrequired Reading

The Best American Science and Nature Writing

The Best American Science Fiction and Fantasy

The Best American Short Stories

The Best American Sports Writing

The Best American Travel Writing

Available in print and e-book wherever books are sold.

Visit our website: hmhbooks.com/series/best-american

EMILY VOIGT
Urban Bird Feeders Are Changing the Course of Evolution. *The Atlantic.*
January 26, 2018.

LIZZIE WADE
The Believer. *Science.* January 19, 2018.

DAVID WALLACE-WELLS
Parenting the Climate Change Generation. *New York Magazine.* December 20,
2018.

BRYAN WASHINGTON
We Go to the Park to Go Somewhere Else: On Houston's Green Havens.
Catapult. February 7, 2018.

CAITY WEAVER
What Is Glitter? *The New York Times.* December 23, 2018.

ED YONG
How to Impregnate a Rhino. *The Atlantic.* May 30, 2018.

ANDREW ZALESKI
The Forever Man. *Popular Science.* May 2018.
The Gray Area of Brain Training. *Popular Science.* March 2018.